MECHANICAL
AND ELECTRICAL
BUILDING CONSTRUCTION

MECHANICAL
AND ELECTRICAL
BUILDING CONSTRUCTION

ROBERT M. HETTEMA

Prentice-Hall, Inc., Englewood Cliffs, New Jersey 07632

Library of Congress Cataloging in Publication Data

Hettema, Robert M.
    Mechanical and electrical building construction.

    Bibliography: p.
    Includes index.
    1. Buildings—Mechanical equipment—Contracts and
specifications.   2. Buildings—Electrical equipment—
Contracts and specifications.   I. Title.
TH6010.H47   1984   696'.068   83-8693
ISBN 0-13-569608-9

Editorial/production supervision and
    interior design: Mary Carnis
Cover design: Diane Saxe
Manufacturing buyer: Anthony Caruso

Printed in the United States of America

10 9 8 7 6 5 4 3 2 1

ISBN 0-13-569608-9

Prentice-Hall International, Inc., *London*
Prentice-Hall of Australia Pty. Limited, *Sydney*
Editora Prentice-Hall do Brasil, Ltda., *Rio de Janeiro*
Prentice-Hall Canada Inc., *Toronto*
Prentice-Hall of India Private Limited, *New Delhi*
Prentice-Hall of Japan, Inc., *Tokyo*
Prentice-Hall of Southeast Asia Pte. Ltd., *Singapore*
Whitehall Books Limited, *Wellington, New Zealand*

Contents

Preface

Our large, tall, complex, and expensive modern buildings challenge constructors not only as to their skill in supervising architectural and structural work, but also as to their ability to manage the intricate process of creating a functional building mechanically and electrically. The modern sizable building project incorporates the best in air conditioning comfort, correct illumination, sufficient power distribution, and efficient communication and transportation systems while maintaining an acoustically correct environment. In addition, these structures contain sophisticated sensors, alarms, controls, and safety devices. Considerable design and construction skill also goes toward producing a building that minimizes energy consumption while meeting all the qualities needed in a truly fine building.

The building construction industry changes and progresses with the passing years. These changes are reflected in new and different materials and methods and greater emphasis on professionally trained managers. This group is not only increasing, but now includes a greater percentage of women engaged in satisfying careers in building construction. Through the means of this text, men and women architects, engineers, and technicians should gain an insight into the broader field of the complete construction process.

The impact of high interest and inflation rates on significant building projects requires professional management to satisfy the demand for rapid construction carried out competently. Through a good understanding of the scope of mechanical and electrical work in building construction, the construction supervisor will have the comprehensive knowledge needed to properly supervise the entire building construction process. This text seeks to broaden the construction manager's perspective beyond the purely architectural and structural, to acquaint him with the content and construction of mechanical and electrical systems, and to give him the ability to plan, schedule, and direct the mechanical and electrical portions of the project in a well-coordinated manner consistent with architectural and structural objectives.

R. M. Hettema
Penn State University

MECHANICAL
AND ELECTRICAL
BUILDING CONSTRUCTION

1

An Overview

1.1 INTRODUCTION

When we think of a familiar building we immediately visualize its external size and appearance. With further thought we might recall various interior spaces such as lobbies, manufacturing areas, and office space. We normally remember the visual image of the finished project. Unless we consciously direct our thoughts to the mechanical and electrical work, most of us in building construction usually think in architectural and structural terms. Perhaps by taking an overview of building construction we can better understand the reason for this mental approach, improve our attitude toward mechanical and electrical installations, and gain a better appreciation for this important part of building construction. Until the construction supervisor completely understands the vital part that mechanical and electrical construction plays in a project, he will not possess full control over the building construction management process. A $50,000 house or a $200 million high-rise can both be tied up by a poorly performing plumber or electrician.

1.2 A DIFFERENT VIEWPOINT

Most building construction supervisors come up through the ranks as carpenter and bricklayer foremen or have a civil engineering degree. They incline toward getting the hole dug and pushing the structural framework; they emphasize enclosing the building and making the roof tight: all important objectives. Unfortunately, many have never had direct mechanical and electrical construction experience and as a result do not understand all that is involved. They also do not feel really confident with mechanical

1

and electrical plans and specifications. They depend heavily on the weight of the obligation the mechanical and electrical contractors have accepted rather than personally assuming a role of firm direction of the coordination and sequencing of this type of work with that of the architectural and structural trades.

It is not difficult to understand this attitude considering the average construction supervisor's background and experience. Consider too the great percentage of mechanical and electrical work hidden behind walls or ceilings or buried under the ground or in a concrete slab. Generally, the layman has no conception of the mass of piping, ductwork, and conduits above a typical hung ceiling. Out of sight is certainly somewhat out of mind. This viewpoint is affected to some degree by the basic division of the design drawings. Usually, the contract documents include (1) structural; (2) architectural; (3) plumbing; (4) heating, ventilating, and air conditioning (HVAC); and (5) electrical drawings. The first two involve numerous trades ranging from excavators and dock builders (piling) to tile setters and painters, while the latter three divisions concentrate employment in only four trades: plumbers, steam fitters, sheet metal workers, and electricians. Mechanical and electrical work means specialization, and typically, construction people tend to defer to the specialists. With their own plans, codes, and specifications, their own union people, and their special status as licensed contractors, they *are* set apart from the rest of construction.

This separation becomes compounded in many areas of the United States in the method of contracting for work. Where separate contracting policy *controls,* such as in public work in Pennsylvania, for example, the work is bid in four parts: general contracting, plumbing, HVAC, and electrical work. Under separate contracting the general contractor loses control over the mechanical and electrical work, whether he wants to or not. Under this system mechanical and electrical work is bound to become something apart from normal building construction.

1.3 MECHANICAL AND ELECTRICAL WORK AND THE CONSTRUCTION DOLLAR

In considering the importance of mechanical and electrical work in building construction we should recognize the large percentage of the cost of the building that go into plumbing, HVAC, electrical work, elevators, and other similar installations. One-fourth of the cost of an apartment building, one-third of the cost of an office building, and 40% of the cost of a hospital go toward mechanical and electrical systems. This large percentage of the cost of a building represents not only a large responsibility but also a significant opportunity to those in charge of construction. The responsibility involves administering large contracts; coordinating the activities of these contractors with the architectural and structural trades; monitoring, together with the mechanical and electrical contractors, purchasing and timely delivery of a multitude of products and equipment; controlling mechanical and electrical progress to blend with overall project progress; and supervising quality so that the completed systems will function properly in accordance with the plans and specifications. This is a big assignment for anyone and deserves intense interest, enthusiasm, lots of energetic work, and much more.

The opportunity, of course, is to perform well as a supervisor responsible for all these things. Mechanical and electrical work offers more of a challenge, however. Since large sums are involved, opportunities exist for cost savings in design suggestions and for purchasing savings as well. Owners, architects, and contractors seeking possible means of reducing the cost of a building invariably look for savings in the large mechanical and electrical installations.

1.4 THE COMPANIES INVOLVED

As the building construction industry has grown and projects have increased in size, mechanical and electrical firms have matched growth with their general contracting counterparts. In the 1978 *Engineering News-Record* "Top 400" issue of April 13,

Fishback & Moore of New York City, a firm doing both mechanical and electrical work, show 1977 contracts in excess of $1 billion. Seven other contractors concentrating on either mechanical or electrical work had acquired $100,000,000 to $250,000,000 in contracts. Obviously, mechanical and electrical work is big business for some companies. Also obviously, many fine career opportunities exist in this line of work for the young construction engineer.

Mechanical and electrical companies generally start out specializing as plumbing, HVAC, electrical, sheet metal, or elevator firms. Many mechanical firms expand their business by doing both plumbing and heating work since the piping work is similar, although both plumbing and steam fitting trades must be used. With growth, particularly in HVAC work, some mechanical companies acquire a sheet metal division. The sheet metal business differs greatly from piping work, involving fabrication facilities and a much larger detailing section, making successful operation in the face of competition from firms specializing in sheet metal work much more difficult. Most mechanical firms prefer to subcontract this type of work.

Six of the seven biggest mechanical and electrical firms are electrical contractors and electrical work accounts for approximately 10% of the building construction dollar. Perhaps with the exception of elevator work, electrical construction is the most specialized in our industry and the least understood by those not directly involved. The electrician works with almost every other trade. Electrical power, controls, and alarms connect to all other mechanical systems and electrical work is built into every interior and many exterior parts of the building. Even electrical plans differ from others in symbols and graphics, but skill in reading electrical drawings can readily be acquired with a small amount of diligent effort. We need not be electrical engineers to manage properly the electrical portion of our projects.

1.5 MECHANICAL AND ELECTRICAL WORK AND PROJECT MANAGEMENT

If as managers we approach this large part of our assignment with real interest, genuine enthusiasm, and proper appreciation we can do a commendable job of construction management. The more proficient we become in understanding the mechanical and electrical plans, specifications, and contracts, the better job we will do. Through experience on many projects we learn the hard way the necessity for truly controlling all aspects of this type of work, and over the years we learn to place greater emphasis on its control. In subsequent chapters we will study how best specifically to control the mechanical and electrical aspects of the project as they apply to job planning, scheduling, progress, cost, and quality control. We will learn what constitutes a complete plumbing, HVAC, electrical, fire protection, and elevatoring system. We will study how best to have mechanical and electrical work installed and where the pitfalls lie. Finally, perhaps we will understand that mechanical and electrical work is different enough to deserve our special attention and alike enough to apply our construction management principles.

2

The Design Process

2.1 INTRODUCTION

The end product of the design process of sketching, rendering, calculations, and preparing models culminates in plans and specifications used to take bids for construction contracts. As the biggest part of the contract documents, they control the construction period graphically through the plans and descriptively through the specifications. Since our interests lie with the mechanical and electrical design, let us look at those portions of the contract documents (plans and specifications) that cover that work: the plumbing, HVAC, elevator, electrical, and special sections. The special sections cover unusual design areas, including conveyor systems, window washing machinery, turntables, and other unique mechanical and electrical items. We must also carefully consider the addenda, which are the corrections and additions sent to the bidders after the initial release of bidding information and which become part of the contract documents.

2.2 THE DESIGN TEAM

The design team is selected by the architect to include, with the artistic designers and structural engineers, experts and specialists who will produce the mechanical and electrical design in the form of plans and specifications. Most prominent architectural firms contract with specialist firms for these services. As their name implies, architectural engineering firms do offer a complete service using their own staff of experts. However, since it is difficult to have all the talent that would be desirable, the use of elevator and acoustical consultants is common. Again the emphasis is on specialization: special designers, codes, and contractors. While utilizing these specialists, the architect or architectural engineering firm has the responsibility to obtain all the necessary design information; make timely decisions; thoroughly coordinate structural, architectural, mechanical, and electrical designs; and insist on good intercommunication of all disciplines. The decisions the team members make and the development of their individual designs greatly affect each other.

2.3   DESIGN DECISIONS

To better appreciate the interrelations, consider, for example, the factors that affect the HVAC designer: the building location, its height and shape, neighboring buildings, the choice of wall materials, the amount of insulation, the amount and kind of glass. In high-rise buildings the location of mechanical equipment rooms and pipe and duct shafts and the placing of cooling towers are critical decisions affecting all other design. Occasionally, the structural design for mechanical floors is not finalized until the mechanical equipment is purchased, because of differing loading caused by manufacturers' configurations.

Consider too the electrical designer who must determine the electrical loads that result from basic HVAC decisions. Power must be brought to pumps, fans, chillers, and so on.

The structural designer, for his part, must provide for heavy roof tank loads, elevator reactions at the top of the building, or design sufficient resistive force into the elevator pits, particularly those that terminate above basement level. Obviously, the team must work closely together, taking turns in advancing the overall design to conclusion. Where they fail to consider each other properly we have, for example, a supermarket roof failure resulting from concentrated mechanical equipment loads.

2.4   FAST TRACKING

Fast tracking is a term applied to phased design and construction. There is nothing new about this method of designing and building except the term and its increasing use. Figure 2.1 shows the time saving that this method offers compared with the conventional method of completing all plans and specifications before asking for bids. The key to fast tracking is the overlapping of design and construction, making the best use of time. The importance of time can be measured in dollars. In a center city property, consider the following time-related factors when a new building replaces an old one:

1. Loss of rental on the old building during demolition and rebuilding
2. Property taxes for the same period
3. Interest on construction loans
4. Losses incurred due to inflation
5. Income the new property will produce

The mechanical and electrical designers usually do not have to prepare plans and specifications for the first bid package. Initially, this gives them a little more time to finalize design details than the structural designer has, without making assumptions leading to overdesign. However, the mechanical and electrical trades must install sleeves, hanger inserts, and so on, as the structural work proceeds, making inclusion of mechanical and electrical work in the second bid package imperative on most projects. This *does* put a burden on the designers to get their work out on time and does lead to making calculated assumptions in some cases. Admittedly, this may cause oversizing of some pieces of equipment, but this policy is economically sound when the financial advantages of fast tracking are considered. During the inflationary periods we experienced in the 1970s, total building cost savings of 5 to 10% annually were readily

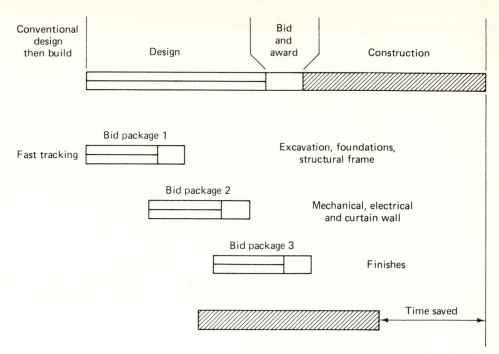

FIGURE 2.1.   Comparison of design and construction methods.

realized with fast tracking. Paying for the excess cost of a few overdesigned fans, motors, or transformers is a small sacrifice to make for the opportunity for such big returns.

2.5 UNDERSTANDING MECHANICAL AND ELECTRICAL DRAWINGS

A good companion to this text to complement the printed word is access to a full set of building drawings. Greater comprehension of mechanical and electrical systems will result from a parallel study of such documents together with this text. We must have confidence in our ability to read all the drawings on a project, not just the architectural and structural sets.

Blueprint-reading courses are certainly of value, but most of us are self-taught in acquiring mechanical and electrical drawing reading skills. Assuming a reasonable ability to read architectural and structural drawings, the neophyte can improve his speed of acquiring this ability by following a few suggestions. Before attempting to understand mechanical and electrical information, architectural and structural drawings should be mastered, at least to the degree that the engineer knows basic locations from the plan views, elevations from the prints, and the structural system from the structural drawings.

In tackling a particular mechanical set such as plumbing, for example, we should think "system." Visualize, for example, getting rid of storm water from roofs, plaza areas, and parking lots. Or imagine a system of waste removal piping, another gravity-flow arrangement. Turn to the plumbing drawings showing schematic diagrams, which are often at the rear of the plumbing set, and study these basic systems shown in three dimensions. Mentally match up the architectural layout with these schematics and you will soon understand where the basic lines will run and how the branch lines connect.

It is essential before going any further to learn the graphical symbols and legends for the plumbing, HVAC, and electrical work. Start with the piping for plumbing, piping and ductwork runs for HVAC, and conduit for electrical. Learn underground and underfloor from exposed piping, and a vertical drop from an upturned riser in ductwork. These runs on the drawings are part of the trunk and branch lines we wish to visualize. The use of small colored-pencil check marks placed on the drawings simplifies following a system, particularly where other systems cross or parallel.

As with architectural and structural plans, engineers present much of their information in schedule form. This is particularly true of HVAC drawings, which show schedules for fans, pumps, diffusers, and air conditioning units. The schedules should be studied so that these pieces of equipment can be located more readily and tied to a specific system.

Mechanical and electrical specifications consist of many pages containing a vast amount of information. They are particularly important in describing the myriad valves, pressure regulators, controls, pumps, motors, and other items that are purchased from manufacturers. They set the standard of quality the designers require in field workmanship and purchased equipment. The specifications should be read and understood as soon as the drawings have been mastered.

2.6 MOCK-UP

A mock-up is the installation of a portion of a building at an early stage, used to evaluate appearance, performance, and practicality. Particularly on larger projects, various requirements for mock-ups greatly improve the likelihood of owner satisfaction. Building construction people consider the mocking up and testing of windows and metal and glass curtain walls very prudent and virtually essential. A mechanical and lighting mock-up of a section of typical office building ceiling or exterior wall, although not essential, is highly recommended, particularly for the larger projects. From the mock-up the owner sees early in the project what he or she will get, and the architects, engineers, and contractors know the owner will not be disappointed later. The mechanical and electrical mock-up also permits minor adjustments for easier installation and large-scale prefabrication, which help reduce costs. Owner and architect approval of the mock-up signals to the contractor that he can release the manufacture and delivery of the mock-up equipment and fabricated items. This helps advance the project.

2.7 RECYCLING

Understandably, new construction appeals greatly to the young architect and engineer. The opportunity to be part of creating something entirely new and (hopefully) meaningful has considerable merit. Considerable satisfaction derives from demolishing dilapidated slum buildings and replacing them with significant new buildings. For most of us, new work certainly constitutes the greatest part of our life's work as building construction industry people. However, a very big part of our industry is devoted to saving and reusing many older excellent homes, factories, hotels, lofts, and office buildings. Many concerned people are sentimentally, emotionally, and practically involved in seeking to retain buildings of architectural and historical worth. New does not mean better. Buildings become a part of people's lives; thus, people often react very keenly to the threat of the loss of buildings they have come to love.

In our capitalistic society property owners are entitled to a fair return on their investment in land and buildings. Our cities contain numerous relatively large old buildings occupying outstanding properties. Their prime location and reasonable management keep them well occupied and financially productive even though they date back to World War I or the Roaring Twenties. In most cases their structural systems are sound, their natural stone exteriors economically irreplaceable, and their energy economics favorable. What they need are revitalized mechanical and electrical systems.

The recycling of our better older buildings challenges our design and construction capabilities. What can we save and what must we rip out? The answers to these questions pose problems to the designer. A reputable specialty contractor can use his broad experience to help advise the mechanical and electrical engineers on what they can still utilize to upgrade the building for another 20 or 30 years of use. In modernizing elevator service we do not dispose of everything and start over. A significant amount is rebuilt and reused.

For the designer it is much more difficult to prepare plans and specifications for recycling than for new work. A survey of the existing building to determine the condition of the various systems makes good sense. The designer faces the dilemma of inadequate plans and specifications which, after bidding, lead to numerous extra charges to the owner or advising the award of cost-plus work to plumbing, HVAC, and electrical contractors. Each architect and owner must make this rather difficult choice of the best method to employ. No matter that alteration or recycling work always costs more per unit than new construction. The end result of recycling work can be just as satisfying as new work as the owner continues to enjoy full space occupancy, good economic returns, and a pleasing new interior appearance to complement the familiar old exterior.

3

Estimating and Bidding

3.1 INTRODUCTION

The estimating department of a large building construction company is a production-oriented operation with a planned work load. One might think that estimating effort concentrates solely on a series of attempts at landing sizable lump-sum contracts—and certainly this is its biggest reason for existence. Estimating services, however, are utilized in many ways during the normal course of activity of a company doing both lump-sum and cost-plus or construction management work. Most estimates are prepared and submitted as a competitive bid to owners requesting lump-sum proposals from two or more contractors. Most public buildings and a large portion of private work is awarded on this basis, the work usually going to the contracting firm that submits the lowest bid. Occasionally, an owner selects a contracting firm first, which then prepares a lump-sum estimate. This method of negotiating a contract from a single lump-sum estimate is fairly common in the construction of expensive custom-built homes for particular clients. A reputation for honest dealing and excellent work is essential to be selected for this type of estimating and contracting.

3.2 NEGOTIATED ESTIMATES

To request lump-sum bids owners must have their architects and engineers prepare a complete set of drawings and specifications. Many owners do not wish to wait a year or more for the entire design before starting their buildings. These owners, thinking in terms of a cost-plus or construction management method of doing business, look for other types of estimating service. The contracting or construction management firm works with its architects and engineers to prepare a series of estimates which culminate in plans, specifications, and costs that satisfy the owner's budget. Owners often initially receive budget estimates from two or three contractors that they are considering for cost-plus work. To be prudent, the choice of contractor should be made on all the qualifications of those being considered; it should not be price oriented in terms of estimates based on sketchy and preliminary information.

3.3 APPROXIMATE ESTIMATES

One further step removed from estimates based on incomplete plans and specifications are approximate estimates, or "guesstimates" where there are no plans at all. To give architects, owners, or developers some concept of costs, quick rules-of-thumb estimates can be made based on so much a room for a motel, so much a bed for a hospital, and so on. More widely used and more reliable are costs per square foot or cubic foot where the complete costs of the building are divided by the gross area of the building. Quick approximate costs for mechanical and electrical work can also readily be obtained. Square-foot costs are available for the plumbing, HVAC, and electrical work for a wide variety of buildings, giving the seeker a fair idea of what that work would cost in the building.

3.4 METHODS OF BIDDING

From an owner's viewpoint the ideal situation exists when the most competent contractor submits the lowest lump-sum bid. *It does happen occasionally.* Private owners improve their chances of success greatly by inviting to bid only those highly qualified, reputable contractors with whom they feel confident in conducting business. The bidders prefer this situation, which puts them in competition with companies of similar qualifications. In public bidding the criteria differ, with the vast majority of contractors in an area "qualifying" for the right to bid, thereby pitting the best qualified contractors against the worst.

Mechanical and electrical contractors bid in essentially two different ways. Most work is bid to a general contracting firm as part of numerous subcontract bids they receive to make up their total bid. The mechanical and electrical contractors interested in a project submit prices to those general contractors that invite them to submit prices and also to any other general contractor whose business they seek. These bids, which combined total as much as 40% of the project, become extremely important to the general contractor. The right combination of attractive plumbing, heating, and electrical subcontract prices can be tremendously helpful in landing the job for the general contractor. Recognizing this facet of bidding, mechanical and electrical contractors will sometimes give a particularly good general contractor an exclusive price which is substantially lower than the broadcast or street price. In giving this special price the mechanical or electrical subcontractor has an understanding with the general contractor that ensures his getting the subcontract work if the general contractor wins the job. Usually, however, these subcontractors submit prices to practically all the general contractors bidding, thereby improving their chances of doing the subcontract work. Whereas the general contractors have only one chance at a project, the subs' opportunities come first from the initial bidding and second from the continued pursuit of the work once it is awarded to a general contractor. Without an exclusive bid they may still be in a very favorable position but cannot be sure of having a subcontract until the general contractor goes through a negotiation period with all of the mechanical and electrical subs that submitted bids. Far too often, bright prospects for a subcontractor on the night the general contractor lands a job evaporate through perfectly ethical competitive price cutting during the subcontract award period that follows. Unfortunately, some contractors not satisfied with honest competition resort to *bid shopping,* a vicious practice in which the subcontractors are unfairly played off against each other and induced into cutting their bid price until all the profit and quality are wrung out of a job.

The more attractive way of bidding for mechanical and electrical subs does not involve a general contractor. Some states and other public bodies request prices separately for plumbing, HVAC, electrical work, and a catchall general contract. Mechanical and electrical contractors lobby in state legislatures for this separate contracting method, claiming it saves the markup the general contractor applies to subcontract prices. Actually, in preparing their bids, general contractors first discount the mechanical and electrical bids by using a reduced figure from the ones received from the subcontractors. Their estimators feel confident that they can "buy out" this work at the lower figure after the general contract has been awarded, and generally do. After this initial cut, which is the part the subs dislike the most, the general contractors add their overhead and profit to the entire material, labor, and subcontract total to cover their contract responsibilities, which include supervision of the mechanical and electrical contracts.

Separate prices are often taken and separate contracts awarded for construction management work. The mechanical and electrical contracting firm should not delude itself that it is free of the general contractor in construction management work. Most construction managers are building contracting firms and have only a slightly less self-centered attitude toward the mechanical and electrical subs on construction management work. Perhaps they may be somewhat less aggressive in seeking price cuts during the negotiation period prior to contract award, but certainly they will demand a first-rate performance in progress and quality control and be tougher about extras than the average separate contract owner.

3.5   AVAILABILITY OF PLANS

Since the mechanical and electrical prices are such an important part of the overall bid, these contractors should be actively encouraged to seek the work. Sufficient copies of the plans and specifications should be procured from the architect and sent to the better subs for estimating purposes. The general contractor can also make these bidding documents available on a well-lighted plan table in his offices. After the bidding period, drawings and specifications should be returned promptly to the architect by the unsuccessful bidders in order to secure the return of deposit money.

3.6   THE BIDDING PERIOD

Owners and architects should choose with care the hour and date for the bidding deadline. They should avoid requesting bids the same week that other owners have attractive projects out for bid. Better competition and better prices will result if they have the attention of the whole industry for their job. The industry will also not appreciate requests for bids on Mondays or following a holiday or at an early hour during the day—any one of which will cause the entire industry to spend night-time and off-time hours on the bid.

The size of the project will dictate the proper amount of time to allow. Depending on the size and complexity of the building two to four weeks is generally sufficient, with three weeks the most popular. The owner and architect should always make sure to allow enough time for estimating because good bids are crucial to the financial success of a project.

3.7   THE DECISION TO BID

The mechanical and electrical contractors make their decision to bid on virtually the same criteria as those of the general contractor. The current work load and the need for an additional contract would be prime considerations. Other factors would be the desirability of the job; its location; the labor and market situation; the availability of estimators to prepare the bid and construction staff to supervise the building; the competition; the type of contract; the quality of the plans and specifications; the owner, particularly where good relations exist; the architectural engineer; and the size of the project. Also, the relations with the general contractor and its business situation at times become very important considerations, and at other times the pressure to bid comes from the owner and architect engineer. Estimating costs money. A construction firm only lands an average of one out of every eight jobs bid. Anything to improve these odds should be considered carefully before deciding on what to bid.

  Mechanical and electrical estimating involves the meticulous measurement and recording of a vast number of items, ranging from ½-in. elbows to large pieces of switchgear or air conditioning equipment. Estimating is a complete study of its own, but perhaps some insight into mechanical and electrical estimating will give the reader a better feeling for this important part of contracting.

3.8 THE STUDY

Mechanical and electrical work consists of systems. For example, in plumbing we have the hot and cold water supply, the sewage removal and venting systems (the air portion that makes it work), the rainwater conductors, and other individual systems. The estimator starts his study of the drawings by learning the various building systems. To do this he would first review the architectural drawings to comprehend the size and spaces of the building and the plan location of specific areas such as toilet rooms and mechanical areas. Next he would look over the structural drawings to understand how his item of work will have to meet the structural requirements, and vice versa. Next he would turn to the trade schematic drawings and gain an understanding of where each system starts, its function, and the areas it serves. With this basic study his overall comprehension of what systems the building contains and how they must fit into and serve the building will be greatly facilitated.

3.9 THE TAKEOFF

Systems are like trees. Starting at the source of each system the estimator would follow the trunk, the principal branches, the smaller branches, and finally the system extremities, taking off smaller and smaller pipes and ducts as he goes along. There are times he must visualize all that is involved based on a plan view of a fixture location, for example. The drawings would show location and pipe sizes, but in the case of plumbing, the estimator will often prepare a simple sketch while estimating to develop all the plumbing fittings needed at that location.

  A lot of estimating mechanical and electrical piping and conduit involves measuring lengths of runs and counting fittings. Forms are utilized for this purpose to simplify the process. Neat check marks using a colored pencil leave a trail on the drawings the estimator can observe to check to make sure the takeoff is complete. Far too often the low bidder is someone whose takeoff is incomplete in one manner or another, including omission of an entire item or leaving out an entire page when totaling up the takeoff sheets. A good estimator exudes mastery of organization—his estimates from initial takeoff to final price are a clearly indicated route.

  In ductwork there are many special fittings, such as elbows, square to round, S-shaped, large to smaller rectangular, and so on. These can be carefully measured and computed in pounds, the unit used for ductwork estimating. This would be somewhat tedious. Estimators utilize quicker and more efficient methods and in this case equivalents of straight runs of duct are used. For example, a 90° elbow might be taken off as an additional three feet of that size ductwork measured along the centerline of the duct.

3.10   EQUIPMENT   A substantial portion of a mechanical and electrical bid would originate from the fixtures and equipment. In plumbing, lavatories, water closets, bathtubs, pumps, tanks, and so on, make up this part of the project. In HVAC, fans, chillers, motor controls, induction units, diffusers, air handlers, coils, cooling towers, pumps, and so on, are in this category. Switchgear, panel boxes, light fixtures, transformers, and so on, make up the electrical part. Many suppliers are involved and their prices can help the mechanical and electrical contractor to win the bid.

The estimator will tabulate all the fixtures and equipment to be priced to assist him. Schedules of the principal pieces of equipment are listed in schedules on the drawings, particularly for HVAC and electrical work. All manner of fixtures, fans, heat exchangers, pumps, transformers, and many other important items are listed, with sizes, capacities, and locations. With this tabulated information the estimator can locate a particular item in a specific location.

3.11   SUBCONTRACTS   Electrical contractors sublet very little, plumbers some, and HVAC a substantial amount. For example, the HVAC contractor might sublet the furnishing and installation of the cooling tower, the fabrication and installation of the ductwork and diffusers, all pipe covering, and air conditioning controls. In preparing the bid, specialty subs submit prices in competition for these special types of work and the mechanical and electrical contractors use the price for which they believe they can eventually purchase this part of the work.

3.12   PRICING   The estimating takeoff produces totals of a great many kinds of labor and materials. An experienced person of sound judgment must apply a price for each one of these work items. This responsibility often falls to the boss or the chief estimator. Based on company records of past performances and utilizing good judgment, the current price for the bid is selected. This is where good cost accounting on past work and work in progress pays big dividends.

3.13   THE AFTERMATH   The last days and hours before a bid goes in are somewhat hectic. Bids that are late should not be accepted, primarily to protect the integrity of those receiving the bids. The bidders in public work must conform to strict procedures generally established by the bid form. In private work the bidders occasionally deviate from the requested form and procedure if they believe they can interest the owners and architects in their ideas. Usually, an alternative is offered which involves a cost saving. This bears out the bidder's secondary hope (its first hope is to be low and be awarded the job) of being called in to discuss the bid. Bidders respond with joy at the sales opportunity to express their interest in the work, their ideas of how the work can be done, and particularly thoughts on savings, a subject dear to the hearts of owners. The latter approach of possible savings coupled with evidence to indicate a good on-the-job organization often convinces the owner, architect, and general contractor and leads to a contract.

Contracts

4.1 CONTRACT LAW Because of the complex nature of the design and construction process for modern buildings, contracts are essential to define clearly what each party must do, its responsibilities, and the conditions under which it must perform. Contract law derives from our basic English-oriented legal system. Construction contract law merits further separate study, but here we only cover essentials, to establish a proper understanding of its importance.

The conditions required for a valid contract are:

1. Mutual agreement
2. Consideration
3. Lawful object
4. Capacity
5. Genuine intent

Very briefly this means:

1. Both parties must agree on the entire contract.
2. Something of value must be exchanged between the parties, if only $1.00.
3. What the contract requires is lawful (no contracts to harm someone or to do anything illegal).

4. The contracting parties are of age and have the mental capacity to act responsibly.
5. The parties must be acting so as not to misrepresent, defraud, or threaten each other.

Although there are valid verbal contracts and times when unsigned contracts become effective, the signatures of both parties constitute proper execution of the contract.

The American Institute of Architects (AIA) has devoted considerable legal time and effort to develop standard contract forms for the various kinds of contracts we discuss in this chapter. These contracts are widely used because of their clarity, fairness, and good performance record. Although more favorable to architects, they are far better for most purposes than long-drawn-out corporation contracts that contain selfish clauses that only create legal bickering.

To better understand contract requirements and avoid legal entanglements, the construction industry conducts numerous legal-awareness seminars. Participants learn the full legal significance of the pertinent clauses and how to protect themselves from legal action arising out of the specific parts of the various contracts. In the larger cities some law firms specialize in the construction industry, giving legal advice to prevent difficulties to their clients or to pursue claims arising out of contracts.

4.2 CONTRACTS REQUIRED

For the design of a building the owner signs a contract with the architect for plans, specifications, and other services, such as accounting and "observance." Somewhat similar to construction contracts, this can be lump sum, cost plus, or a time card hourly basis contract. The architect, in turn, signs contracts with supporting plumbing, HVAC, electrical, elevatoring, and acoustical engineers and consultants who contribute to the design. When an owner contracts with an architectural engineering firm that has a complete engineering staff, such secondary contracts are unnecessary.

The owner also contracts for the complete construction project in three basic ways:

1. With a general contractor to do the entire project
2. With a general contractor, the plumber, the electrician, and HVAC contractor, called separate contracting
3. Under 20 to 30 separate contracts, with or without a construction manager, also called separate contracting

He would also sign separate contracts for various testing and inspection services with firms specializing in this type of work.

4.3 THE OWNER'S OPTIONS

The conditions under which the owner builds, and his or her knowledge of, and attitude toward construction contracts will largely dictate the owner's choice of contract. Unlike his counterpart in governmental work, the private builder can choose from many different forms of contract. Public organizations must conform to the local law controlling construction contracting, which usually dictates a lump-sum contract. In some states the mechanical and electrical interests have had laws passed requiring the separate contract system listed above, where work is let lump sum in one general and three or more mechanical, elevator, and electrical contracts. The public official, although much more inhibited than his private counterpart, still has some opportunity of contract selection. The U.S. government frequently approves cost-plus contracts in times of national emergency. During the past decade the General Services Administration has utilized construction management for hundreds of millions of dollars of work.

4.4   THE LUMP-SUM CONCEPT   The popularity of the lump-sum contract can easily be appreciated. People want to know what something will cost before they agree to pay for it. An owner who has gone to the time, effort, and expense of having an architect prepare plans and specifications can easily proceed to request and receive bids for the construction work. Using qualified contractors, possessing a good set of plans and specifications, assuming normal market conditions and a design familiar to the contractors, the owner is in a favorable position to receive good competitive firm prices for a project.

The conditions required for lump-sum bidding bear examination. The lump-sum contracting method requires a complete set of drawings before prices are taken. Often a year is consumed during this design process. The year spent in design costs the owner 5 to 10% of the construction dollar based on the average national inflation rate. The quality of documents is also important. Low-quality plans and specifications can cause two adverse lump-sum conditions:

1. Inflated prices containing hidden contingencies to cover situations that worry the contractors
2. Low prices as the basis of a contract that later results in numerous costly extras

The degree of competition also becomes a very important consideration. At a midwestern airport one general contractor had all the work—more than he could properly handle. Lump-sum bidding on additional fingers and satellite lounges produced two bids, both very high. Competition was lacking because the on-site contractor dominated the scene. Another reputable general contractor was employed on a cost-plus contract, resulting in large savings and a strong performance.

Finally, the conditions under which a project will be built greatly affect price. Contractors must place large contingency sums in their bids to cover the risks of an unusual design or unusual working conditions, such as rebuilding a store without shutting down business. If these sums are not needed, the contractor pockets them as profit.

Human nature being what it is, lump-sum contracting can readily produce an adversary relationship between the owner and architect on one side and the general contractor and his subs on the other. Mediocre contract documents offer happy hunting for an aggressive contractor looking for extras. The lump-sum contractor first considers profit, with the quality of his work and reputation of secondary importance. When the lump-sum contractor assumes all the risks, his interests are not necessarily compatible with the owner, particularly as to progress. With highly productive labor in short supply, he will sacrifice progress rather than properly staff the work. Faced with poor weather, he will stop work instead of spending money to provide weather protection. Generally, his selfish interests dominate.

We generally think of competition and bidding when we refer to lump-sum contracting; however, negotiating a lump-sum price can have considerable merit. When you want the best home builder in town to build your house, you negotiate with him. The process consists of requesting an estimate based on your plans and specifications and then proceeding with discussions until reaching agreement on price and contract form. In a tight market when all the HVAC contractors have lots of work and HVAC prices are coming in high, it sometimes works well to negotiate with one company having a good reputation and a sound organization. By selecting a worthy contractor who appreciates saving the cost of competitive estimating, he can be expected to submit an estimate containing a reasonable profit for himself and a fair price for the owner. After all—estimating lump sum produces only one job in eight or ten attempts. Having been selected, the fortunate contractor generally responds by giving a superior performance in quality and progress.

Even when an owner awards a cost-plus or construction management contract, 80 to 85% of the work ends up under a lump-sum specialty contract. This occurs because the general contractor buys out most of the subcontracts for the entire project, including the electrical, plumbing, HVAC, elevator, architectural, and structural trades. The contractor clarifies the architects, and engineers trade documents and spells out who will provide the necessary job overhead items, such as hoisting, cleaning, and so on. The cost-plus contractor convinces the owner that with his construction experience he can buy better, control more, and get the owner the most for his money. He sells his management ability.

4.5   THE COST-PLUS CONCEPT

For those not familiar with cost-plus contracting and particularly for those who are psychologically adverse to the cost-plus contracting idea, the great amount of building construction performed under cost-plus contracts would probably be surprising. Reading down the list of the nation's leading building firms, one comes across the names of numerous contractors who have built strong organizations based on their ability to satisfy clients, improve their reputations, increase their business, and make money while successfully executing both lump-sum and cost-plus contracts. Many of these firms consistently obtain more than 50% of their work on a cost-plus basis.

Owners sign cost-plus contracts with companies with excellent reputations, strong construction and management organizations, depth of cost-plus experience, and top financial standing. The size of the job need not be a deterrent to letting a cost plus contract, although admittedly the bigger and more difficult the project, the easier it is to convince the owner that the cost-plus way will work best.

Utilizing a cost-plus contract, the owner pays all of the direct and indirect on-site costs, which are clearly spelled out in the section defining the cost of the work. In addition to all labor and materials, the costs include equipment rental, supplies, temporary facilities, salary of job personnel, and other job overhead items. Main office expenses, particularly the salaries of the company's officers, are not generally accepted as legitimate costs of the work, and this is spelled out in a section defining costs not to be reimbursed.

The kind of cost-plus contract the owner chooses will depend on the conditions existing at the time the project will be constructed. Essentially, the objectives of the owner determine the type of cost-plus contract selected. We will discuss them starting with the cost-plus-percentage fee type, which contains the greatest risk to the owner while offering maximum control, and proceed to the more financially conservative types.

Below are reproduced Articles 4, "Cost of the Work," and Article 5, "Costs Not to Be Reimbursed," from a typical cost-plus contract.*

ARTICLE 4.   COST OF THE WORK.   The Owner shall pay to the Contractor, or directly, the cost of the Work, which shall include costs necessarily incurred in connection therewith (except those costs specified in this Agreement to be borne by the Contractor); such costs to include the following items:

(a)   Wages of labor directly on the Contractor's payroll.
(b)   Salaries of Contractor's employees stationed at the field office.

---

* Articles 4 and 5 are reprinted by permission of Turner Construction Company.

Expeditors, General Superintendents and Supervising Accountants, whose part-time services are required for the Work, shall be considered as stationed at the field office and their salaries paid for such part of their time as may be devoted to the Work.

When the preparation or analysis of schedules, material lists, shop drawings, working details, periodic cost studies and similar services necessary to define the Work, and control its cost and progress, are performed by employees located in the Contractor's main or regularly established branch offices, the salaries of employees directly engaged in such duties shall be charged for such part of their time as may be devoted to the Work, together with an allowance equal to fifty percentum (50%) of such salaries to cover office expenses incident thereto.

(c)   Payroll taxes and contributions for Federal Old Age Benefits, Unemployment Insurance or other employee benefits required by law, Welfare Funds and Pension Funds with respect to salaries and wages chargeable to the cost of the Work.

(d)   Traveling expenses of representatives of the Contractor incurred in the discharge of duties connected with the Work.

(e)   Expenses incurred, with the approval of the Architect, for the transportation and board and lodging of the personnel required for the performance of the Work, in case it is necessary to secure any of such personnel at a distance from the place in which the Work is located.

(f)   All materials, whether for permanent or temporary use, including cost of inspection, testing, transportation, storage and handling.

(g)   Supplies of whatever nature; tools and equipment required for the work; cost of water, power and fuel; cost of telephone service, telegrams, postage, blueprints, photographs, field office supplies, stationery, and similar items; cost of surveys, soil and other investigations; protection and altering of public utilities; royalties; protection and repairs of adjoining property; rental of property for storage, job office or other purposes; Federal, State, Municipal or other taxes based upon labor performed and material furnished, including among others the Sales and Use tax, Gross Receipts and Occupational or License tax; fees for permits and licenses; surety bond premiums; premiums on insurance carried by the Contractor; cost of discharging liens; attorneys fees and expenses growing out of the conduct of the Work.

(h)   All subcontracts let in connection with the Work.

(i)   Rentals of tools and equipment or parts thereof. Transportation of said tools or equipment, costs of loading and unloading, cost of installing, dismantling and removal thereof and repairs and replacements made necessary by use on the Work. Rental charges for tools and equipment belonging to the Contractor shall be in accordance with a schedule of rentals approved by the Architect.

(j)   Losses, expenses or damages, to the extent not compensated by insurance or otherwise (including settlements made with the approval of the Owner), except to the extent any such loss or expense is caused by the failure on the part of the corporate officers of the Contractor, or its other representatives charged with the supervision or direction of the operation as a whole, to exercise good faith or the standard of care normally exercised in the conduct of the business of the Contractor.

Wherever the Contract Documents state that the Contractor shall perform any work or incur any expense, it shall be understood to mean, in the absence of specific language to the contrary in this Agreement, that the cost thereof shall be included in the cost of the Work payable by the Owner.

All charges for the services of the Architect are to be paid directly by the Owner and shall not be considered a part of the cost of the Work.

ARTICLE 5.   COSTS NOT TO BE REIMBURSED.   The cost of the Work as defined in Article 4 shall not include the following costs which shall be borne by the Contractor:

(a)   Salaries of the officers of the Contractor.

(b)   Services of Contractor's regularly established Main or Branch Office Purchasing and Estimating Departments.

(c)   Administrative or general overhead expense of the Contractor's Main or regularly established Branch Offices, including salaries of all persons located in such offices whose time is devoted to the general conduct of the Contractor's business.

(d)   Interest on capital employed, unless payments are not made in accordance with Article 8.

4.6   COST PLUS PERCENTAGE FEE

Prior to World War II the U.S. Navy needed a massive accelerating effort to build land-based facilities in the Pacific Ocean. Starting in Hawaii with three joint-venture companies, Hawaiian Dredging, Raymond Concrete Pile Company, and Turner Construction Company, and $15,000,000 in contracts, work spread all over the Pacific and eventually amounted to $400,000,000, close to $3 billion in 1978 dollars. The joint venture expanded to eight companies called Contractors Pacific Naval Air Bases, and the history of their work was published in a book called *Builders for Battle*. All of the work was performed on a cost-plus-percentage basis, with the contractors responsible for their own inspection and quality control. All of the best construction cost control, purchasing, personnel, engineering, management, and construction procedures were utilized particularly prior to the attack on Pearl Harbor. A giant step in the war effort resulted through a contract that fit the need. Large projects of this type often have a descending percentage of fees. Rapid industrial expansion can be handled this way, with the fee dropping, say, from 5% to 3% as the project expands and goes from $20,000,000 to $60,000,000. However, use of this type of contract also applies to all sizes of jobs, such as rewiring part of a store destroyed by fire or quickly replacing the HVAC portion of a larger complex. Urgency and difficulty in determining the eventual extent of the work (how much of the store building will have to be rebuilt?) are prime reasons for choosing the percentage fee contract. The use of this contract should not preclude, wherever possible, estimating, labor cost accounting, good purchasing practice, scheduling, and control of labor. It is even essential at times for the general contractor to impress on the owner, who exercises so much control, that the conduct of the owner's staff can be helpful in holding down costs and eliminating waste, particularly where overtime programs are being administered.

4.7   COST PLUS FIXED FEE

The fixed-fee contract appeals to owners by offering the protection of a limit on the profit and off-site overhead the contractor will receive. Theoretically, with his potential earnings established by the fixed fee, the contractor should concentrate on speedy contract completion consistent with high quality standards. The amount of the fixed fee is based on a defined construction program. There are usually preliminary plans indicating the size and scope of the project from which a rough estimate of the building can be made, which goes a long way toward establishing the fixed fee. The contractor also sets his request for fee based on the complexity of the work and the length of the construction period. Fixed fees are set by dollar volume, the number of contractor's staff required, and the length of time they must work on the project. Remember, contractors offer management ability as represented by experienced personnel. Some people call these experienced construction personnel *money-makers*.

Some owners add complexity and dollars to their project without increasing the fixed fee. They do this by writing the cost-plus contract with a clause stating that the fixed fee can be increased only if the square footage of the project becomes greater. The owner, on the other hand, depends on the integrity and skill of the contractor to control costs. Aside from maintaining his reputation and increasing his chances for another contract, no other strong incentive exists for the contractor.

4.8 GUARANTEED MAXIMUMS AND SAVINGS SHARING

A large percentage of the cost-plus contracting mentioned earlier is performed under the cost plus fixed fee guaranteed maximum saving sharing contract. Many view this as the ideal method of building contracting, since it has attractive features for both parties. The owner still only pays the defined costs, the fee is fixed, and further protection extends to the owner with the establishing of a maximum amount to cover costs and fee. For budgeting purposes, particularly to receive board of directors' approval for the project or to obtain construction and long-term mortgage funding, the owner has a definite figure that limits the amount to be paid for a particular project. Since the guaranteed maximum figure limits the sum of the costs plus the fixed fee, excessive costs reduce the fixed fee paid to the contractor.

Knowing that he or she pays only the actual costs, satisfied that the fee represents a fair payment for the services the contractor will render including a reasonable profit, the owner now has the assurance that the ultimate expenditure has been determined. Many contracts are concluded on this basis and the project progresses happily to completion for all.

However, all through the sales and negotiating period the contractor has emphasized that with this type of contract the owner, architect, and contractor will work as a team to achieve the maximum quality in the shortest time while operating in the most efficient and economical manner. As an incentive to both the owner and the contractor to operate efficiently, work together cooperatively, and produce savings, they agree that if the project costs less than the guaranteed maximum, 75% of these savings, for example, will be retained by the owner and 25% of them will be added to the contractors fixed fee. Savings can be generated by good labor cost control, purchasing of subcontracts and materials, changes in specifications that save money without jeopardizing quality, and even by unusually mild or sunny weather. The architect should also recommend the use of this contract, since he enjoys the benefit of having the owner and contractor working harmoniously together.

Establishing the guaranteed maximum usually occurs before work starts, but there are cases where the poor state of the plans and specifications dictate the establishment of the guaranteed maximum at a later stage when the plans and specifications are more acceptable for this purpose. Many owners willingly start their projects based on reliable budget estimates under a fixed-fee contract, which later converts to the full guaranteed maximum savings sharing type.

4.9 COST PLUS AND MECHANICAL-ELECTRICAL WORK

There are many fine old buildings on choice properties in the United States that are being or should be recycled. We tend to notice the new buildings and ignore the old. The Woolworth Building in New York City, although a designated landmark, still operates profitably while enjoying a high rate of occupancy. Most recycling centers on modernizing plumbing, installing or improving the air conditioning, increasing light intensities, and redoing the elevators. This work, involving both retaining and replacing whole systems or parts of systems, lends itself readily to some form of cost plus contract.

When mechanical and electrical contractors work cost plus on these buildings, they provide a significant additional service not offered by the general contractors. All tools, vices, pipe benders, cutting and welding machines, band saws, and other tools and machines of the trade are excluded from the cost of the work. Mechanical and electrical contractors establish their fixed fees to cover the very real cost of supplying these items.

When deciding which mechanical and electrical companies to approach for cost-plus work, only those contractors who have a demonstrated record for good productivity and labor relations should be considered. Firms with recognized superior engineering staffs should receive preference since the experience of contracting firms is valuable to the designers in making recycling decisions. Finally, since the owner depends on reliable budget estimates, only those firms known to be capable of producing dependable figures should receive cost-plus consideration.

4.10 COST PLUS AND FAST TRACKING

The cost-plus contract gives the project team a great deal of flexibility of operation. Any one of the three (owner, architect, general contractor) can propose an action such as changing to high-early-strength concrete to speed construction. The owners approval triggers immediate response. Construction can start with excavation alone as a simple design package of plans and specifications suitable for bidding. In like manner, with preliminary steel drawings to use for bidding, structural steel can be awarded on a price per ton basis so that mill material can be ordered and detailing started. In this way the architect releases the plans and specifications in bid packages in time for the general contractor to take bids and have the material detailed, fabricated, and delivered—all to meet the project schedule (Fig. 4.1). Although it is not imperative, architects prefer to assemble documents in groups. In the first grouping they usually release excavation, caissons, concrete foundations, and structural steel for construction at the same time. Of course, there should be enough mechanical and electrical information included so that pipe sleeves and other foundation penetrations can be formed in the foundation wall.

This procedure, now known as fast tracking, has been used extensively during this century and undoubtedly was used by many of the great builders in the past. The greatest value of fast tracking derives from the efficient use of time. The architect does

FIGURE 4.1. Comparison between traditional and fast-track concepts.

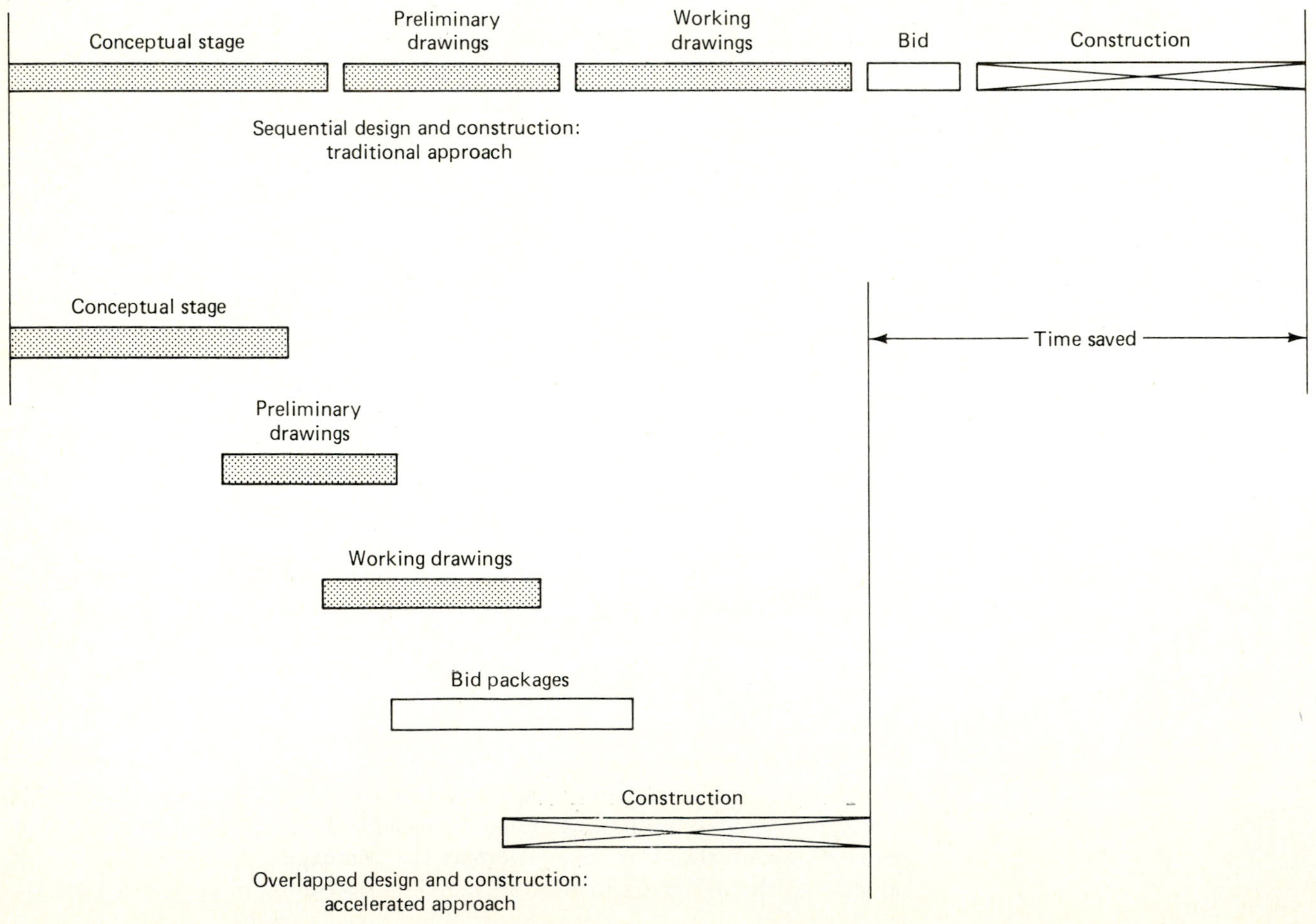

not *design faster* nor the contractor *reduce construction time,* but by overlapping the two, perhaps as much as one-third of the total design and building time can be saved. As mentioned under the lump-sum contract discussion, in this age of inflation an earlier start on construction saves construction dollars. Reducing the overall design–construction time with fast tracking also permits the owner to occupy a property considerably earlier, thereby increasing investment earnings.

4.11 CONSTRUCTION MANAGEMENT

Up to now in our discussion of lump-sum and cost-plus contracts, we have dealt with legal instruments in which the contractor agrees to complete a building in accordance with the plans and specifications. He contracts *to build* a building. In construction management he contracts *to see* that a building gets built. Acting as an agent for the owner, he is paid a fee for the professional services which benefit the owner. The owner, in turn, accepts legal responsibility for the actions of the agent. As with all contract law, the agent–principal relationships are clearly spelled out and may be of interest for further study.

Contract management extends the trend of less and less direct involvement in the construction process by the general contracting firm. Where general contractors formerly poured their own concrete, placed their own rebars, and laid their own brick, they now subcontract for all phases of the work except hoisting, cleaning, safety, and supervision. This trend continues in construction management, where the construction manager operates one step further away from direct performance. In some cases a lump-sum general contractor even works under the supervision of the construction manager. At other times the owner signs 20 to 30 separate contracts for the work, which are all administered by the construction manager.

Unfortunately, there are innumerable interpretations of construction management. Firms doing contracting work agree to a "construction management contract" to humor an owner and get the job, knowing full well that they really have a form of cost-plus contract. The Associated General Contractors of America in March 1972 spelled out their interpretation of construction management in their "Construction Management Guidelines" (see Appendix A). The real significance of the Guidelines derives from directing attention to the services required of the construction manager during the various stages of the project. *Controlling the budget* from start to completion and performing *value engineering* services during the various phases of design to get the maximum construction per dollar do set construction management somewhat apart. The astute owner will put a construction manager to work at the earliest possible time, asking for budget figures, savings suggestions, purchasing expertise, and construction advice.

With construction management the mechanical and electrical work are usually lump-sum contracts. Prior to their award the construction manager will probably request mechanical and electrical budget assistance. This very valuable assistance by the mechanical and electrical contractors should be clearly recognized by the construction manager, the owners, and the architects. Those performing these services without payment should receive special consideration when lump-sum plumbing, HVAC, and electrical contracts are awarded.

Scheduling

5.1 CONTROL OF TIME   In the construction industry our clients expect a good-quality building built in a reasonable time at an economical cost. We say they want their job "good, fast, and cheap" and kiddingly wonder which two of the three they expect to get. Occasionally, everything blends perfectly and we do succeed in reaching that splendid plateau. More often, when speed becomes paramount, costs accelerate and quality suffers. Completing a construction project when required depends on the proper use of available time. Progress control (a subject we will discuss later) and scheduling go hand in glove. Within the confines of straight-time working hours and the local qualified labor supply, efficient management and high productivity are limited as to what they can accomplish. If sufficient time for the work has not been scheduled, either a delay in completion or extra expenditure on overtime will result. Skillful scheduling allots the proper amount of time for every part of the project program. When properly done it starts with the owner's planning, includes design time, and ends with construction.

Scheduling controls time. Time can be very important as a *fixed date,* such as the opening of a store, starting up a manufacturing plant, or getting a stadium ready for the first game. It can also be financially critical in meeting commitments for moving out of existing space, and starting the flow of rents in a new building. The amount of time consumed or the *duration* of a project critically affects the amount of money spent on construction funding interest, taxes, and income lost on property in preparation for rental. Owners recognize the importance of time by stating in their construction contracts that *time is of the essence,* a tough legal statement. The careful expenditure of the allotted time requires skillful project scheduling.

5.2 SCHEDULING ATTITUDES

General contractors and construction managers develop most project schedules. As a result, their lack of understanding and appreciation for the importance of mechanical and electrical work shows up in their schedules. The most widely used schedule is the *bar* or *Gantt chart* (Fig. 5.1). Most of these show plumbing, HVAC, and electrical work as long bars that start when sleeves are needed for the foundation walls and underground piping for the basement slabs, and end on the last day before occupancy. This type of scheduling, although frequently used, contributes very little to good project management. It merely shows that the general contractor somehow will get the mechanical and electrical work done with the rest of the work (see Fig. 5.1).

Office buildings, hospitals, schools, and similar structures have most of their mechanical and electrical work out of sight enclosed in shafts or above hung ceilings. This requires proper trade sequencing and scheduling of shaft and toilet room work so that walls and finished ceilings can proceed in an orderly and efficient manner. Other areas, such as kitchens and cafeterias, mechanical equipment rooms, computer spaces, and fancy lobbies, because of their mechanical and electrical complexity, require special attention from the scheduler. Elevator programs, because of their importance in providing temporary service during the construction period and essential service upon occupancy, deserve special care and skill on the part of the scheduler.

FIGURE 5.1. Bar chart.

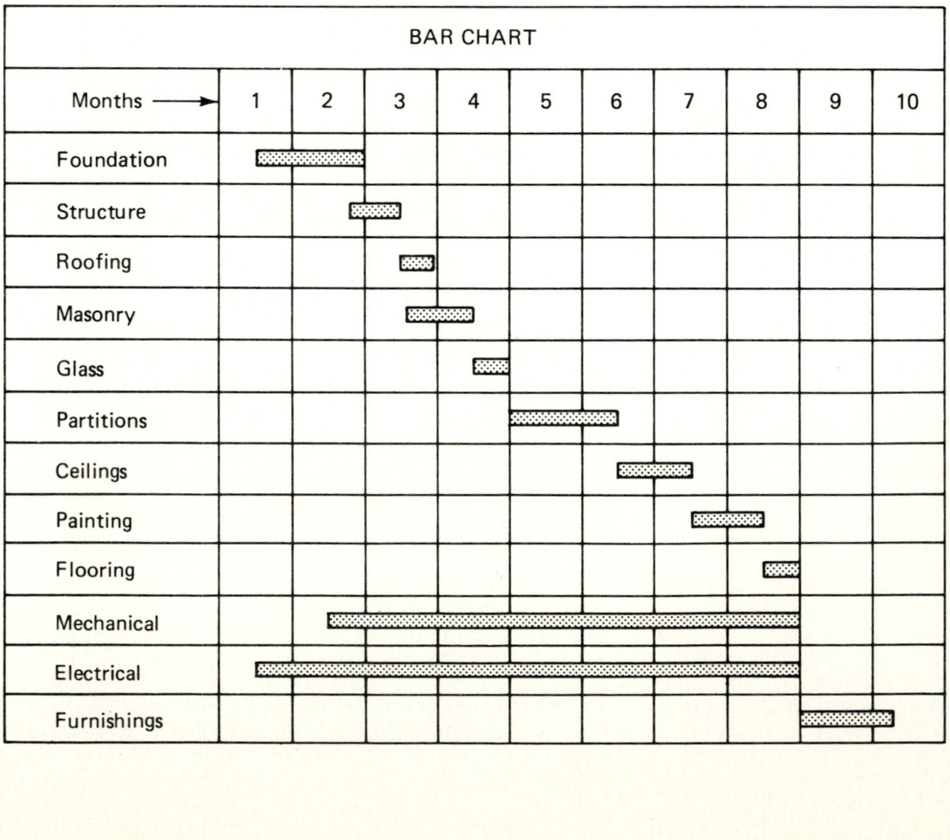

BAR CHART										
Months →	1	2	3	4	5	6	7	8	9	10
Foundation	▦	▦								
Structure		▦								
Roofing			▦							
Masonry			▦	▦						
Glass				▦						
Partitions					▦	▦				
Ceilings						▦	▦			
Painting							▦	▦		
Flooring								▦		
Mechanical			▦	▦	▦	▦	▦	▦		
Electrical		▦	▦	▦	▦	▦	▦	▦		
Furnishings									▦	▦

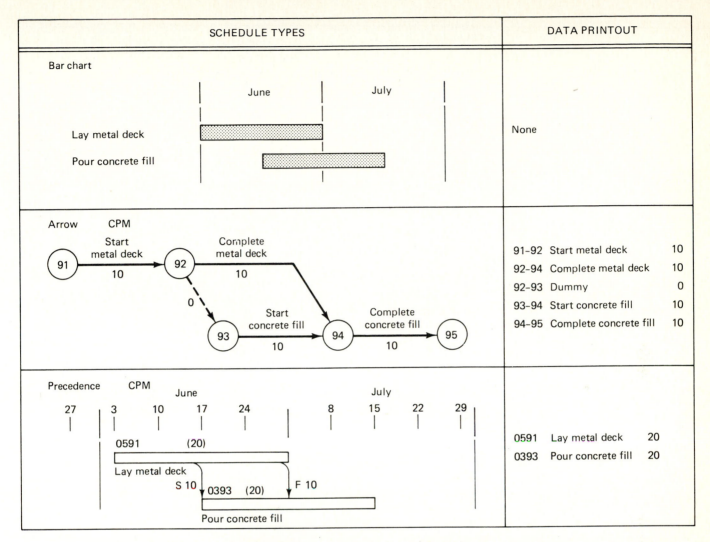

SCHEDULE TYPES	DATA PRINTOUT

Bar chart

	June	July	
Lay metal deck	▓▓▓▓▓▓▓		None
Pour concrete fill		▓▓▓▓▓▓▓	

Arrow   CPM

91–92 Start metal deck	10
92–94 Complete metal deck	10
92–93 Dummy	0
93–94 Start concrete fill	10
94–95 Complete concrete fill	10

Precedence   CPM

0591 Lay metal deck	20
0393 Pour concrete fill	20

FIGURE 5.2. Schedule types. (Courtesy of Turner Construction Company.)

5.3 USE OF SCHEDULES

For scheduling purposes, a building project should be considered as one all-inclusive program which includes the owner deciding the building program objectives, the architect conceiving the building that meets the owner's requirements, and the contractor and his subcontractors executing the developed drawings and specifications. Too often the owner and architect fritter away months of valuable time and then try to recoup this loss by demanding a tight schedule for the construction phase. An important New York City civic project spent incredibly large sums on overtime to recover from this kind of gross misuse of decision and design time. Comprehensive scheduling, which includes time allotted for the conceptual, preliminary, and development design as well as procurement and field construction, usually produces big dividends in time and money saved.

This early scheduling should allow *time* for the owner to decide his or her needs and communicate them to the architect, *time* for environmental impact statements and building department review, *time* for creation of ideas and their development by the architect, *time* for choosing materials and approvals, and *time* for bidding. Where a construction manager provides early services such as construction feasibility studies and budget estimates, he should also be assigned the overall scheduling responsibility. When dates are established for completion of working drawings and specifications, the balance of scheduling information can readily be determined. This would include the bidding period, contract award date, shop drawing preparation and approval time, the period required for fabrication and delivery, and finally the construction erection duration.

5.4 COMPREHENSIVE SCHEDULING

Whether or not the best use of time has been made prior to start of field work, all construction activities should be carefully scheduled for the time remaining. A skillfully prepared schedule properly used helps ensure that each subcontractor will start and finish on time, that all materials and equipment will be delivered when required, and that specific areas or facilities are completed when needed. The initial schedule often originates with the estimator during the estimating and bidding period. His schedule essentially determines project duration. When the project becomes a reality the construction scheduler develops a complete comprehensive schedule which the construction managers can use to control progress.

Too frequently, scheduling stops after the preparation of the *project schedule.* Scheduling should start with this overall schedule and develop from there. A large contractor building a nuclear plant, for example, has a *five-year* schedule, a *three-week* schedule, and a *one-week* schedule. The overall schedule is a project management tool and should be revised periodically if events make the original schedule unrealistic. We do not throw a road map away because we run into a detour. Neither should scheduling be abandoned because of an unexpected delay. The original schedule can always be filed away as part of the project's records, to be used later to determine facts concerning delays.

Sound construction management will insist on and encourage the development of numerous area and interim schedules. The assistant superintendent in charge of structural work should be directed to prepare a schedule for the foundation and superstructure part of the project. The supervisor of mechanical work should schedule in detail piping and ductwork in shafts, work in mechanical equipment rooms, the condenser water and cooling tower systems, and mechanical requirements for hung-ceiling areas. He should also contribute to special area schedules, such as kitchens and cafeterias, lobbies, and executive offices. The young engineer directed to prepare a detailed schedule will know the requirements of that part of the job much better and as a result should do a superior job of supervision. At the same time, the engineer will be acquiring skill as a scheduler, which will improve his or her ability as a project manager.

5.5 INPUT

In developing a schedule the scheduler uses numerous kinds of input. Basically, all field construction scheduling depends on when an operation can start, whether the necessary material and equipment are available, and the number of *worker-days* required for field work. Sometimes the *quantity of work* is more meaningful in determining an operations duration. In the case of excavation we would think in terms of *cubic yards of earth or rock* to be excavated per day. In high-rise work we schedule in terms of *stories:* for example, one tier (two stories) of structural steel every six days or a floor of branch ductwork per week. While assembling the input data we should use whatever method will give us the most accurate duration of time.

During the period when subcontract bids are being taken, much valuable schedule information can be obtained from interested bidders. In addition to the time required for installation, questions can be asked concerning equipment to be used, construction methods, and crew sizes.

Having the subcontractor state the time required to perform his contract, works particularly well for excavation and structural steel, where their part of the work does not depend on others. In mechanical work this would apply to cooling towers, kitchen equipment, conveyor systems, elevators and moving stairs, and refrigeration machines. Where mechanical and electrical progress depends on others, the mechanical and electrical input must be obtained and worked into the combined schedule.

The purchasing process leading to the award of subcontracts should also determine initial delivery dates and off-site production rates. The latter becomes very important on large buildings where the arrival of the first light fixture is no more important than the quantity of fixtures to be received each week. The delivery dates for switchgear, boilers, cooling towers, and refrigeration machines, on the other hand, would determine specific starting dates for that portion of the work and often have an important impact on the overall schedule.

5.6 REQUIREMENTS FOR EFFECTIVE SCHEDULING

General contractors visualize the construction schedule in terms of four objectives:

1. Excavation and foundation
2. Superstructure
3. Enclosing the building with exterior walls and roof
4. Interior finish work

The completion of each of the first three phases makes possible the start of many following operations. While the scheduler determines durations from worker-days and other data, he also decides the appropriate time for the next operation to begin. For example, in a high-rise structure the metal deck follows structural steel erection, and the electrician in turn installs trench header ducts and access devices immediately behind the metal deck contractor. When these two have completed their work on a floor, the all-important concrete floor can be poured. This dependency of one operation on another exercises considerable control over the entire schedule. Not only does it control the start of operations, but their completion. No matter what happens, the metal deck cannot be completed until after the last steel is in place and the steel contractor's equipment has been removed.

As you can see, scheduling requires sound logic, complete knowledge of the work, and intelligent judgment. This is particularly true in deciding the duration of an operation. Here the scheduler faces the problem of establishing the probable work force. For example, if it takes *20 worker-days* to erect the sheet metal ductwork on an office floor, should he decide on *2 workers for 10 days* or opt for the optimistic *4 workers for 5 days,* which the scheduler would prefer? One of the primary benefits of having the ultimate supervisor schedule his own program derives from the experience he will receive in trying to keep to the schedule he conceived. Four workers sounds very modest when making the schedule, but keeping them there is another story when they are needed on other parts of the project or, worse, are shifted to someone else's building.

Perhaps we can summarize these thoughts on scheduling by saying:

1. Assemble all the data for each operation that can be obtained.
2. Try to determine duration by more than one means.
3. Determine the proper amount of work that must be accomplished before a following operation can start and finish. Think in terms of these dependencies.

4. Duration is determined from the worker-days of work and the size of the work force.

5. Scheduling your own work and living up to your own schedule is very valuable experience.

5.7 KINDS OF SCHEDULES

The building construction engineer has many different scheduling methods from which he can select the one best suited for the operation he wishes to control. More important than the scheduling method is the skill and care used in its preparation and the intelligent use made of the schedule. In this chapter we discuss these kinds of scheduling methods.

1. The list of dates
2. The bar or Gantt chart (Fig. 5.2)
3. The critical path method (Fig. 5.2)
4. The precedence method (Fig. 5.2)

5.8 THE LIST OF DATES

This method consists of determining the start and completion of the operations and listing them. The simplicity of this method makes it very useful for subschedules or short-duration scheduling. For example, a schedule for installing a package boiler might look like this:

Boiler installation	Start	Complete
1. Rig and set boiler	June 1	June 3
2. Flue connections	June 4	June 9
3. Fuel supply piping	June 4	June 10
4. Water connection	June 6	June 12
5. Steam supply piping	June 15	June 19
6. Electrical hookup	June 15	June 19
7. Initial startup and testing	June 22	June 26

This schedule can quickly be reissued by a typist and copies can be carried in one's pocket to be used right where the boiler work takes place. The use of this simple scheduling method is recommended to the young construction engineer responsible for small areas or parts of a project.

5.9 THE BAR OR GANTT CHART

By all odds the most widely used and most popular scheduling method is the bar or Gantt chart. In its simplest form a list of dates is shown graphically as duration times (see Fig. 5.1). Construction supervisors have always readily accepted the bar chart as a good easily understood scheduling tool. The recording of progress periodically by appropriate markings quickly brings the chart up to date. Some construction companies use a prepared blank form in which the headings are typed in and the bars are indicated by various types of plastic tape which a draftsman can quickly secure in place. These can be revised by the scheduler by marking up the existing schedule and turning it over to a draftsman and typist for retyping and retaping as necessary. Figure 5.3 shows a simple bar chart for the installation of the refrigeration machine for central air conditioning in a large building.

Progress on a bar chart can be recorded in several different ways. The examples below record identical progress. In Fig. 5.4 work performed is recorded according to

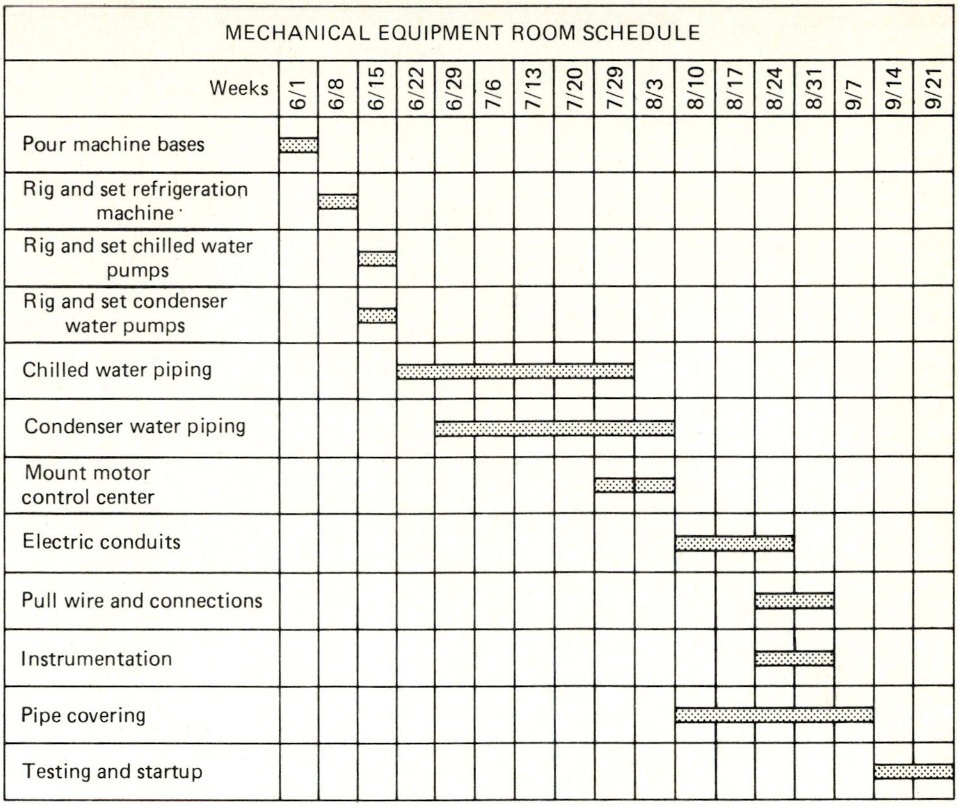

MECHANICAL EQUIPMENT ROOM SCHEDULE																		
Weeks	6/1	6/8	6/15	6/22	6/29	7/6	7/13	7/20	7/29	8/3	8/10	8/17	8/24	8/31	9/7	9/14	9/21
Pour machine bases	▨																
Rig and set refrigeration machine		▨															
Rig and set chilled water pumps			▨														
Rig and set condenser water pumps			▨														
Chilled water piping					▨▨▨▨▨												
Condenser water piping					▨▨▨▨												
Mount motor control center									▨▨								
Electric conduits											▨▨						
Pull wire and connections													▨▨				
Instrumentation													▨▨				
Pipe covering											▨▨▨▨						
Testing and startup																▨▨	

FIGURE 5.3. Mechanical equipment room schedule.

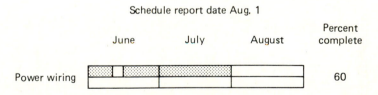

Schedule report date Aug. 1

	June	July	August	Percent complete
Power wiring	▨ ▨▨▨▨	▨▨▨▨		60

FIGURE 5.4. Bar chart on which work performed is recorded without regard to percentage completed.

periods of activity without regard to percentage completed. Here the work was interrupted in June. Percent complete is shown in a separate column with a reporting date. Time is 65% used up but *only* 60% of the work is complete. In Fig. 5.5 the chart is marked up with required completion percentages. The reporting date is still August 1, but *only* 60% of the work has been completed of the *required* 70%, so the progress bar falls short of August 1.

FIGURE 5.5. Bar chart showing required completion percentages.

	June	July	August
		30	70	
Power wiring	▨▨▨▨	▨▨▨	

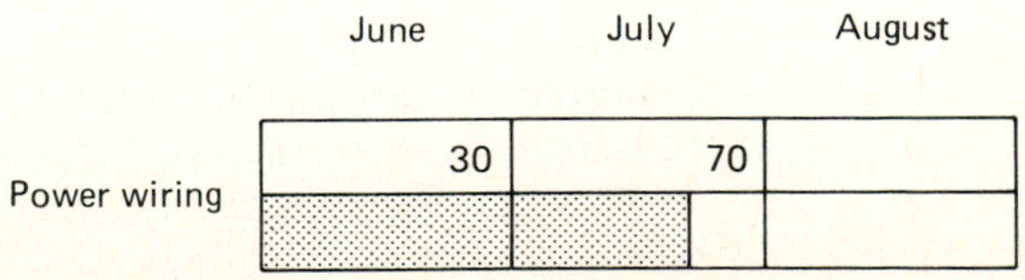

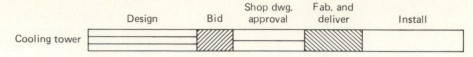

FIGURE 5.6.   Modified bar chart.

The Gantt chart can be modified further to show the following:

1. Design time
2. Bidding period
3. Shop drawing preparation and approval
4. Fabrication and delivery

A bar for a large cooling tower might look like Fig. 5.6.

Money and manpower curves can also be plotted on the Gantt chart as shown in Fig. 5.7, with the estimated and actual curves shown for comparison.

5.10   THE CRITICAL PATH METHOD

During the late 1950s the DuPont Company and the U.S. Navy, at about the same time, independently developed and used with great success separate network analysis systems of scheduling utilizing computer techniques. These systems, which have since been modified and improved, generally are now identified as the *critical path method* (CPM). The critical path is one of the routes through a network diagram of all of the important operations that make up a schedule and, as its name implies, it is the critical one—the one route that takes the longest to accomplish.

The CPM has not received universal acceptance after two decades despite its successful use by many firms, enthusiastic initial reception by the construction industry, specification by owners and architects, and recognition of its worth by highly competent schedulers. The serious-minded scheduler, however, will benefit greatly from learning the CPM of scheduling and becoming acquainted with its advantages and disadvantages. This method of scheduling basically resembles similar methods utilizing computer techniques which originated from the CPM.

The critical path method of scheduling utilizes a disciplined procedure in which the scheduler must completely study and understand every operation required to construct a project. The scheduler must also know how these activities relate and depend on each other. Every step is spelled out. Every step is evaluated for the time it will take to accomplish. The objective becomes the creation of a network diagram forming a complete study of all operations of a project, which can be entered into a computer program for recording, updating, and reporting.

Computerized applications of the CPM consist of three phases: planning, scheduling, and control–monitoring.

The Planning Phase

A considerable amount of time and effort go into the planning phase of the CPM. As indicated, each activity is identified, studied as to its relation and dependency to other activities, and its duration established. Naturally, the scheduler forced to examine in detail everything that must occur on a project will want to draw from all the sources available, including company records, company experienced personnel, and subcontractor judgments. The study required in the project planning and analysis produces one of the real advantages to the CPM. From it comes a complete understanding of every facet of the job which will be very valuable during the management of the project.

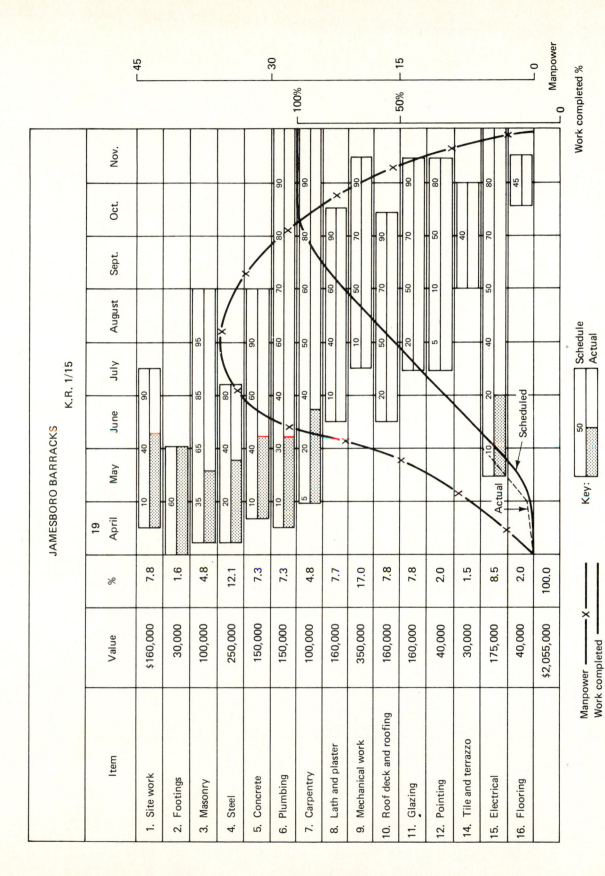

FIGURE 5.7. Modified Gantt chart. (From King Royer, *The Construction Manager,* © 1974, p. 218. Reprinted by permission of Prentice-Hall, Inc., Englewood Cliffs, N.J.)

35

Creating the Network Diagram

In this study of scheduling the intent is toward familiarization rather than direct training. The Associated General Contractors have prepared and published a manual for CPM training called "CPM in Construction" which has been very helpful to those wishing to become proficient CPM schedulers. However, let us look at a simple CPM schedule to become acquainted with the nomenclature and some basic rules. Figure 5.8 shows a network diagram for the installation of two unit heaters by pipe fitters and electricians.

The solid arrow from 1 to 2 is an *activity:*

$$① \xrightarrow{\text{Deliver Unit Heaters}} ②$$

An *event* is the starting point of an activity:

$$① \xrightarrow{\text{Deliver Unit Heaters}}$$

Dependency is the term used to check on what an activity is dependent on.

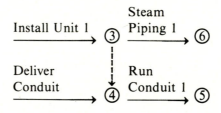

2–3 depends on 1–2.

A *dummy* does not represent work and has a zero duration time. Dummies show dependency and are shown as an arrow with a dashed-line shaft:

FIGURE 5.8.   Simple CPM diagram for installing two steam unit heaters.

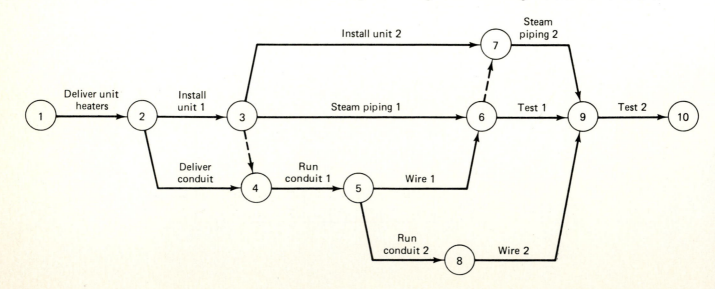

The numbering of events progresses left to right with the number at the arrowhead greater than the one at the start of the activity.

Duration is the estimated time for an activity:

$$\underset{\text{2 days}}{\underbrace{\text{Run Conduit}}} \longrightarrow ⑤$$

In Fig. 5.8 time is allotted for installing conduit.

During the planning phase the network diagram and duration estimates per activity are required. Using these as a basis, the following optional information can be produced:

1. Cost estimates per activity—for cost monitoring and cash flow requirements
2. Resource estimates—for resource requirement calculations
3. Trade (responsibility) indicators—for activity grouping

The Scheduling Phase

With the compilation of the planning phase data, the computer begins to list all the activities, compute total time, and identify the critical path. This phase produces the following information:

"1. A *Schedule* for the activities described in the arrow diagram, indicating:

   a. Each activity's STATUS—which are the critical activities.
   b. The EARLIEST START DATE for each activity.
   c. The EARLIEST FINISH DATE for each activity.
   d. The LATEST START DATE for each activity.
   e. The LATEST FINISH DATE for each activity.
   f. The amount of extra time (FLOAT) available to an activity

2. A *Bar Chart* for the project, showing time during which each activity is under way, without the percentage of completion at any particular time and without a comparison of the schedule to date.
3. A *Resource Analysis*—indicating how many of any kind of resource, i.e., manpower, equipment, etc., will be required for each day of the project.
4. A *Cash Requirement Prediction*—indicating how much will be required to pay for the tasks in the project and the amount of money coming in as a result of the activities completed"*

The Control–Monitor Phase

As in bar chart and other scheduling methods, effective monitoring of the CPM calls for periodic review of the project's status. In the CPM the original expertise must again be utilized for evaluation and updating and although much information can be produced, the skill of the scheduler conducting the updating process again determines the worth of the information produced.

This process of collecting and evaluating utilizes the following job information:

"1. *Additions to the arrow diagram*—new activities caused by change orders or by new approaches to the project.

---

*This section is taken from *CPM in Construction*, copyright 1965 Associated General Contractors of America.

2. *Deletions from the arrow diagram*—same reasons as 1.
3. *Changes*—one might decide to change an activity's duration, description, trade indicator or cost.
4. *Activity Actual Start Dates*—this is used by CPM to revise the schedule for the activities remaining.
5. *Activity Actual Finish Dates*—or notification of an activity's completion; used to revise the schedule for those activities remaining."*

The outputs available from the Control-Monitor Phase consist of the following items:

"1. *Time Status Reports*—the difference between the project's finishing date in the revised schedule (based on the actual date described above) and the original schedule's end date.
2. *Revised Schedules*—based on the new schedule reports.
3. *Revised Bar Charts*—same as 2.
4. *Revised Resource Analysis*—same as 2.
5. *Revised Cash Flow Predictions*—same as 2.
6. *Cost Status Reports*—compares actual cost incurred to a point in time for a project against estimated cost expected. This is expressed both as a percentage and a dollar difference."*

Developing a CPM schedule for a portion of a larger project provides excellent training for a young engineer–manager. Large mechanical equipment rooms, electric vaults and switchgear rooms, and complicated kitchens and cafeterias make ideal minischedule subjects. The skill required can be utilized for more complicated CPMs and other computer methods such as the Precedence method.

Advantages and Disadvantages*

Advantages:

1. The discipline required in identifying every activity, determining its duration (2 workers for 10 days or 4 workers for 5 days), and understanding the requirements of each activity and its relations to others. This acquired knowledge is very valuable to the later managing of construction.
2. The ability to utilize computer techniques.
3. The variety of studies and reports available.

Disadvantages:

1. The amount of time and effort required to produce the original project CPM schedule.
2. The cost of a CPM schedule.
3. The degree of skill, judgment, and experience needed to create and intelligently update the CPM schedule. Many companies prefer to utilize this type of talent to solve project problems rather than merely identifying them.

---

*This section is taken from *CPM in Construction*, copyright 1965 Associated General Contractors of America.

4. The lack of acceptance of the CPM below the managerial ranks.
5. The misuse of the CPM contrary to the user's original intentions. Contractors utilize the owner's CPM schedule to justify delays and pursue claims.
6. Excessive detail, leading to inflexible, overcomplicated networks and voluminous computer printouts which inhibit use by job management.

5.11   THE PRECEDENCE METHOD

Dissatisfaction with the arrow network diagram critical path method has led to the development of other forms of CPM scheduling. Recognizing the benefits of the older system while seeking a comprehensive easy-to-use system has led to the use of the precedence system by the Turner Construction Company.

This system provides the user with an efficient technique for scheduling, monitoring, and controlling work, all within the defined resources of time, available money, manpower, materials, and seasonal impacts. By implementing this system, other widely used scheduling methods, such as bar charts, work-in-place analysis, and manpower control, have been complemented.

Objectives

The three primary objectives of the precedence system are:

1. Creation of a well-defined realistic schedule plan.
2. Visually conveying this plan to all levels of project management by using timely updating and monitoring procedures.
3. Developing compatible feedback information to ensure the best results, including firmly holding original completion dates. The latter is referred to by the Turner organization as "recovery planning."

To satisfy the visual output requirement so important to this kind of scheduling, a computer-plotted, time-scaled bar chart was developed which forms a critical path network illustrating activities and their logic interrelation. The network plot includes all information describing an activity, such as work item number, activity description, duration, and interdependencies. The key to the acceptance of the precedence method is the plotted network, but users also draw from the same kinds of computer printout reports sorted out by performer, time frame, and milestones as in other computer scheduling methods.

The system's success stems from the ease with which the current schedule can be updated and corrective action recognized and applied. The more familiar-looking graphic display readily lends itself to the recording of current field progress at the field level, ensuring up-to-date schedule maintenance and monitoring. Where owners and others require a formal updating, the simplification built into the system makes it possible to submit this kind of update in a few days. With the simplicity of this system, a small number of logic changes in a given time period may even make a manual update more realistic.

All schedule systems cost money to operate, and perhaps the lack of a schedule system may be the most costly of all. Despite the use of special bar chart computer plotting equipment, operational costs compare favorably with other computer scheduling systems. Where costs of construction are substantial or early occupancy produces big returns, proper application of the system is a small cost compared with the results produced.

Some Project Control Concepts

Earlier approaches to CPM scheduling perhaps drew undue attention to "float," that time in a schedule in which an activity did not have to start. In utilizing the precedence method an initial concept is to concentrate on "early start." Job managers gear their thinking to how soon an activity *can start* rather than when it *must start*.

Many initial enthusiasts of CPM were "turned off" by the complexity of arrow network diagrams. Another precedence concept is directed at drastically reducing detail and creating a network containing the significant control activities. The nature of the project dictates the degree of detail *necessary*. The objective is the creation of a schedule containing the *minimum* level of detail that can still provide effective monitoring and control of project operations.

A third precept recognizes the importance of involving the personnel who will use the schedule in its creation. The superintendent who has helped prepare a schedule will try harder to hold to that schedule. This philosophy increases the continuing interest of the user throughout the life of the project.

Recovery planning is the final concept. Once the short- and long-term objectives have been agreed upon, the project completion date can be established and *locked in*. As the precedence monitor recognizes slippage, the management team formalizes specific recovery plans to return the project to the locked-in date. Corrective action takes the form of logic reassessment, super-expediting of shift work, and so on.

5.12   THE PROJECT SCHEDULING CYCLE*

The project scheduling cycle is the process by which the project schedule is developed, reviewed, and monitored. The cycle is illustrated in (Fig. 5.9), and a detailed discussion of each stage follows.

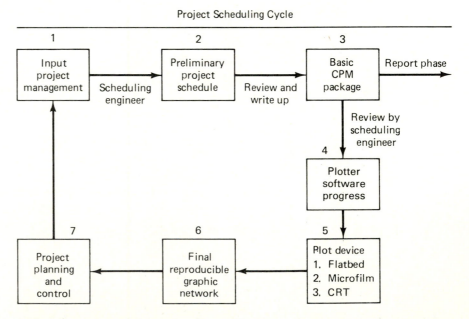

FIGURE 5.9. Project scheduling cycle. (Courtesy of Turner Construction Company.)

---

*Section 5.12 is taken from Garrett Thompson, *The Precedence Method* (instructional material), Turner Construction Company, New York, 1975.

Stage 1: The Input Phase

At this stage, the project manager and/or the project superintendent, along with a scheduling engineer and members of the project management team, organize the information necessary for the development of a preliminary project schedule. This initial involvement of the project staff is vital to the establishment and support of the computerized project schedule. It becomes the responsibility of a scheduling engineer to stimulate and encourage discussion during the planning phase. Activities should be forced to start as early as possible, thus supporting an "early-start philosophy." However, realistic constraints to the start and finish of activities should be recognized, such as: manpower, costs, logistics, flow of work, and so on.

A key to the success of any project schedule is the degree of detail which is developed. Prior to drawing the network diagram, consideration should be given to the level of detail. It is important to recognize that there are two distinct concepts involving the level of detail. The first concept deals with the number of different contract items to be delineated on the schedule. The second concept deals with the dissection of each of the contract items.

As an example of the first concept of level of detail, consider the scheduling of every item of work handled by each trade or subcontractor. This has been done in the past using CPM software packages and has met with varying degrees of acceptance and success. A second approach is to focus planning efforts on identifying those contract items which most often control the flow of work on a project. In general, this approach has resulted in the production of a much more effective and more manageable schedule.

As an example of the second concept of level of detail, consider the scheduling of two projects similar in construction but different in size. The larger project may require a greater dissection of the construction activities to provide the same control as the smaller project. There may be little difference in the type of work activities scheduled, but there will be a significant variation in the number of monitorable activities scheduled.

The most important schedule planning tool is a project master schedule. This schedule should be developed at the beginning of the project and used to set the framework for all detailed schedules.

The master schedule is made up of all the applicable important items in the project delivery process—i.e., land acquisition, budget approvals, demolition, clearing, surveys, borings, all major design phases, and other key milestones. In order to obtain this information the project manager must communicate with all of the parties who are involved (client, architect, and engineer, and so on), visit the site, analyze the construction market, note the status of related activities such as site acquisition or fund allotments, and, in general, gain an overview of the major tasks which must be accomplished from the time of initial development to completion of the project.

Stage 2: The Preliminary Project Schedule

Once project management has organized the information which will define the preliminary schedule, the scheduling engineer will begin drafting the working time-scaled network. It is suggested that the working network diagram be time-scaled in order that manpower distribution, flow of work, and seasonal effects may be considered.

The diagramming technique used is a form of CPM network diagramming termed "the Precedence Method." Through the use of various overlapping types, logic relationships between activities are described (see Fig. 5.10). There are three types of overlap: the "E" type or end-to-start, the "DS" type or delayed start, and the "DF"

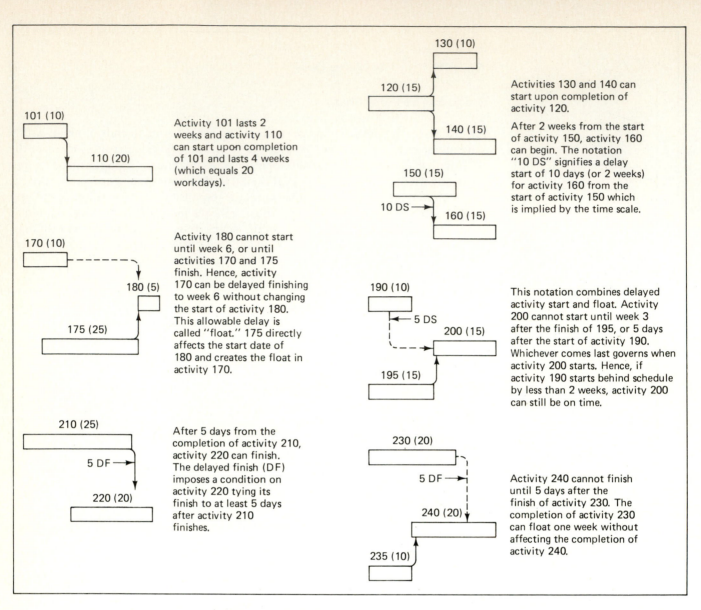

FIGURE 5.10. Logic relationships of the network plot based on a scale of five-day weeks. DS, delayed start; DF, delayed finish. (Courtesy of Turner Construction Company.)

type or delayed finish. By being able to overlap it is possible to produce a more realistic and workable relationship between activities and reduce the number of total project activities required. (A project scheduling system showing the DS and DF types of overlap is shown in Fig. 5.11.)

Stage 3: The Basic CPM Package

After the hand-drawn network has been reviewed by project management, the schedule is transferred to data processing format and is processed by the CPM software program. Estimated dates are calculated as the program makes a forward pass through the activities based on when activities *can* start and finish. Required dates are calculated as the program makes a backward pass through the activities based on when activities *must* start and finish to meet an established project completion date. The program is run in a time-sharing mode, with terminals located at some 12 territory offices. Off-line data entry is provided by means of a tape cassette device.

Once all format and data input errors have been corrected, the scheduling engineer will initiate the printing of a work report titled, SWI Report (succeeding

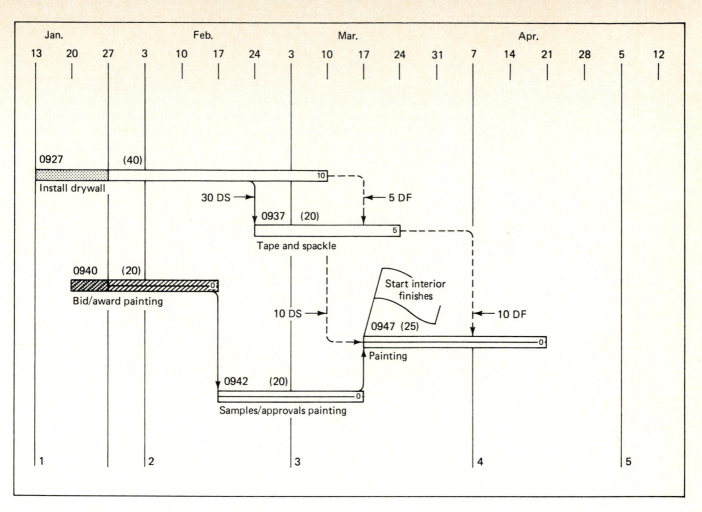

FIGURE 5.11. Project scheduling system. (Courtesy of Turner Construction Company.)

work item report). All items of information shown on the hand-drawn network diagram plus calculated data are incorporated in this report: i.e., preceding work items (PWIS), succeeding work items (SWIS), estimated and required start and finish dates, total float, total and remaining duration, binding codes, critical path designation, work item numbers, work item descriptions, and milestone critical activities.

Stage 4: The Plotter Software Program

Step four is the plotter software program phase and is really the point which separates the Turner system's output from that of more conventional systems. After input data are checked for errors by the scheduling engineer, he can either initiate a report or a plot of the schedule or both. Within the software plotting program, the user has a high degree of flexibility, with many available options. Some of these options are: (1) banding groups of activities; (2) plotting only specific activities, such as mechanical or electrical; (3) plotting a specified time period; (4) graphically identifying the critical path; and (5) designation of procurement activities. This plotter software program was developed specifically for a large building contractor, is linked internally with that of the CPM program, and provides the user optimum turnaround time and ease of operation and handling.

Stage 5: The Plotter Device Phase

Following the selection of system options, a decision is made regarding the use of one of the several alternative plotting methods for producing the final graphic network. A drum or flat bed computer-driven plotting device is generally used to produce the graphic display. Or, a microfilm plotting procedure can be used to reproduce the network on 35 mm microfilm. A third alternative is available which utilizes a cathode ray tube unit to produce an image of the network. This method has been used successfully in presentations and internal training to demonstrate the capabilities of the system.

Stage 6: The Reproducible Graphic Network

Once the plotting of the schedule has been completed, the plot is sent to a specified address via a predetermined designated delivery method. In general, delivery time of plot revisions is within a few days from the time the data are received at the center.

Stage 7: The Project Planning and Control Phase

During this phase the schedule is implemented by project management. A key component of the implementation of the schedule is periodic review and update which generally identifies problem areas and stimulates the investigation of alternative solutions. This phase closes the loop to the cycle and integrates a continuous chain of action and interaction.

The monitoring process is, of course, nothing more than determining on a periodic basis where the project actually stands when compared with the plan. In addition to consideration of actual versus planned time expenditures, effective monitoring should evaluate physical progress, value of work in place, and manpower expended against the plan. Obviously, the real task of controlling a job with regard to schedule only begins with the creation of a formal plan and monitoring actual progress against that plan. The difficult part is to evaluate the update or monitoring data and to reanalyze the logic of the original plan based upon this and other new information to develop the plan to complete the project in the most expeditious manner economically consistent with the quality desired. In effect, this may mean replanning the entire job on some periodic basis. Included in this process is the communication and leadership required to put the new plan into action.

5.13 CONCLUSION* The contractor feels that the degree of flexibility along with the components of the graphic display make the CPM scheduling system an effective and responsive management tool. The fact that the contractor is using this scheduling system does not mean the elimination of more traditional tools. The system is not a cure-all; it is only one of several procedures such as statistical modeling, cost control systems, job minutes, manpower analysis, and so on. Thus far, however, it has been an extremely effective tool resulting from the contractor's philosophy of adopting, developing or modifying systems which will support the firm's method of operations, rather than subject personnel to "canned" programs which may adversely affect the management process itself.

While most management tools identify where a project is at any given time, the CPM project scheduling system is an attempt to define where the project is *going*. It is a *forward planning* tool as well as a *monitoring* tool. The distinctive graphic display is the tool provided to help managers plan for completing the construction project on time . . . and on budget.

---

* Section 5.13 is taken from Garrett Thompson, *The Precedence Method* (instructional material), Turner Construction Company, New York, 1975.

6

Project Management

6.1 PROGRESS CONTROL
The general contractor and construction managers responsible for controlling progress on a modern construction project are faced with a much more difficult assignment than were their predecessors. Two things have occurred to cause this condition. The design of buildings has steadily grown more complex, as exemplified by computer areas, curtain wall construction, exotic lighting, underfloor distribution systems, solar energy, and energy conservation. While buildings have become more difficult to build from a design standpoint, construction has year by year become more specialized and fragmented, to the point where we have subcontractors for every conceivable operation from cement finishing to air balancing. The modern manager must coordinate, sequence, expedite, and otherwise seek to control the progress generated by 20 to 30 separate organizations.

The construction supervisor at a very early date should determine the acceptance of the reality of the progress schedule. Some schedules unfortunately are unduly optimistic. Sometimes the construction supervisor has not been involved with the preparation of the schedule handed to him as he is freed from one project and shifted to another already under way. If he is to perform well, the person in charge of construction must be convinced that he can meet the established schedule as the work proceeds under normal conditions of time allotted, using efficient crew sizes, and economical methods. Unless he has faith in the schedule he will want the authority to rephase work, employ inefficient numbers of workers, or work overtime hours to keep up to date.

The other participants, subcontractor supervisors and trade foremen, must have confidence that work goals are normally attainable. Psychologically, their input or their company's input toward schedule presentation commits them to the time allowed for that portion of the work. Wherever possible while taking subcontract prices, the mechanical and electrical subs should be required to commit themselves to a schedule for their part of the work.

When attempting to make a schedule work, recognize that there are a great number of variables. Mathematically a schedule originates from a contract amounting to *X amount of dollars* containing *Y amount of labor* which converts into *Z number of worker-days.* It all looks as simple as XYZ. However, while the days slip by, weather, strikes, material and equipment delays, availability of labor, disruptive changes, quality of management, supervision, and productivity all affect the construction period. Realistic or not, the person in charge must do his utmost to meet the commitment made by his firm. To do this he should break down the overall schedule into more digestible parts. Some firms doing nuclear power plant work have a *five-year,* a *three-week,* and a *one-week* schedule. In a high-rise structure, for example, we can break down the HVAC schedule into pipe and duct risers, basement and top-floor mechanical equipment rooms, cooling towers, floor development (tenant work), and controls. Executive, computer, and kitchen areas, which are the most complicated, should also receive special schedule attention.

6.2 PURCHASING LONG-LEAD-TIME ITEMS

It must be difficult for the inexperienced to realize the market cycles that occur in construction. Since 1970, depending on where you were in the United States, there have been wide swings in the availability of labor, materials, and equipment. During this period electric cable producers were quoting as much as 72 months for delivery on certain types of cable. Many items were listed available in excess of 12 months. No matter when you build, long-term deliveries should be determined and action taken to see that these items are purchased, expedited, and delivered to meet schedule requirements. Electrical switchgear usually falls into this category. When imperative, owners and architects often make these purchases prior to the awarding of the construction contract. Later they assign this purchase to the appropriate contractor.

6.3 MECHANICAL AND ELECTRICAL COORDINATION

In certain parts of the United States, in order to reduce architects' and engineers' fees, the mechanical and electric working drawings are developed only to the point where the architects and mechanical engineers are relatively confident that all the pipes, ducts, electric conduits, sprinklers, light fixtures, conveyors, pneumatic tubes, and so on, can somehow be made to fit into the space provided by the established ceiling height (Fig. 6.1).

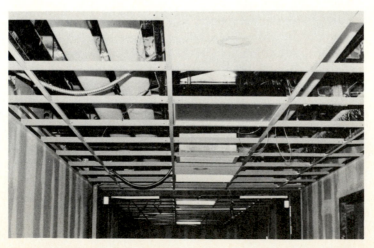

FIGURE 6.1. Corridor ceiling reflects coordination problems.

The contract documents in these instances provide that the mechanical and electrical subcontractors will develop their trade shop drawings to be used by their own plumbing, electrical, HVAC, and sprinkler engineers, who will meet with the other trades to develop coordinated shop drawings. Since each subcontractor has his own interests at heart, either the mechanical designers, the construction manager, or the general contractor should be present to control, encourage, mollify, and expedite the subs in determining the best overall solutions. Certain rules must prevail: 4-in. electrical conduits are not spaghetti to be hashed about; soil, steam supply, return, and sprinkler lines must pitch; the bulkiness of sheet metal ductwork should receive proper consideration. Good judgment, fairness, firmness, and the ability to get others to cooperate are attributes required of the engineer controlling the coordination process.

The older traditional design method leaves coordination in the hands of the design team. The contractor deviates from the locations assigned under these coordinated conditions at his own peril. In either case, the final coordinated locations are subject to architect and engineer approval. From the owner's standpoint it is doubtful that one method costs less than another since the subcontractors provide for this engineering coordination time in their bid prices. The general contractor or construction manager should recognize that before he can do very much in the field mechanically and electrically, drawings must be coordinated. If the contract documents and drawings require coordination, he must start the coordination process as soon as the necessary design drawings are available.

6.4 SHOP DRAWINGS, CUTS, AND SAMPLES

Mechanical and electrical contractors must submit a formidable amount of shop drawings, samples, and manufacturers' brochure material called "cuts." The approval of these items is vital to progress since material and equipment should not be manufactured or shipped without architects' and engineers' approval. For standard shelf items, manufacturers of fans, grilles, plumbing fixtures, lights, and similar items have prepared sheets or "cuts" showing the specified item with dimensions, capacities, and other pertinent information listed for all sizes of this item that they manufacture (Fig. 6.2). The submission is a simple matter of indicating by arrow the code designation with its sizes, dimensions, and capacities.

Piping and duct systems require shop drawings, with the key to progress generally the sheet metal drawings. Mechanical equipment rooms require generous amounts of detailing time to satisfy operating personnel requirements and to "shoehorn" huge fans and ducts into minimal spaces. On major buildings the preparation of these drawings, and their submission, correction, resubmission, and approval, requires the time and effort of numerous subcontractors, general contractor, architect, and design engineer personnel. For Madison Square Garden, estimates of the number of shop drawings were obtained from all subcontractors and this information was shared with the architect and engineers so that they could prepare and organize for a very large amount of approval work. For example, Bethlehem Steel Company prepared and had approved *2500 shop drawings* for their portion of the work. When you wish to control progress, certainly the preparation and approval of shop drawings should be a major concern of the general contractor or construction manager.

Samples of pipe, hangers, light fixtures, switch plates, pipe covering, and numerous other items are also submitted and after approval are kept on the job for ready on-site comparison. This process generally does not cause problems unless the owner or architect delays color selection or expresses dissatisfaction with metal or other finishes.

The foregoing does not happen overnight, but fortunately for mechanical and electrical contractors they have an excellent opportunity to complete these preliminary tasks while excavation, foundations, and structural frame proceed. On

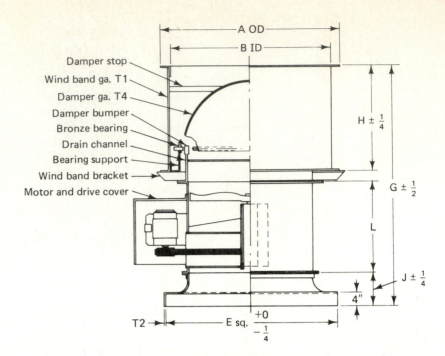

FIGURE 6.2. Data sheet for a belt-driven fan. (Courtesy of Aerovent Inc.)

SIZE	A	B	E	G	H	J	L	STEEL			ALUMINUM		
								T1	T2	T4	T1	T2	T4
915mm 36	1162mm 45¾"	1092mm 43"	1264mm 49¾"	1559mm 61⅜"	610mm 24"	249mm 9¹³/₁₆"	660mm 26"	1.9mm .075"	1.9mm .075"	.9mm .036"	2.0mm .080"	3.2mm .125"	1.3mm .050"
1066mm 42	1314mm 51¾"	1245mm 49"	1416mm 55¾"	1781mm 70⅛"	686mm 27"	244mm 9⅝"	813mm 32"	1.9mm .075"	1.9mm .075"	.9mm .036"	2.0mm .080"	3.2mm .125"	1.3mm .050"
1467mm 57¾"	1397mm 55"	1568mm 61¾"	1959mm 77⅛"	762mm 30"	244mm 9⅝"	914mm 36"	1.9mm .075"	1.9mm .075"	.9mm .036"	2.0mm .080"	3.2mm .125"	1.3mm .050"	
1372mm 54	1626mm 64"	1549mm 61"	1721mm 67¾"	2007mm 79"	813mm 32"	241mm 9½"	914mm 36"	1.5mm .060"	2.7mm .105"	.9mm .036"	2.0mm .080"	3.2mm .125"	1.3mm .050"
1525mm 60	1778mm 70"	1702mm 67"	1899mm 74¾"	2115mm 83¼"	914mm 36"	248mm 9¾"	914mm 36"	1.5mm .060"	2.7mm .105"	.9mm .036"	2.0mm .080"	3.2mm .125"	1.3mm .050"
1830mm 72	2235mm 88"	2159mm 85"	2254mm 88¾"	2375mm 93½"	1067mm 42"	302mm 11⅞"	965mm 38"	1.5mm .060"	2.7mm .105"	.9mm .036"	2.0mm .080"	3.2mm .125"	1.3mm .050"
2135mm 84	2540mm 100"	2464mm 97"	2559mm 100¾"	2658mm 104⅝"	1219mm 48"	330mm 13"	1067mm 42"	1.5mm .060"	3.4mm .135"	.9mm .036"	2.0mm .080"	4.1mm .160"	1.3mm .050"
2440mm 96	2845mm 112"	2769mm 109"	2864mm 112¾"	2962mm 116⅝"	1372mm 54"	330mm 13"	1219mm 48"	1.5mm .060"	3.4mm .135"	.9mm .036"	2.0mm .080"	4.1mm .160"	1.3mm .050"

many of the larger projects during this period the architect and construction manager are involved with a mock-up of part of the exterior wall of the building. This affords the HVAC contractor the opportunity to set up window convector units, working out pipe bends and valve locations. From this mock-up information he can establish an on-site copper piping shop and prefabricate large quantities of the perimeter piping system.

6.5 PROGRESS MEETINGS

The best managed projects operate under a system of progress meetings. There are three types of meetings generally conducted for these large projects. The types of meetings held will depend on the wishes of the owner, architect, and the general contractor/construction manager. Since owner and architects must meet frequently over design decisions, they usually continue to be represented as the owner meets with the *contractor or construction manager* to monitor progress. The construction manager or general contractor in turn controls progress through meetings with his *trade foremen and subcontractors.* A very effective third type of meeting is the *internal type* conducted by the project executive or general superintendent with his *project manager, engineer, and purchasing agent.*

48

6.6 OWNER'S MEETINGS

Meeting with the architects, engineers, and contractors, the owner determines the status of the project through required reports. An updated progress chart or computer report from the contractors shows the status of the various operations, plus any extension of time required in part or overall. The bars on the chart should show the original schedule and above or below them current progress to the report date. The percentage of work completed on all items should be recorded and any necessary extension to the original bar added. Figure 5.4 illustrated an update of a progress report. The owner's meeting chairman, in addition, should determine the following:

1. Information or decision affecting progress for which the owner is responsible
2. Plans, specifications, changes, or approvals required from the architect causing delay or required to prevent delays
3. The contractor's or construction manager's requests for decisions, information, or approvals from owner or architect
4. Items of work behind schedule
5. Reasons for work being behind schedule
6. Opportunities for recovery
7. Corrective action the general contractor/construction manager will institute
8. Options available to owner if contractors fail to take action

The minutes of these meetings and of all progress meetings should be *promptly* completed and distributed since they lose half their value weeks later and become mere historical records. The owner's designation of a respected chairman with the reputation for being fair-minded and adaptable to changing situations will inevitably produce favorable results. A person with these qualifications usually ranks high in his company and it is not unusual for him or her to rise to top executive status. In one instance the chairman for the headquarters project became the president of a very large New York insurance firm.

6.7 CONTRACTOR-SUBCONTRACTOR MEETINGS

Meetings called by the general contractor with his subcontractors are used extensively for progress control. When properly run with good control, proper discipline, and a cooperative attitude on everyone's part, they can be very effective. The individual separate reports of each trade require beneficial thought and preparation. The mechanical and electrical contractors, in particular, must submit a *detailed purchasing report* which is updated for each meeting, covering all essential data concerning order numbers, supplier, shop drawings, fabrication, promised delivery date, and so on, for all major equipment, fixtures, and material (Fig. 6.3). Necessary action on the part of the specialty contractor for improvement of unsatisfactory dates or potential expediting help from the construction manager or owner can then be flagged. The specialty contractor should be prepared to indicate a constructive program of correction to recover lost time. It is best for affected contractors to accept his intention to recover lost time and not assume that they can let down. If information or expediting help are required, the subcontractor can then make these needs known. The general contractor has an excellent opportunity in these meetings to outline the overall program and sequences for the next few weeks. Individual contractors can be alerted to prepare to start work in a new area or add or shift workers to quicken the tempo of current operations.

    Most construction people have a heavy schedule of energetic activity and this type of meeting *wastes many people's time.* The subcontractor meeting requires the electrician to sit and listen to all manner of information on windows and other items that do not concern him in any way. Whenever possible, separate mechanical and electrical submeetings should be held. Let's face it—meetings requiring the attendance of many subs are inefficient. Second, it becomes *more difficult* in a peer-group contractors' situation to *reveal completely the real job situation.* Most human beings

NO.	ITEM	QUANT.	MANUFACTURER	VENDOR	ORDER NO.	ORDER DATE	SHOP DRAWINGS DRAWING NO.	SHOP DRAWINGS SUBMIT.	SHOP DRAWINGS RELEASE	REQ'D.	DELIVERY PROM.	DELIVERY REC'D.
1	Cooling tower	3	Baltimore AirCoil	Newton Eng.	5630	9/29/78	632-1	10/8/78	10/12/78	10/31/78	10/27/78	10/25/78
2	Refrigeration mach.	3	York		5591	9/14/78	SEE CUTS					
3	Air handling units	ALL	American Std.	American Std.	5617	9/20/78	→			11/17/78	11/15/78	
4	Fans, Motors	ALL	"		"	"				"	"	
5	Unit htrs and cab htrs	ALL	"		"	"				"	"	
6	Convectors and fin tube	ALL	ITT Nesbitt Inc.	ITT Nesbitt Inc.	5623	9/27/78				11/15/78	11/4/78	10/27/78
7	Coils	ALL	American Std.	American Std.	5617	9/20/78				11/15/78	11/15/78	
8	Pumps	ALL	Worthington	Worthington	5609	10/2/78				11/10/78	11/8/78	
9	Heat exchanger	ALL	Patterson Kelley	Patterson Kelley	5611	10/16/78				"	11/15/78	
10	Condensate pumps	ALL	Worthington	Worthington	5609	10/2/78				"	"	
11	Filters	ALL	American Air Filt.	Assoc. Thermal Prod.	5619	10/16/78				11/15/78	11/4/78	
12	Strainers	ALL	Cutler Hammer	Cutler Hammer	5627	10/19/78				11/30/78	11/29/78	
13	Air outlets	ALL	Titus	Albert Weiss Acco.	5624	10/19/78				12/1/78	11/29/78	
14	Water treatment equip.	ALL	Water Service Labs	Water Service Labs	5632	10/19/78				12/1/78	11/8/78	
15	Pressure red. valves	ALL										
16	Tanks											
17	Ball joints											
18	Relief valves											
19	Exhaust heads											
20	Thermostats											
21	Gauges											
22	Electric unit htrs											

PH CONSTRUCTION CO. CONTRACT # 1383 — EQUIPMENT AND MATERIAL SCHEDULE

BUILDING PLAZA CENTER - CHICAGO

SUBCONTRACT # 238 HVAC

SUBCONTRACTOR HEAT AIR INC.

FIGURE 6.3. Equipment and material schedule.

50

are thin-skinned enough to tend to prevaricate when bad news must be reported in public. Overoptimism usually prevails. Third, intertrade rivalry leads to one-upmanship, grandstanding, or bombast, all *detrimental to cooperative attitudes* among the subs. The general contractor's representative must have the ability to run a tightly controlled, expedited, fact-finding, no-nonsense meeting. Few are very good at doing this.

6.8   INTRACOMPANY MEETINGS

One of the most effective progress control methods uses as its basic method a periodic intracompany meeting. Conducted by the person with maximum responsibility, such as a construction vice-president or general superintendent, the meeting examines *each item* of construction from every operational viewpoint. Purchasing, engineering coordination, and field supervision each report in turn on the current status and explain the next action step to be taken. For example, after full reporting on the current status of an item, the minutes would state: "Jones follows up for approval of pump room plumbing shop drawing" or "Smith to set up meeting with electrical contractor and fixture manufacturer." Figure 6.4 illustrates a portion of this type of minutes.

FIGURE 6.4.   Minutes of a job meeting.

	MEETING NO. 3 PAGE 10
MINUTES OF JOB MEETING	DATE October 26, 1978
	CONTRACT NO. 383

43.   Elevators   Upsome Elevator Co.
          Shop Drawing Submitted — 12
          Approved to Date   —   6
Mr. Black (Engineer) follows up with sub for balance of shop drawings and architects approval. Mr. Green (Purchasing Agent) follows up for prices on new type cab selected by owners. Mr. Blue (Superintendent) follows up with sub to obtain more elevator constructors and improve progress. Rail installation just started.

44.   Plumbing   Soil Lines Inc.
          Shop Drawings Submitted — 17
          Cuts Submitted   — 85
          Approved to date — Dwgs   — 20
          Cuts   — 75
Mr. Black follows up for balance of shop drawings and estimate on new toilet room 8A.

Toilets roughed to third floor, fire standpipe run to top floor. Water meter and incoming water service installed. Mr. Blue follows up for temporary water service.

45.   Electrical Work   Doe Contracting Co.
          Shop Drawings Submitted — 20
          Cuts Submitted   — 24
          Approved to Date — Dwgs — 10
          Approved to Date — Cuts — 18
Mr. Black follows up for balance of shop drawings, cuts and their approval. The sample fixture was rejected. Doe to obtain proper fixture. Mr. Black, please expedite.
Electric risers installed to third floor. Incoming service work started, progress satisfactory. Mr. Blue follows up for temporary electrical installation.

46.   HVAC   Heat Air Inc.
          Shop Drawings Submitted — 32
          Cuts Submitted   — 39
          Approved to Date — Dwgs — 10
          Approved to Date — Cuts — 10
The failure to approve shop drawings on time is delaying the work. Messrs. Black and Blue to meet with architects and engineers to arrange for quicker approval of drawings. Mr. Black to expedite machine foundation shop drawings so that concrete pour can be made.

No sheet metal ductwork received to date.

Mr. Green will visit shop to expedite risers basement to third floor.

These meetings are very effective at establishing the current status of each portion of work and distributing the action load to advance every necessary part of the project. Because each next step of every item is spelled out with the responsible person designated, such meetings promote action and go a long way toward eliminating excuses.

Distribution of the minutes to company department heads or the company boss alerts these persons of problem areas and brings to bear a greater supervisory pressure.

6.9 SWEAT SESSIONS

All of the meetings discussed above, which are general in nature, hopefully will uncover only minor, easily correctable matters. The serious problems of *fabricated material* and *equipment supply* or *lack of field progress* call for stronger measures. As an initial step the project management personnel should summon the subcontractor's personnel to the site to develop problem solutions. If satisfactory or sufficient action does not develop from the sub following this special on-site meeting, a *"sweat session"* should be set up in the general contractor's main office with the general contractor's or construction manager's construction vice-president and the top-rated person responsible for subcontract awards present. The mechanical or electrical sub in question, while not satisfying job requirements, may be turning flip-flops seeking other work from these same people. Construction people, being very pragmatic, require results on the current work before entertaining discussion of additional contracts.

Confronted with this degree of pressure the subcontractor usually develops a reasonably satisfactory recovery program, particularly if the construction manager maintains a firm, constructive, and understanding attitude.

6.10 THE TRUTH AND NOTHING BUT THE TRUTH

Perfectly honest contractors often produce inaccurate job status reports through overoptimism or by careless evaluation. There have been painful instances where otherwise good mechanical and electrical contractors have, through sloppy cost control procedures, deluded themselves and the construction manager as to progress status. Without a good cost reporting system from which to evaluate productivity, otherwise competent mechanical and electrical firms have been known to report that since they have spent *X amount* of their estimated labor they must have *(estimated − X) amount* of work to complete. In two such cases the construction manager's own studies eventually revealed the true amount of work to complete, whereupon stringent measures were applied and some lost time was recovered. Electricians can initially appear to be keeping up with the general progress of the work while continually falling further behind with their electrical progress. They concentrate on conduit installation while letting wire pulling, panel board connecting, and other work slide. Eventually, the project is confronted with one electrical progress crisis after another until the project ends. On one large project the electrical contractor had to employ 50 workers for five months of catching-up work after the building had been turned over to the owner for use. Beware the sub who is fooling himself. Carefully review the progress of the sub who is consistently overoptimistic or is inclined to give hasty reports without careful checking. A touch of pessimism can be very helpful at times.

6.11 EXPEDITING HELP

We all like to solve our own problems and through pride are reluctant to seek outside help. However, *many people share the responsibility* for a construction project completed on time. Mechanical and electrical subcontractors should energetically expedite their own fans, motors, fixtures, and other equipment whenever possible as one of their contract requirements. General contractors and construction managers can also greatly assist when imperative, depending on their reputation and purchasing clout. Occasionally, through banking interests or director relationships, owners can be helpful in critical situations. However, as with anyone drowning, the call for help should come in time to prevent disaster. A good ploy is to alert others by saying: "Here's my situation; if it gets worse, I may need your help." That way everyone is alerted.

Telephone expediting *generally* produces satisfactory results. Proper preparation by the expediter in obtaining shop drawing status, order numbers, job requirements, and so on, before he picks up the phone improves his performance. Through a series of well-thought-out questions he should be able to determine accurately the status of the work. Unfortunately, some suppliers and manufacturers, acting immaturely, deliberately deceive expediters, hoping to buy some time and somehow scrape through a bad situation. The exposure to being deceived is the biggest drawback to telephone expediting.

There is no substitute for a plant visit by a well-prepared job representative. Evaluation of work completed, work in progress, and rate of progress plus on-the-spot commitments obtained for the remaining portions of work make these plant visits very worthwhile. Both parties benefit by establishing closer business ties through interpersonal relationships.

6.12 GETTING RESULTS

The simplest structure can face serious delays through lack of coordination, failure of key materials to arrive on time, changes by the owner, and refusal of a key subcontractor, such as the plumber or electrician, to perform on schedule. To keep a project on schedule the general contractor–construction manager conducts a running battle with the weather, the material and equipment situation, and numerous other changing circumstances. Each of the participants can contribute to the progress of the job by doing his part. For example, the owner can:

1. Make timely decisions.
2. Utilize restraint on changes.
3. Exercise good judgment and a cooperative attitude, as evidenced by realistic requests.
4. Promptly pay monthly requisitions and release retained money when prudent.

The architect can:

1. Conscientiously review drawings and specifications to eliminate the need for later corrective changes.
2. Promptly reply to requests for needed information.
3. Expedite approval of samples, shop drawings, and color selections.
4. Punctually process payment requisitions and change-order requests.

The general contractor or construction manager can:

1. Carefully review schedules to make the best use of time to purchase, detail, fabricate, deliver, and erect.
2. Provide qualified and sufficient supervision for subcontract and own work.
3. Utilize quality subcontractors (it only takes one really bad subcontractor to create progress havoc).
4. Back up field forces with appropriate coordinating engineering, purchasing, and accounting talent.
5. Periodically update schedules followed by indicated corrective action.
6. Determine to stick to the schedule.

7. Provide intelligent leadership and enthusiasm for maintaining the progress objectives.
8. Be sincerely interested in the welfare of the staff, workers, and sub-contractors while pursuing company policy.

Subcontractors can:

1. Be responsive to the plan and directives of the general contractor–construction manager and be willing to take corrective steps to improve performance.
2. Adopt a long-term attitude whereby project performance becomes more important than maximum short-term profits.
3. Show a disposition to set a good pace on his type of work.
4. Provide good supervision and utilization of well-supported, productive labor.
5. Follow policies of early purchasing and intelligent expediting.

Progress depends on the good performance of many and the adequate performance of all. Despite all our energetic and intelligent efforts we are still vulnerable to severe weather, strikes or material problems at suppliers' plants, and shortages of productive labor. Not every job will be completed on time, but our clients deserve our maximum effort to increase the percentage of projects brought in on schedule.

7

Budget and Cost Control

7.1 COST CONTROL: A NECESSITY

General contracting and mechanical and electrical work differ somewhat in their approach to cost control. The general contractor purchases large quantities of materials such as cinder blocks, brick, 2×4's, plywood, concrete, and so on, which after being cut, fitted, and joined become part of the structure. A certain amount of material is always wasted and it is common to order extra material to ensure having enough. This may account for the absence of good cost control on the part of many general contractors starting in business. A neophyte plumbing or electrical contractor, on the other hand, by the nature of his work should be much more conscious of the necessity for accounting for material and fixtures than his general contracting friend. The plumber's work entails a considerable amount of distinctly different kinds of pipe, pipe fittings, valves, fixtures, and so on, all to be ordered and charged to a specific job. Electrical work requires a similar range of different outlets, switches, fixtures, motor controls, lamps, and so on, chargeable to distinctly different parts of the work.

It stands to reason that someone starting a mechanical and electrical contracting business which requires accounting for pipe, fittings, fixtures, and equipment would also institute a labor cost control system. Certainly the beginning contractor should want to know how much it costs him in labor to install the various kinds and sizes of piping and conduit and his unit cost for lavatories, water closets, panel boxes, and transformers. Monitoring productivity is essential to financial survival. A company that does not know every two weeks where it stands on labor costs (productivity) usually ends up disappointed or bankrupt when it does find out. Accounting for labor, material, fixtures, equipment, rental charges, subcontracts, and job indirect costs is one more part of proper construction management that separates the successful from those who go bankrupt.

7.2 SETTING UP A COST CONTROL SYSTEM

A contracting company's accounting system should contain all of the accounts general to most businesses: assets and liabilities, net worth, income, and expense. For *each project* a separate ledger should record every expense chargeable to that particular project. These would include those items that become *directly* part of the project, such as:

1. Labor
2. Material and project equipment
3. Rental on cranes and other construction equipment
4. Subcontracts (insulation, rigging, and so on)

They would also include *indirect* (overhead) items for that job, such as:

1. Taxes
2. Insurance
3. Temporary services (electric, water, heat)
4. Salaries of supervisors
5. Engineers' and expediters' charges
6. Cleaning
7. Watchmen
8. Tools and supplies
9. Blue printing
10. Temporary buildings and installations, and so on

In this manner each project stands alone to be monitored while in progress and smiled at or mourned when completed.

Looking over the foregoing lists we see that the charges to most of these accounts would be prepared from some kind of paper record: delivery ticket, rental charge invoice, and so on. However, labor charges—perhaps the biggest single item and usually the most important—are lost permanently unless recorded on a daily basis.

Proper business management, therefore, requires creating a system that breaks labor down into easily recognizable and readily described classifications that are convenient to use. The system should be created by or with the guidance of those company personnel most interested in cost control and estimating. A good balance between complexity and simplification will produce the best results. An overly complicated system produces a myriad of isolated information that does not seem to match up with anything, and too simple a system does not define the cost picture with sufficient detail.

Prudence might dictate that the system should be compatible with an eventual move toward computer usage. As computers become smaller and less expensive, computer manufacturers are counting on many small businesses converting to a computer system.

An example of part of a labor classification code numbering and activity description for a plumbing contractor is shown in Fig. 7.1.

Code	Description	Unit
2.0	Pipe, cast iron, single hub, 2 to 6 in.	L.F.
2.01	Pipe, cast iron single hub, 8 to 12 in.	L.F.
3.0	Pipe, copper Type L tubing, ¼ to ⅝ in.	L.F.
3.01	Pipe, copper Type L tubing, ¾ to 1½ in.	L.F.
3.02	Pipe, copper Type L tubing, 2 to 4 in.	L.F.
4.0	Pipe, plastic fiberglass reinforced, 1 to 2 in.	L.F.
4.01	Pipe, plastic fiberglass reinforced, 3 to 8 in.	L.F.
4.1	Pipe, plastic PVC, ¼ to ¾ in.	L.F.
4.11	Pipe, plastic PVC, 1 to 3 in.	L.F.

FIGURE 7.1. Example of labor code numbering for a plumbing contractor.

Figure 7.2 shows a similar example for an electrical contractor.

Code	Description	Unit
6.0	Conduit, aluminum, ½ to 1 in.	L.F.
6.01	Conduit, aluminum, 1¼ to 2 in.	L.F.
6.02	Conduit, aluminum, 2½ to 4 in.	L.F.
6.1	Conduit, galvanized steel, ½ to 1 in.	L.F.
6.10	Conduit, galvanized steel, 1¼ to 2 in.	L.F.
6.11	Conduit, galvanized steel, 2¼ to 4 in.	L.F.
7.0	Pull wire, copper No. 14 solid	L.F.
7.01	Pull wire, copper No. 12 solid	L.F.

FIGURE 7.2. Example of labor code numbering for an electrical contractor.

7.3 LABOR

Implementing the System

Ideally, trained personnel, such as timekeepers, should record the amount of labor going into each labor classification. Since they do not have to meet a production goal, they should be able very objectively to charge the labor to the proper account. However, the majority of companies depend on their foremen to classify and record work performance. It is essential that the foremen accurately charge out the labor as it is performed, being careful not to "fudge" accounts to sugar-coat poor performance. Deliberate or careless misclassification distorts not only the current project but future company records as well.

Each week all the labor charges are summarized and entered into the labor sections of the job ledger. Since they are recorded by payroll week date, any one week or combination of weeks can be used to prepare a labor cost report.

Measuring the Work Performed

The best foreman will know at the end of a day, in many instances, his unit labor cost. Suppose, for example, that his crew of plumbers installed 43 ft of 4-in. plastic pipe. The timebook would show something like this: under classification 4.9 (Installing 4-in. plastic pipe on hangers):

Name	Job classification	Hourly rate	Labor classification 4.9
Joe Jones	Foreman	$16.00	2
John Smith	Journeyman	12.00	7
Tim Tinker	Helper	8.00	7

This would compute out as:

Jones	$16.00 \times 2 =$	$ 32.00	
Smith	$12.00 \times 7 =$	84.00	$\dfrac{\$172.00}{43 \text{ ft}} = \$4.00$ per foot
Tinker	$8.00 \times 7 =$	56.00	
		$172.00	

Along with keeping track of labor expended on this item, the foreman also measured the amount of work performed that day. With many crews working, the task of reporting the amount of work completed becomes more difficult and a system of reporting this item must be developed. A weekly summary from a well-trained foreman or superintendent is commonly used. On the larger projects the demands on the superintendent's time become so great that a cost engineer reports the amount of work accomplished under each classification. By touring the project weekly or biweekly he would prepare marked-up sketches to be used with the project drawings to compute quantities of work accomplished.

Fringe Benefits

The decision as to how a company wishes to account for various monies that are added to the basic hourly wage should be made by the accountants, cost people, and estimator. Tacked on to the hourly rate are pension, health plan, vacation, disability, apprentice, travel, subsistence, industry promotion, social security, and workmen's compensation payments. Some companies work with the basic wage and consider all other payments as indirect job costs. Others apply union fringe benefits, travel, and subsistence payments and keep their records in this fashion. Whatever the decision the system should be clearly understood and consistently applied.

Preparing the Labor Cost Report

During early and late stages of a project frequent cost reporting diminishes in importance. However, during periods of intense activity involving many workers, periodic labor cost reports are very important. An experienced foreman or superintendent should be cost-conscious enough to be aware of items or areas of poor productivity prior to the release of the labor cost report. The report confirms what he anticipated and alerts his supervisor, who hopefully can assist in correcting problem areas. The labor report concentrates on three things: (1) the estimate allowance, (2) the performance for the period since the preceding report, and (3) the labor cost unit to date. To prepare the labor cost report, therefore, we require:

1. A labor classification system
2. Hours worked under each classification
3. An accurate record of work completed

   a. During the period under study
   b. Since the start of the project

4. Target costs to be met

In the examples shown in Fig. 7.3, the estimate allowance is the target amount of $5100. The estimator measured 1200 ft of 4-in. plastic pipe from the plans and believes $4.25 a foot is the correct labor unit for this work. On the report, after the code number and work description, we see "Pd 600 ft." "Pd" means the labor cost period, which in

LABOR COST REPORT NO. _____3_____ PAGE __1 of 1__

SUPT. __D.D. Doe_____ CONTRACT __#204 H.V. Corp.____

DATE __9/19/80__ PERIOD __9/1 to 9/14__ LOCATION __Eric, Pa._____

ESTIMATE ALLOWANCES		CLASSIFICATION	QUANTITY	MECHANICS		HELPERS		TOTAL	
				UNIT	AMOUNT	UNIT	AMOUNT	UNIT	AMOUNT
1200 ft @	4.9	Install 4"	Pd 600 ft.	2^{70}	1620	1^{40}	840	4^{10}	2460
4^{25} = 5100		plast. pipe	TD 900 ft.	2^{80}	2520	1^{35}	1215	4^{15}	3735

FIGURE 7.3. Labor cost report.

this case is the two weeks from 9/1 to 9/14. The period performance based on 600 ft of pipe installed at a labor cost of $2460.00 follows, showing $4.10 against a $4.25 target. The next line, "TD 900 ft," is the quantity of pipe installed to date. The total labor of 3735 divided by 900 gives the combined labor unit of $4.15/ft.

In estimating and cost work, to avoid decimal point errors amounts less than $1.00 are shown as, for example, $45^{\mathbb{c}}$. Dollar values are shown without dollar signs or cents in amounts over $10.00.

$$\text{Example:} \quad 2340 = \$2340.00$$

Amounts from $1 to $10 are shown as $6^{\underline{45}}$, which equals $6.45. This widely accepted practice ensures accuracy, particularly for manually created reports and estimates.

Using the Labor Cost Report

Labor cost reports should be prepared periodically and freely discussed with those responsible for productivity. Operations proceeding poorly will get the most attention, but since the overall labor performance is more important than any single item, operations proceeding well should be reviewed for further improvement. Company philosophies differ as to what extent the foremen and superintendents are made aware of estimate allowances. Some companies do not let their supervisors know what labor units are in the estimate. Other companies completely involve their foremen and superintendents in knowing and beating target objectives. In some cases, the best foremen and superintendents are even consulted when the estimate allowance is being established. As in sports, both the "tough" approach and the "team involvement" concept produce satisfactory results provided that good company management exists.

7.4 MATERIAL AND EQUIPMENT

A mechanical and electrical firm without a good material and equipment ordering, delivering, checking, and issuing system will probably fail in a short time. Paperwork starts with ordering and the record grows with the checking of deliveries. The material portion of a cost report can readily be prepared from the accountant's records. A significant material shortage should signal that something is wrong. Perhaps the amount of material was underestimated. In such a case the labor allowance may also turn out to be grossly insufficient.

Rental Versus Company Ownership

Some building construction companies of considerable size own very little construction equipment. They consider the problems of maintenance, yard space, utilization, and ownership disadvantages to the extent that they prefer to utilize their capital in better ways. Renting equipment follows the easy and ceaseless trend toward specialization. Cost keeping involves simply applying the rental rate to the labor classifications on which the rented machine works.

Many contractors prefer to own their own cranes, backhoes, and bulldozers. Good business practice makes it imperative that they charge the project using a machine at a *rate* that reflects all the *owning* and *operating* costs. Since money must be invested in a company-owned machine:

"Money invested in plant and equipment should produce an income, the same as any other investment. The amount of income that should be produced varies in the market, but obviously, to show a *profit,* the amount should be greater than the premium the company has to pay to borrow money from the bank."*

Owning costs include:

1. Investment costs (interest, insurance, taxes, storage)
2. Maintenance (major repairs and replacement)
3. Depreciation (loss of value by wear and obsolescence)

The hourly rate includes the more obvious *operating costs:*

1. Fuel and lubrication
2. Minor repairs and small parts
3. In and out charges (transportation)
4. Operators' wages, including fringe benefits

7.5   PROJECT OUTCOME REPORTS

Many companies do not prepare true project outcome reports. They recognize potential labor losses only to the extent that they may have occurred. They do not attempt to *analyze* future performance. A true project outcome report makes use of good judgment to estimate the *final result.* Using the performance report to that date, quantities of work remaining and the labor unit applicable are evaluated. The estimated quantity of work is reexamined since there may be more or less work to perform upon reevaluation. *High*-labor-cost units when reviewed might indicate that the balance of the work will be simple and easier to perform, justifying using a *lower* unit for the remaining work. Where the balance of the work appears more difficult, higher units for the remainder of the work would be prudent. Obviously, some training, good judgment, and consultation with project personnel will produce the best project outcome reports.

In Fig. 7.4 we use the labor cost information from Fig. 7.3. The person preparing the report knows that the work remaining is actually 325 ft and must be run through an area much more difficult for the workers. In his judgment the unit cost for this remaining work will be $4.40 per foot.

A project outcome report also includes a *material, subcontract,* and *indirect cost* study. The savings and overruns are summarized for all categories and anticipated project earnings are compared with the original estimate. With these reports company officers know the status of individual projects as well as company-wide performance summarized for all contracts. Executive action based on these reports might result in changes in method, workers, or supervisory personnel.

*Keith Collier, *Fundamentals of Construction Estimating and Cost Accounting,* Prentice-Hall, Inc., Englewood Cliffs, N.J., 1974.

LABOR SECTION

CONTRACT #204	H.V. Corp	LOCATION	Erie, Pa.	DATE	Sept. 30, 1980

CODE	DESCRIPTION	ESTIMATE	INDICATED COST	SAVING	OVERRUN
4.9	Inst. 4″ pl. pipe	1200 ft @ 4^{25} = 5100	TD 900 ft @ 4^{15} = 3735 Bal. 325 @ 4^{40} = $\dfrac{1430}{5165}$		65

FIGURE 7.4.   Project outcome report 1.

7.6   BUDGET REPORTS TO OWNERS

On cost-plus and construction management contracts, the owner receives periodic budget reports. The total estimated cost of the job is broken down into the principal divisions of the work, such as excavation, foundations, structural steel, plumbing, HVAC, electrical, and so on. The report contains a comparison of these trade estimates compared against subcontract awards for that portion of the work. The status of changes to the work is shown according to the degree to which they have been processed:

First—approximate change orders

Second—pending change orders

Third—approved change orders

Approximate change orders designates those being estimated by the various contractors. *Pending change orders* designates those being evaluated and reviewed by the general contractor, architects, and engineers. *Approved change orders* designates those that have been accepted by the architect and the owner. Since changes to the mechanical and electrical systems generally amount to over 50% of the dollar value of change orders, these contractors should expedite submitting approximate and processed change-order figures. This in turn will keep to a minimum the amount of change-order work they must carry on their books while change orders are being processed.

7.7   COST CONTROL AND ESTIMATING

The relationship between cost control and estimating can be shown in a circular or triangular diagram such as that of Fig. 7.5.

Accurate estimating and intelligent bidding leads to a contract award. Sound cost control ensures meeting estimate objectives. The cost records from completed projects are added to the estimator's permanent cost data. Each leg of the triangle helps improve the integrity of the whole triangle. The better the cost control, the better the estimating data, and finally better estimates—which can also be stated: "Nothing succeeds like success."

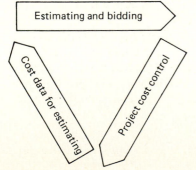

FIGURE 7.5.   Cost data estimating cycle.

8

Quality Control

8.1 INTRODUCTION Quality control in mechanical and electrical construction has a somewhat different character from the architectural trades with which we usually associate the term. Although appearance and performance in brick laying, tile setting, plastering, painting, cement finishing, and so on, relate strongly to each other in producing a quality product, appearance receives the strongest emphasis. In structural, mechanical, and electrical work, on the other hand, the concentration is on performance. Due to the nature of mechanical and electrical work, appearance largely results from an off-site manufacturing operation. Plumbing, for example, terminates in fire hose cabinets and toilet room fixtures. All most people see of an HVAC system are the diffusers, window induction units, and thermostats—all manufactured items. Elevator, moving stair, and electrical work make similar use of factory items for finish appearance. Final appearance depends mostly on the equipment and fixtures specified and the color and finish selected by the architect.

High quality in mechanical and electrical work relates to things such as dependability, long life, freedom from maintenance, and quiet operation. As we study these trades in succeeding chapters and the functions their various systems perform, we will become acquainted with the control over quality required of the construction supervisor. We will learn how the various codes and licensing requirements establish a standard which helps ensure quality. We will see how apprentice training programs, industry training programs (elevators, HVAC controls, and so on) and certification (welding) are all directed toward producing trade proficiency and high-quality workmanship.

8.2 TESTING Because our work in mechanical and electrical contracting concerns operating systems, testing assumes much greater importance than in the architectural parts of buildings. Public health and fire safety dictate many of the tests performed. Much of our testing evaluates the integrity of the system with x-ray and ultrasonic testing of welds and liquid testing of plumbing, heating, fire, and cooling lines. Specific tests are conducted along safety lines for elevators, moving stairs, sprinkler systems, fire dampers, and alarm and electric systems.

The prudent construction manager recognizes the importance of mechanical and electrical testing procedures and the interruption to everyday production that testing entails. Work on a system must cease, test apparatus rigged, and the test successfully conducted and inspected before architectural work covering the system can proceed. Testing must be thought out and made part of the project schedule to minimize inconvenience and delay to following trades.

8.3 QUALITY AND THE DESIGNER Because elevators, chillers, fans, faucets, pumps, light fixtures, motors, valves, and so on, *operate,* the quality specified makes a world of difference. This quality factor the workers and contractor on the job do not control. The architect and mechanical engineer determine capacity, size, weight, energy consumption, degree of maintenance, vibration, noise, and other important factors when they specify mechanical and electrical equipment. Low initial cost may have been dictated by the owner to the detriment of energy consumption, quiet operation, and minimum maintenance. What the laymen sees in the exposed equipment, diffusers, fixtures, and so on, has been selected by the designers from technical sales information. The finished appearance conforms to a great extent to a standard set by the manufacturer and accepted and selected by the designer. The workers and contractor essentially can concentrate on proper installation and protection from damage by other trades provided that the specified material and equipment have been delivered.

8.4 INSPECTION BY GOVERNMENT AUTHORITIES Quality does not solely lie in the hands of those who design and construct buildings. Depending on the local code, inspection of mechanical and electrical work is the function of municipal and other government inspectors. Boilers, elevators, underground sewage lines, plumbing systems, fire damper controls on ventilation systems, fire standpipe, fire alarm, and electric systems all fall within the inspection jurisdiction of the local authorities. Their inspections cover proper materials, adequate support, workmanlike joining of materials, and satisfactory operation.

8.5 CONCLUSION Despite the effort made to ensure quality by the architects, mechanical and electrical engineers, general and trade contractors, manufacturers, municipal inspectors, and proficient workers, quality problems arise with this part of building work. Bearings burn out, air volumes or temperatures are excessive or insufficient, machinery vibrates, and plumbing makes noise. With such a large volume of the construction dollar involved and the operational nature of the work, it is only natural that some problems do arise early in the occupancy of a building. It has been the author's experience that most of these irritating problems are expeditiously solved with far less adverse owner's reaction to the contractors than we experience from a leaky roof or curtain wall. Good quality is an inherent characteristic of mechanical and electrical construction.

Plumbing

9.1 PLUMBING CONTRACTING   The plumbing contracting industry consists of numerous small and medium-size firms mostly confining activities to a restricted area. Whereas some of the manufacturers of fixtures tend to dominate their field, plumbing contractors generally neither monopolize the business in their part of the country nor grow to a very large size compared to their counterparts in HVAC and electrical construction. Four factors probably cause this condition:

1. Plumbing generally is the smallest of the mechanical and electrical contracts.
2. Most of the work consists of on-site assembly of standard materials installed in an established manner, leaving little or no opportunity for innovation.
3. Very little of the contract consists of large expensive equipment or substantial subcontracts offering significant purchasing possibilities.
4. These three factors combine to create a condition of strong competition, limiting dramatic growth of any one contractor.

Plumbing's importance stems more from its function than from the size of its cut of the construction pie. This can be seen from Table 9.1, which shows plumbing as a percentage of the total cost of certain general types of buildings.

TABLE 9.1   PLUMBING AS A PERCENTAGE OF TOTAL COSTS

Type building	Median %
Apartments	8.7
Churches	4.8
College classrooms and administration	6.5
Factories	6.1
Hospitals	9.1
Offices	5.3
Schools—senior high	7.6
Supermarkets	6.0
Warehouses	4.6
Average	6.5

Source: This information is copyrighted by Robert Snow Means Co., Inc., and is reproduced from the 1982 edition of *Building Construction Cost Data* with permission, pages 276–285.

9.2   THE IMPORTANCE OF PLUMBING

Plumbing ranks very high as one of the elements of a building that we operate and actively use. Unlike static items, such as exterior walls and foundations, the plumbing systems operate continuously to satisfy the needs of the building occupants. Our sanitation, drinking water, cleanliness, cooking, and fire safety depend on plumbing. Proper functioning of a building's plumbing system is essential by law to the occupancy of a building.

9.3   REGULATION

Licensing

To safeguard the well-being of the public and the building occupants a three-part program is widely used throughout the nation, consisting of a code, licensing, and an inspection program. Because of its importance to health, most organizations in authority require the licensing of plumbing contractors from single entrepreneurs to large-scale firms. "This is the first in the list of trades that require licensing almost everywhere. This license is required before any contractor can install or alter any plumbing system, or gas piping. The granting of license is done by an appointed state or local plumbing board. An applicant usually must have had 10 years of experience in the design and installation of plumbing systems. An oral or written examination is required, and sometimes a practical examination is required. An applicant for a plumbing license must be familiar with all the codes and rules and regulations referring to the installation of piping systems to supply potable water and for the safe elimination of all wastes."*

Codes

A plumbing code is that part of the law which covers the design and installation of building plumbing systems. States and cities have their own plumbing codes. To assist in creating and maintaining good codes, the Building Officials and Code Administrators International (BOCA) every three years publishes "The BOCA Basic Plumbing Code," which they believe is "the most complete and up-to-date plumbing

---

*Laurence E. Reiner, *Handbook of Construction Management,* Prentice-Hall, Inc., Englewood Cliffs, N.J., 1972, pp. 182, 183.

code available." These codes cover the design and installation of water supply and distribution systems, storm water drainage, waste removal, venting systems, and all other aspects of plumbing. Plumbing design engineers must possess a complete knowledge of the local code governing the building on which they are working. The plumber is also looked on as someone who should know the code well, and although not responsible for the plumbing design as explained in AIA Document A201, General Conditions of the Contract for Construction, if he notices that any of the contract documents are at variance with applicable laws, statutes, building codes, and regulations, he is required to notify the architect.* "Further, if the contractor performs any work knowing it to be contrary to such laws, ordinances, rules and regulations, and without such notice to the architect, he shall assume full responsibility therefore and shall bear all costs attributable thereto."†

Because of their expertise we refer to subcontractors as specialty contractors since they are recognized as having special knowledge in their field. We expect them to help keep us—owners, architects, engineers, and general contractors—out of trouble. This should be a reasonable assumption to rely on since the law is specific, and the contractor has been duly examined and carries a license.

Permits. A proper code will require that before any plumbing construction can proceed, a plumbing permit must be obtained. Permits are not required for maintenance, repairs, or removal of stoppages, but are required for alterations to an existing system.

Most authorities issue permits based on a review of the plumbing plans to ensure compliance with the code. In most localities plumbers pay a modest fee for the necessary permit.

Inspection

The third part of a sound regulatory program requires on-site inspection. Qualified organizations making installations in accordance with accepted practice backed up by an inspection program produce safe and sanitary plumbing systems. The nature of these tests for water supply and waste removal lines will be covered under specific headings to follow.

9.4   WATER SUPPLY AND DISTRIBUTION

Historians, sociologists, and other experts of ancient studies traditionally use the progress shown in the use of water as a measure of the degree of civilization development. Ancient Egypt and Syria had extensive canal systems and we recognize the Romans for their superb acqueducts and lovely bath facilities. The renewed interest in the quality of our own drinking water and the extensive current federal program for cleaning up our rivers and lakes demonstrate the public's determination to have pure clean water for drinking, cleanliness, and recreation. Authorities with fire safety responsibilities must also concern themselves with providing a large capacity of water at an adequate pressure for firefighting purposes. In addition, many manufacturing processes require large quantities of high-quality water.

Most owners need only tap into the existing water service in the street for their water needs. However, many private homes and manufacturing plants have found it advantageous or necessary to drill their own wells (Figs. 9.1 and 9.2). The development of the full potential of a water resource requires the services of water supply engineers and well drilling specialists. In sandy locations without municipal water supply or

---

* Section 4.7.1.

† Section 4.7.2.

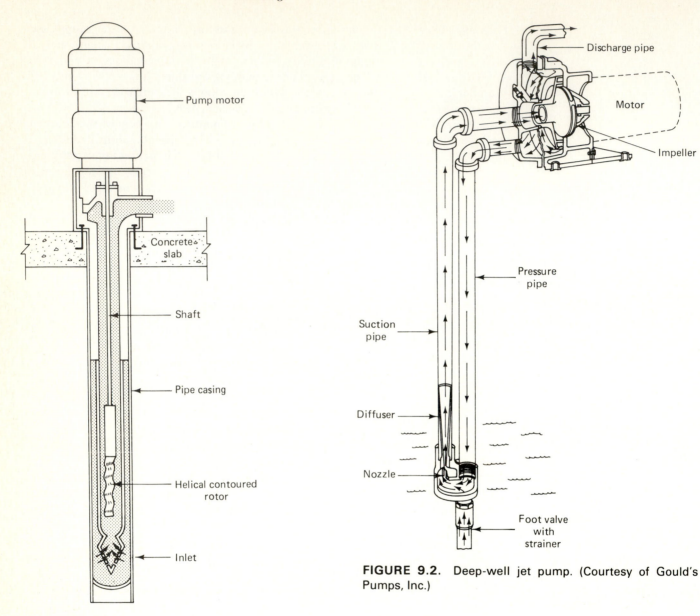

FIGURE 9.1.   Deep-well pump.

FIGURE 9.2.   Deep-well jet pump. (Courtesy of Gould's Pumps, Inc.)

sanitary sewers, by common practice the well is often located in the front yard and waste disposal by septic tank and leaching field occurs in the back. By code the water service must be separated from any possible sources of contamination, such as sewer lines, septic tanks, cesspools, and drainage fields. To ensure against corrosive action, or rupture due to settlement, clearance must be provided as the water service passes through the foundation wall. This is very important. This type of rupture occurred in a large New York City bank building due to subsidence of the water main in the street, resulting in serious flooding of five basement levels, all elevator pits, and a large computer area.

While installing the potable water supply system as required by the design drawings the plumbing contractor must be aware of the basic requirements for these systems. Separation of the system, regulation of pressures, elimination of water hammer, isolation of the parts of the system for control and ease of maintenance, and provision of adequate supply constitute a complete system.

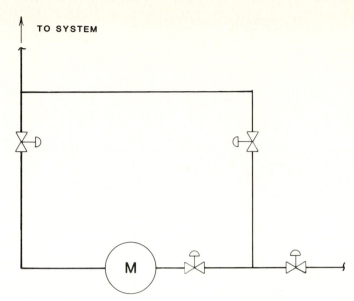

↑ TO SYSTEM

M

FIGURE 9.3. Water meter with valves and bypass loop.

Valves and Meters

Codes require a shutoff valve at the curb outside the building and near the point of entry inside the building. Valves are required at the meter and its bypass. Figure 9.3 shows a meter and valve assembly for incoming water service. The plumber prefabricates this in his main shop, utilizing his fabrication specialists to reduce costs and eliminate errors. On the site the assembly is connected to the incoming service line. Valves are required at the bottom of upfeed and at the top of downfeed risers. In addition, owners and designers add valving to divide the system into smaller parts for the convenience of maintenance workers, building occupants, and tests required of the plumber. Some typical valves are shown in Fig. 9.4. Whenever possible, the plumber should install all valves in readily accessible well-marked locations.

FIGURE 9.4. Typical valves. (Courtesy of Crane Co.)

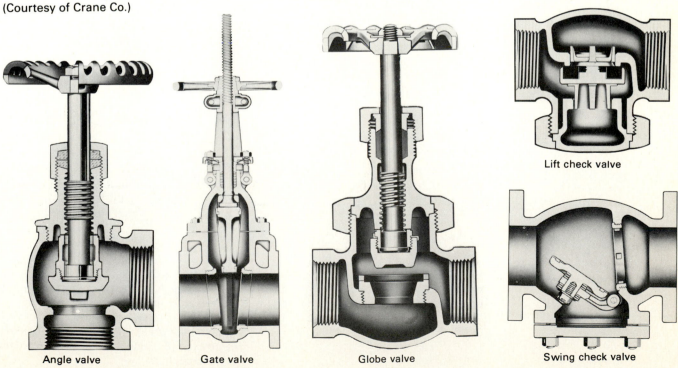

Angle valve   Gate valve   Globe valve

Lift check valve

Swing check valve

Vacuum Breakers, Check Valves, Air Gaps, and Backflow Preventers

Our interest in isolating the potable water supply derives from our concern with waterborne diseases. Not many years ago the Holy Cross football team contracted hepatitis and did not play for the entire season. The players had drunk from a drinking fountain contaminated by an underground lawn sprinkling system. Unfortunately, water can flow in more than one direction in a water supply system due to differential pressures; we prevent this occurrence with vacuum breakers, check valves, air gaps, and blackflow preventers. A urinal flushometer contains a vacuum breaker which prevents water from being syphoned into the flushing water supply line.

    The check valves shown in Fig. 9.5 rely solely on the sealing ability of the valves for protection. They permit flow in only one direction, thereby protecting the potable water supply. Figure 9.6 shows the use of an air gap in a faucet. Other influences, such as the effective supply opening area and its relationship to a side wall, are illustrated in Fig. 9.7. Figure 9.8 shows the provision the code makes for the possibility of having the overflow freeze or clog up by providing an air gap equal to twice the diameter of the supply pipe. The best means of preventing backflow utilizes differential pressure, check valves, and an air gap, as shown in Fig. 9.9.

FIGURE 9.5.   Double check valves. (Photo courtesy of Hersey Products.)

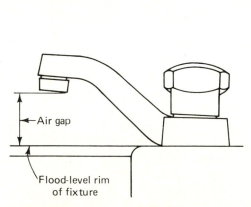

FIGURE 9.6.   Air gap and an effective opening. (Courtesy of Building Officials & Code Administrators International, Inc.)

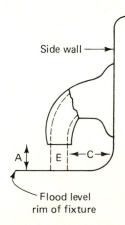

FIGURE 9.7.   Near-wall influence on an air gap. A, air gap; C, distance from the side wall to the effective opening; E, effective opening (minimum cross-sectional area). (Courtesy of Building Officials & Code Administrators International, Inc.)

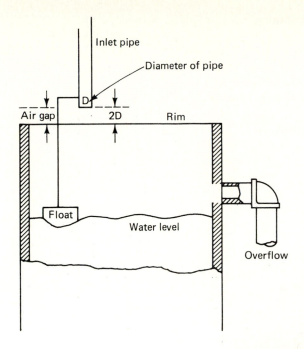

FIGURE 9.8. Air gap in an open tank with overflow. The air gap should not be less than two times the diameter of the inlet pipe. (Courtesy of Building Officials & Code Administrators International, Inc.)

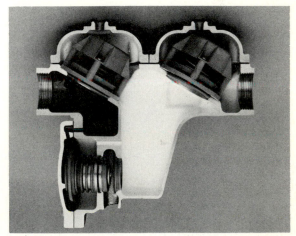

FIGURE 9.9. Cutaway of a small backflow preventer. (Photo courtesy of Hersey Products.)

Air Chambers and Shock Absorbers

Many younger readers brought up in modern homes and schools may not have experienced the annoyance of water hammer. The term "hammer" is very descriptive. In old-fashioned steam heating systems the arrival of heat was accompanied by noisy pounding of water in the system. The rapid closing of a water valve creates a heavy shock, but since trapped air compresses readily, air chambers (Fig. 9.10) installed in the proper location prevent water hammer.

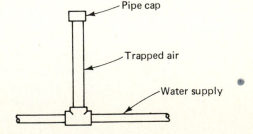

FIGURE 9.10. Job-installed shock absorber.

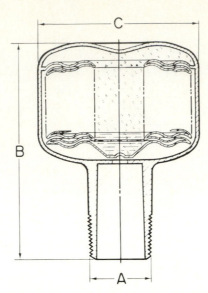

FIGURE 9.11. Stainless steel shock absorber. (Courtesy of Tyler Corp.)

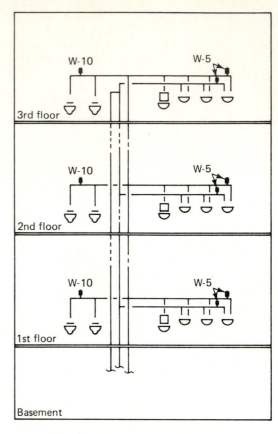

FIGURE 9.12. Proper location for shock absorbers. (Courtesy of Tyler Corp.)

Figure 9.11 shows a shock absorber utilizing a stainless steel bellows filled with nitrogen gas. These compact devices run little risk of malfunction and need no maintenance. Figure 9.12 illustrates the location and use of various-size absorbers.

9.5 WATER QUANTITY AND PRESSURE

Firefighting requirements and potable water use are compatible concerns in municipal installations. As a result, most American cities provide adequate pressure and quantity for the vast majority of water users. In tall buildings and those located on higher elevations where water service pressure is inadequate, water needs can be provided for in three ways: (1) direct pumping, (2) utilizing hydropneumatic tanks, and (3) the use of gravity tanks.

For direct pumping a minimum of two pumps are installed. They are selected by the designer to produce the quantity and pressure required at that location. Most users wish them to be programmed to pump alternately increasing the longevity of the individual pumps (see Fig. 9.13). In very tall buildings this pumping chore is shared by pumps in the basement and a second set located in an upper-story mechanical equipment room.

The hydropneumatic system has gained popularity with homeowners using their own water supply. The pump partly fills the tank with water until the air trapped at the top compresses to the required pressure. The pump does not have to operate again until water usage reduces the pressure to the low cut-in level. Tanks and pumps are sized for large installations as well as individual homes.

The third method utilizes gravity or house tanks located at the top of a building or, in the case of buildings over 40 stories high, tanks located on the mechanical equipment floor levels. In the latter case the building has separate zones serviced by these tanks (see Fig. 9.14). A float switch controls the pumping.

FIGURE 9.13. Water supply pump.

FIGURE 9.14. Building water supply system.

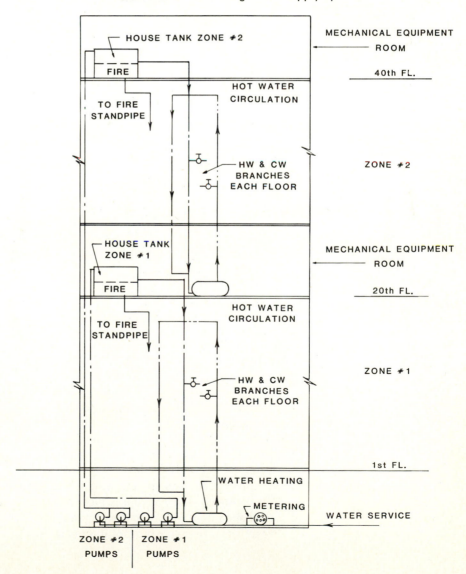

Even with a power failure and no emergency electrical pumping, the piping and tank design ensures a minimum of at least 50% of the capacity for firefighting with a gravity tank, although under normal conditions the full capacity would be available. To keep a building operational during an electrical blackout, at least one house pump should be hooked into the emergency electrical generating system.

Even in these advanced times, gravity tanks are constructed of wood with steel rods or bands. Since the tank should always be full of water, it is considered fire safe. Designers generally specify cylindrical and cube-shaped steel tanks for use inside the building enclosure.

Too much pressure also presents problems; therefore, the code calls for pressure-reducing valves where pressures exceed 80 psi. This rule applies to incoming service and pressure at any given fixture. A pressure-reducing valve is shown in Fig. 9.15.

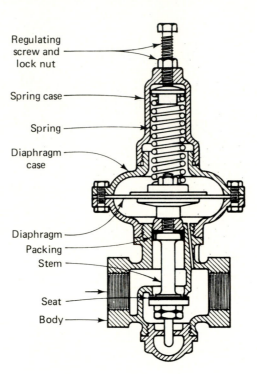

Regulating screw and lock nut

Spring case

Spring

Diaphragm case

Diaphragm

Packing

Stem

Seat

Body

FIGURE 9.15. Pressure-reducing valve. (Courtesy of ITT Hoffman Specialty.)

9.6 WATER SUPPLY INSTALLATIONS

Materials

Although the vast majority of homes and commercial buildings require only a short water service run, many installations for industrial plants, shopping malls, or campus-type office complexes require medium-size underground water supply installations. Cast iron pipe with bell and spigot joints for this kind of service has an excellent record for low maintenance cost and dependable performance. The pipe is joined by installing the spigot end into the bell portion and properly aligning the pipe. Oakum in rope form is inserted between bell and spigot and tightly compacted using a yarning iron (Fig. 9.16a). An asbestos runner wrapped around the joint retains the molten lead being poured into the joint (Fig. 9.16b). The solidified lead is then wedged tightly against the bell and spigot, using inside and outside caulking irons (Fig. 9.16c).

No-hub cast iron pipe, which is easier to manufacture and ship, joins by inserting the ends into a neoprene shield. Stainless steel clamps fasten pipe and shield tightly together to form a leakproof joint (Fig. 9.17).

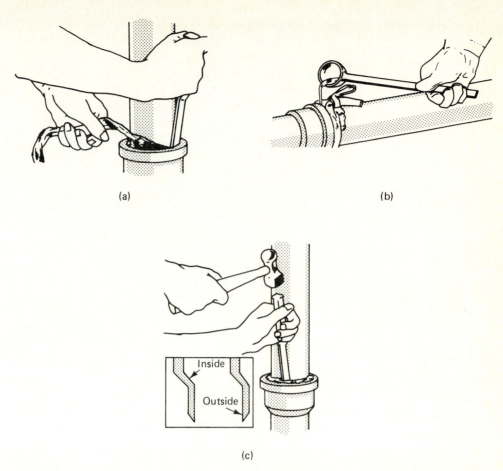

FIGURE 9.16. (a) Compacting oakum; (b) Pouring lead; (c) Caulking lead. (From T. Philbin, *Basic Plumbing*, Reston Publishing Company, Inc., Reston, Va., 1977.)

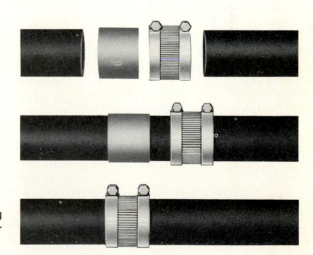

FIGURE 9.17. Steps in joining no-hub piping. (Courtesy of Tyler Corp.)

Polyvinyl chloride water service pipe, competitive in price with cast iron, has good corrosion-resistant properties. The pipe can be easily sawed and surfaces smoothed for joining. One type has a fiberglass epoxy covering and bell and spigot joint (Fig. 9.18). A lubricant facilitates inserting the spigot end into the bell and the rubber ring makes a tight seal (Fig. 9.19).

FIGURE 9.18. Joining PVC pipe. (Courtesy of Johns-Manville Sales Corporation.)

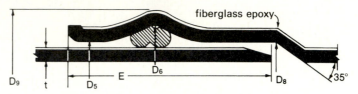

Size	D5/D8	D6	D9	OD	t	E
4″	4.85	5.70	5.80	4.80	.142	3.75
6″	6.96	7.94	8.06	6.90	.183	4.50
8″	9.12	10.29	10.45	9.05	.216	5.25
10″	11.19	12.54	13.21	11.10	.270	6.14
12″	13.30	14.79	15.64	13.20	.325	6.73

FIGURE 9.19. Cross section through a bell, spigot, and rubber seal PVC joint. (Courtesy of Johns-Manville Sales Corporation.)

Asbestos cement water service pipe also has superior corrosion resistance over ferrous materials under certain soil and electrolytic conditions. Its low material and installed cost make it attractive to designers, particularly for larger installations. Although perhaps not as foolproof or resistant to damage as other materials, asbestos cement has gained wide acceptance.

Joints are made by utilizing pipes with tapered ends which are forced into a coupling containing rubber gaskets seated in grooves in the coupling. The wedging

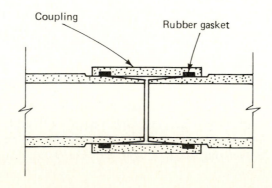

FIGURE 9.20. Asbestos cement pipe joint.

action compresses the rubber gasket to form a tight joint (Fig. 9.20). Since copper is used extensively for water supply inside the building 2- to 4-in. incoming services of copper are common. Copper is joined by soldering as shown in Fig. 9.32.

Methods

In building construction installations, most water services are located 5 to 10 ft below ground level, which usually does not involve a substantial hazard to workers or problem to the contractor. A backhoe, also called a pull shovel, excavates only enough trench to lay (Fig. 9.21) one joint (length) of pipe at a time. This prevents cave-ins and permits the machine to assist with the laying of the pipe. Depending on the type of soil, a shield (Fig. 9.22) may be required to protect the workers from

FIGURE 9.21. Laying cast iron pipe. (Courtesy of American Cast Iron Pipe Co.)

FIGURE 9.22. Use of a steel protective shield for pipe laying. (Courtesy of American Cast Iron Pipe Co.)

cave-ins. In unstable soil, banks are held in place with solid wood sheeting or sheet steel piling (Fig. 9.23) or with shoring consisting of vertical wood planks and horizontal wood timbers or screw jacks spaced 2 ft or more apart according to the soil conditions. Folding metal shoring drops into place quickly and safely, as shown in Fig. 9.24. Activation of the hydraulic pistons retains the earth's lateral pressure. Products like these can often be rented, saving the contractor's capital for other uses. Where rock occurs, drilling and blasting using light charges and millisecond-delay explosive discharge give good fragmentation with little throw for easy excavation. Rerouting to avoid rock makes good sense since trench excavation in rock costs 15 times that of similar earth excavation.

  High groundwater levels mean dewatering and considerable extra expense to the contractor. In earth and clay, pumping with a light 2- or 3-in. centrifugal pump supplemented by a bedding of small crushed stone usually puts the pipe "in the dry" (Fig. 9.25). Since water accumulates overnight the pump operator should report in early to pump the trench dry to avoid having an expensive machine and crew stand idle while the water recedes. In sandy soil a well point system is imperative since it is impossible to work in wet, runny sand. From a central pumping point a long header

FIGURE 9.23.   Pipe laying using sheet steel piling in unstable soil. (Courtesy of Ductile Iron Pipe Research Association.)

FIGURE 9.24.   Folding shoring, installed folded. The jacks should be at right angles to the channels to take the Earth's thrust.

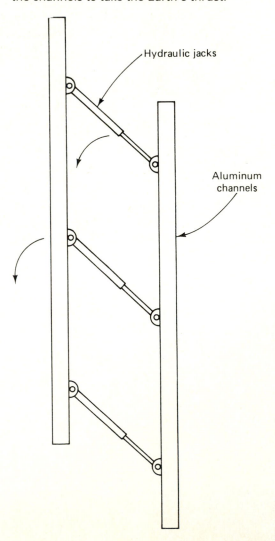

Hydraulic jacks

Aluminum channels

Type 4

FIGURE 9.25. Laying condition: pipe bedded in sand, gravel, or crushed rock to a depth of ⅛ pipe diameter, 4 in. minimum. Backfill compacted to top of pipe approximately 80% Standard Procter. (Courtesy of Ductile Iron Pipe Research Association.)

pipe parallels the section to be laid. Well points are installed at regular intervals along the header perhaps 16 ft apart, depending on conditions, sufficient to lower the water table while the pipe is being installed and tested. Well point systems must usually operate 24 hours a day, 7 days a week, although most contractors will at least *try* reducing these hours. The large amount of pump, pipe, and well point equipment, fuel or electric energy, and labor make this method of dewatering expensive, essential as it may be.

Although the joining of the pipe has been made simple and dependable, other factors are required to produce proper underground installations. The firm support of the invert of the pipe is of utmost importance. Because machine pipe laying operations disturb the soil, the proper pipe support is usually created *after* the joining of the pipe by compacting good supporting soil or fine crushed stone under the lower third of the pipe.

In marshland where some of our airports have been built, water lines are supported on concrete cradles or on a thick wide bed of crushed stone (Fig. 9.26). Sometimes piling supports the line from below when outside the buildings (Fig. 9.27) or the pipe hangs from pile-supported slabs (Fig. 9.28) inside the structure.

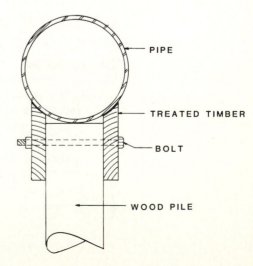

FIGURE 9.26. Pipe supported on concrete cradles. (Courtesy of Ductile Iron Pipe Research Association.)

FIGURE 9.27. Pile support for pipe.

PIPE

TREATED TIMBER

BOLT

WOOD PILE

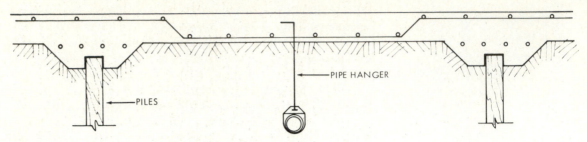

FIGURE 9.28.   Underground pipe support in unstable soil.

9.7 TESTING

Domestic water service lines are tested to meet code requirements and in general should withstand a pressure of at least 1.5 times the working pressure. The line is left with joints exposed until the test has been completed, but may be partly backfilled to hold the pipe in place. By bleeding all air from the section under test the line fills completely with water, and testing apparatus equipped with gauges applies the required pressure. Inspectors look for any drop in pressure and any visual leaking of the joints during the test period. On satisfactory completion the line is carefully backfilled with clean fill. Fill installed in 6-in. layers is compacted with hand-operated compactors until 18 in. of compacted material covers the top of the pipe. Properly installed, water service piping performs satisfactorily for decades on end. Most leaks and breaks stem from improper support of the lower third of the pipe and careless, overambitious backfilling.

9.8 FLUSHING AND DISINFECTING

Before placing a new supply line in service, thorough flushing and disinfecting are required. Authorities recommend a flushing rate of 2.5 ft/sec for supply lines to commercial buildings.

"Disinfection of mains should be accomplished only by workmen who have had experience with chlorine or other disinfecting agents. Chlorine gas and water solutions are fed into the main being disinfected to a concentration of at least 50 parts per million available chlorine. This chlorinated water solution should remain in the pipe for at least 24 hours, at the end of which period the chlorine concentration should be at least 25 parts per million. If this is achieved final flushing can then be accomplished and chlorine residuals checked to determine that the heavily chlorinated water has been removed from the pipeline."*

9.9 COORDINATION OF UNDERGROUND PLUMBING LINES

During the construction period of most buildings, such as offices, schools, apartments, or manufacturing units, ample time exists for installing the many underground services. Besides the water service the plumber runs in the gas line and connects the sanitary and storm water drain lines. In addition, the electrical service and steam supply lines, where central steam exists, require coordination with the general contractor and other subs to minimize the inconvenience of installing those particular services. Underground work should be scheduled to avoid tieing up ground space needed for steel erection, concrete, forming, reinforcing steel, concrete pouring, and erection of precast concrete panels. Particularly in a city, all trades use the area around a building for receiving material and for access for cranes and hoists. Connecting to services should be expedited to reduce the inconvenience to all.

---

* A Guide for the Installation of Ductile Iron Pipe, Cast Iron Pipe Research Association, Oak Park, Ill., 1972.

Particularly on one-story industrial work, expediting the underground work avoids delays in pouring the slab on ground—an important milestone in this type of construction. Close liaison with the construction manager results in a well-planned program benefiting both the plumber and the other trades.

Inside the building in basements, mechanical equipment rooms, and industrial buildings, water supply lines usually run exposed, hung from the slab or roof structure above. Vertical lines (risers) generally are hidden and run in pipe chases, preferably enclosed in masonry. Where horizontal lines are supported by poured concrete slabs, the plumber prepares shop drawings showing dimensioned location of hangers. As soon as the concrete contractor has the metal or plywood slab forms in place, the plumber lays out horizontal pipe runs and hanger locations and fastens metal inserts to the forms for future support of pipe (see Fig. 9.29). Examples of hanger inserts available for casting into concrete are shown in Fig. 9.30. Figure 9.31 shows the method of supporting the vertical riser load of water and piping at various floor levels. Support is specified based on type and size of piping.

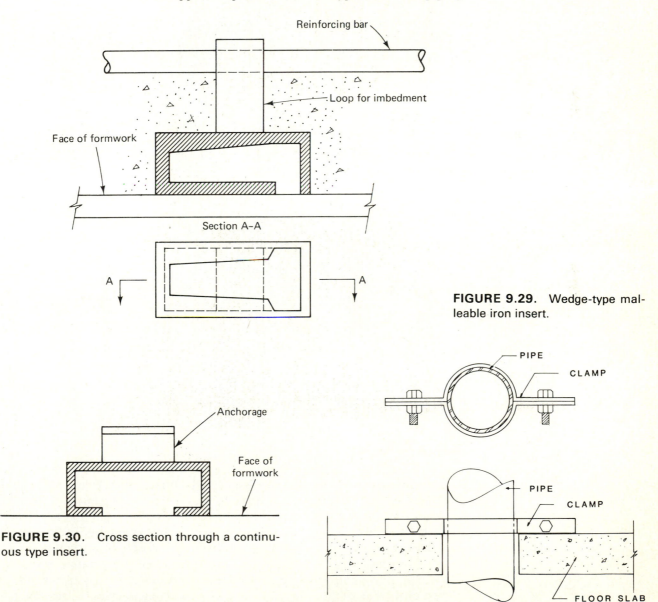

FIGURE 9.29. Wedge-type malleable iron insert.

FIGURE 9.30. Cross section through a continuous type insert.

FIGURE 9.31. Method of supporting a pipe riser at floor level.

Inside the building a water distribution system can be designed using brass, copper, copper tubing, galvanized iron, galvanized steel, lead, or plastic piping, depending on the local code. Some codes prohibit the use of plastic piping for commercial buildings while permitting its use in one- and two-story residential construction. Copper tubing ½ to 4 in. and galvanized steel 4 in. and larger are the most widely used materials due to low installed costs. The copper tubing cuts quickly and readily lends itself to preassembly (Fig. 9.32). An entire assembly can be checked for fit and alignment before joining the piping. Copper also will not rust, which can occur with galvanized steel pipe. Galvanized steel pipe is usually joined with threaded connections. Welding of steel pipe and the use of Victaulic connections utilizing grooved pipe is economical for the larger, 6- to 10-in. pipe (see Section 13.9).

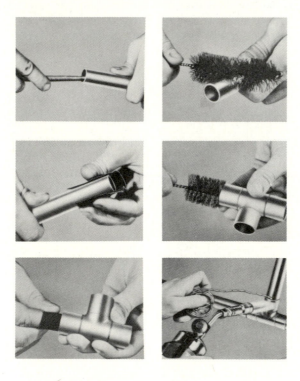

FIGURE 9.32 Steps in making a solder joint. (Courtesy of Crane Co.)

Except in very large buildings, water distribution pipe is readily handled by workers with a minimum of equipment. A come-along or chain-falls helps raise the pipe and hold it in proper position for joining. Adjustable hangers hold and help position the pipe for horizontal runs. The plumbers work from metal rolling scaffold, which puts them in good working position.

9.10 HOT WATER SERVICE

The heating of water has received a great amount of attention from designers, inventors, and users since the oil crises of the 1970s. In our effort to conserve energy we are dramatically increasing the efficiency of our water heating systems. Heat recovery plays a big part in this approach, where heat from refrigeration systems, manufacturing processes, and flue gases from direct consumption of fuels is recaptured. Water and air are the two principal media for this type of heat recovery.

In many installations the incoming water supply temperature is increased by solar or heat recovery methods and the basic hot water heating equipment brings it the rest of the way to the required heat level. In electing to invest in the piping, tanks, controls, equipment, and labor to make these important energy savings, the designer makes a cost-effectiveness study to justify the additional initial costs, interest on that sum for the study period selected (say 20 years), additional operating and maintenance costs, salvage, and so on. Few owners are willing to invest in energy saving unless they can make a true monetary saving in the time period chosen.

Most residential houses have an electric or gas hot water heater containing storage capacity independent of the home heating system (see Fig. 9.33). In commercial applications water is often heated in conjunction with building heating boilers. Where a central steam system supplies the building, a converter heats the water to the required temperature (see Fig. 13.5). Where electricity is cheap, such as in hydroelectric areas, large buildings utilize electric heaters (see Figs. 9.34 and 9.35).

FIGURE 9.33. Typical gas hot water heater. (Courtesy of Building Officials & Code Administrators International, Inc.)

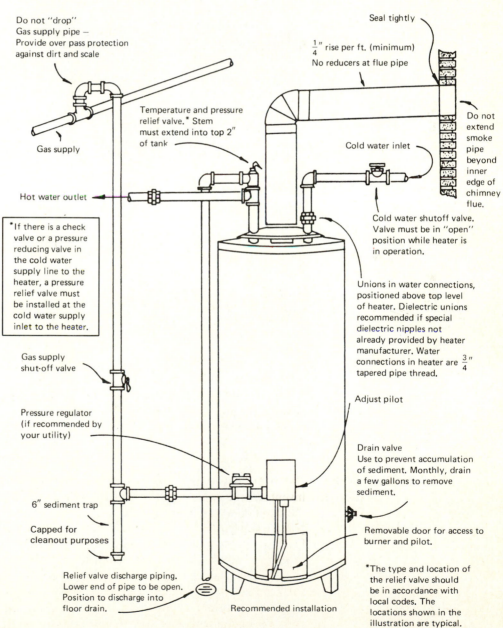

Do not "drop"
Gas supply pipe —
Provide over pass protection
against dirt and scale

Gas supply

Temperature and pressure
relief valve.* Stem
must extend into top 2"
of tank

Hot water outlet

*If there is a check
valve or a pressure
reducing valve in
the cold water
supply line to the
heater, a pressure
relief valve must
be installed at the
cold water supply
inlet to the heater.

Gas supply
shut-off valve

Pressure regulator
(if recommended by
your utility)

6" sediment trap

Capped for
cleanout purposes

Relief valve discharge piping.
Lower end of pipe to be open.
Position to discharge into
floor drain.

Seal tightly

$\frac{1}{4}$" rise per ft. (minimum)
No reducers at flue pipe

Cold water inlet

Do not
extend
smoke
pipe
beyond
inner
edge of
chimney
flue.

Cold water shutoff valve.
Valve must be in "open"
position while heater is
in operation.

Unions in water connections,
positioned above top level
of heater. Dielectric unions
recommended if special
dielectric nipples not
already provided by heater
manufacturer. Water
connections in heater are $\frac{3}{4}$"
tapered pipe thread.

Adjust pilot

Drain valve
Use to prevent accumulation
of sediment. Monthly, drain
a few gallons to remove
sediment.

Removable door for access to
burner and pilot.

*The type and location of
the relief valve should
be in accordance with
local codes. The
locations shown in the
illustration are typical.

Recommended installation

FIGURE 9.34. Electric hot water boiler. (Courtesy of CAM Industries Inc.)

FIGURE 9.35. Cutaway view of an electric hot water boiler. (Courtesy of CAM Industries Inc.)

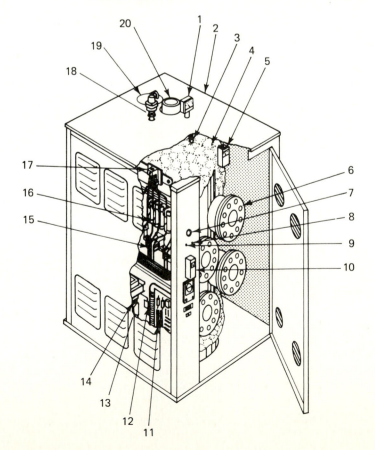

1. Pressure-temperature gauge
2. Boiler enclosure
3. Temperature control sensor
4. Low-water cutoff probe
5. High-temperature cutoff
6. Heating element assembly
7. Pilot switch: control circuit on-off
8. Pilot light: control circuit power
9. Pilot light: limits satisfied
10. Temperature control
11. Solid-state step control
12. Terminal block: external interlocks
13. Control transformer primary fusing
14. Control circuit transformer: 120 V
15. Magnetic contractors
16. Fuse blocks: element circuits
17. Main supply lugs
18. A.S.M.E. pressure relief valve
19. Inspection opening
20. Outlet

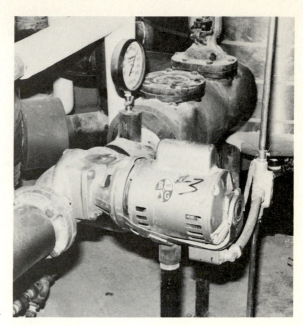

FIGURE 9.36. Circulating pump.

In vertical buildings the circulation of hot water takes place in a loop system. The heated water expands and rises in the supply leg while the cooler water moves toward the heat source. In horizontal buildings a small pump provides the necessary hot water circulation. Figure 9.36 shows a pump used for hot water circulation.

9.11 INSTALLATION AND CONSTRUCTION SEQUENCING

In horizontal construction much of the plumbers' work is underground or hung exposed from the structural steel. Vertical construction, particularly tall apartment and office buildings, differ radically and require closer sequencing of plumbing installations with the work of the architectural trades. Water risers run vertically 20 floors at a time. The architectural trades, such as the installers of hollow metal door frames, masonry and drywall partitions, work floor by floor, starting on a low floor and completing the bulk of their work before going to the next floor. Since the risers and piping for the toilet rooms, slop sinks, and wet columns* must be tested and covered before being enclosed, the plumber should divide a building for work purposes into convenient groupings of floors. For example, he should put in the water lines on the 7th to 12th floors, installing, testing, and covering to coordinate the plumbing with the other trades, who are working one floor at a time. This is particularly important to the wet trades, especially the bricklayers and the plasterers, who bring a large quantity of scaffolding and materials to the floor they are constructing. We can readily understand the piping contractor's reluctance to test frequently since all work stops on a given system in preparation for a test. All open piping must be capped off and time spent in filling, performing the test, and draining down the system. However, it is essential that the mechanical trades think "horizontal" even though their work is vertical in nature.

9.12 TEMPORARY WATER

Fortunately for modern-day owners and construction managers, designers specify less and less masonry, plaster, and wet tile work each year. These wet trades load the building with moisture which must be dissipated before wood paneling and paint can be completed. However, the absolutely essential spray fireproofing of structural steel

---

* In large rental office buildings water and waste removal risers are installed at two or four columns in open areas for connections for tenant private toilets and sinks that may be needed in the future.

uses large quantities of water throughout the structure. The installation of a temporary water system servicing at least every other floor is necessary for the work of this trade.

Temporary water is also required for sanitation. Some codes, labor law, and union contracts require running-water toilet facilities for the workers on large high-rise work. Meeting the requirements for the trades and sanitation calls for a weather-protected installation. A pump, and in tall buildings an upper-story relay pump, connected to a 2-in. black iron line with hose bibs every second floor, will generally provide the required water. A temporary storage tank is sometimes installed on the top floor to meet the heavy requirements of the spray fireproofer. In addition, 55-gallon drums at each hose bib draining into the permanent waste stack will catch most spillage and help keep the floors dry. Temporary toilets of plywood construction with some old water closets and a sheet metal urinal handle the sanitation. Usually these systems function satisfactorily except in cold weather. Despite insulation, electric trace lines, and permitting the water to run, supply lines freeze and so do waste lines if the building has not been enclosed. Basically, temporary water supply and wet sanitation are a nuisance to the construction supervisor on a large building. See Appendix B for a temporary plumbing specification for a 44-story office building.

In horizontal work, particularly a big industrial building, a temporary water line of black iron can be laid underground with shutoffs and drain-down below the frost line. The trend is away from elaborate installations and toward supplying only temporary water at central locations.

10

Drain, Waste, and Vent Systems

10.1 INTRODUCTION

During the course of a project the general contractor has differing plumbing objectives depending on the type of building being constructed. In hospitals and apartment buildings, which have a higher percentage of plumbing than offices and industrial buildings, work on the drain, waste, and vent (DWV) system would receive special attention for two important reasons: first, the amount of work involved, plumbing being 9 to 10% of the cost of the building, and second, the amount of architectural, HVAC, and electrical work that can progress after the plumbing installation.

In office buildings over three stories, the toilet rooms usually are stacked one above the other in the core of the building. The general contractor considers this an important part of the *core work,* which includes stairways, elevators, telephone and electric services, ductwork, and HVAC pipe shafts as well as toilet rooms. This is a distinctly different part of the operations he plans, schedules, and controls from the building perimeter or open office space. Concerted effort on the part of the plumber to install the built-in portion of the DWV system accelerates the overall core program. This includes the disposal lines carrying human, washing, and kitchen wastes, and the venting system that makes the waste system work properly.

In industrial plant construction, objectives differ. The roof and the floor are of paramount importance. The general contractor wants all mechanical and electrical underground lines inside the building installed expeditiously to accelerate pouring the concrete slab on ground. He looks to the plumber to dig and lay pipe for the sanitary sewer and storm water drains. He also wants the rainwater conductors (leaders) run up to roof level and connected to the roof drains as the roofing progresses.

LEGEND

——   DRAIN LINE
— —   VENT LINE
CO   CLEAN OUT
FD   FLOOR DRAIN
P1   WATER CLOSET
P2   URINAL
P3   LAVATORY

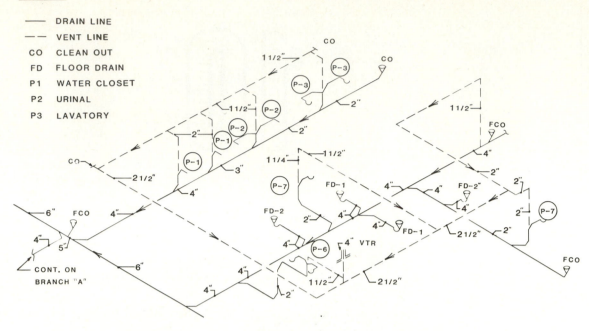

FIGURE 10.1.   DWV system.

The DWV system for a building can best be understood by studying the three-dimensional sanitary waste and vent riser diagram of the project's plumbing drawings (Fig. 10.1). However, before anyone attempts to comprehend any of the mechanical and electrical systems the architectural and structural plans should first be mastered. Time, effort, and frustration can be saved by learning the building in which the systems operate. A short but sufficient plan perusal will locate plumbing spaces, shafts, mechanical equipment rooms, basements, and speed up the process of understanding how the plumbing fits into and serves the building.

The DWV system consists of piping, fittings (Fig. 10.2), and fixtures that act together *hydraulically* and *pneumatically* to function properly. Proper functioning means:

1. Disposal of waste quickly, cleanly, and completely
2. Sealing of the system from odors originating in the system and from the exterior sanitary sewer system
3. Venting to the outside air to ensure smooth hydraulic flow and to carry off unpleasant odors

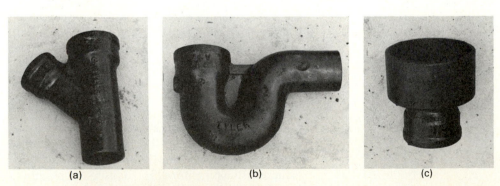

FIGURE 10.2.   Basic DWV fittings: (a) Y; (b) trap; (c) reducer.

The various municipal and national codes specify the requirements that will carry out these functions. Some of these are still needlessly restrictive, particularly with certain materials, but they do produce the desired result. The water portion of these systems are sized and joined and pitched to quickly carry off the effluent from all the fixtures connected to that particular system. The code recognizes that theoretically, numerous water closets could be flushed simultaneously on a series of floors and provides for this possibility.

10.2 TRAPS

A very important part of the system are the *traps* (Fig. 10.3). Traps, by retaining the last of the flushing or washing water, form a constant water barrier against the system's foul and unsanitary air. Certain floor traps (Fig. 10.4) are subject to evaporation, which would destroy the seal. A connection to the water supply adds water to keep these traps functional. The house trap, located at the point where the soil line leaves

FIGURE 10.3.   Cast iron trap used between a vertical drop and a horizontal run.

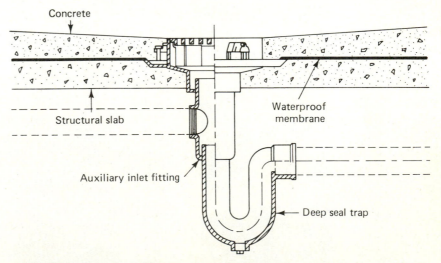

FIGURE 10.4.   Floor drain and trap. In this cutaway section an auxiliary inlet fitting is provided for priming the trap from the water supply system to keep the trap full. (Courtesy of Tyler Corp.)

FIGURE 10.5. House traps for a large office building.

the building, prevents street sewer gases from entering the building (Fig. 10.5). The trapping device can be either a separate fitting installed in the line or an integral part of a fixture. To the plumber it is another part of the *roughing* he must install to make way for architectural and structural trades. The roughing includes piping, supports, and covering for the DWV and the water supply systems. Roughing is a general term used in reference to mechanical and electrical work which is installed in advance of architectural finishes. It is generally considered as the part of the work that eventually gets built in or covered up.

10.3   CLEAN-OUTS   Adults and children occasionally put cloth and plastic objects into the waste system, causing paper and solid wastes to catch on them and plug up the system. Clean-outs (Fig. 10.6) are pipe fittings with a removable threaded plug and are located strategically at the ends of horizontal runs and where piping changes direction for easy access for the plumber to clear any obstructions. They are particularly (Fig. 10.7) important where the waste lines join together and where a vertical and horizontal line meet.

FIGURE 10.6. Tee-fitting clean-out for plastic piping.

FIGURE 10.7. Clean-out adjacent to two Y fittings on a horizontal run of piping.

10.4 VENTING
The venting portion of a DWV system (Fig. 10.8) is a system of interconnected piping open to the *outside air,* which connects with the *liquid parts* of the system (Fig. 10.9). Just as loosening the top cap on a gasoline can allows the gasoline to flow freely, the venting system performs the same function. In addition, by breaking the *syphoning action,* water remains in the various traps. Foul air, in turn, can exhaust to the outside atmosphere (Fig. 10.10). Vent connections from the fixtures join to branches, which in turn connect to the main vent. Each fixture must be vented by the system, but the codes have been developed with many exceptions which ensure proper functioning while encouraging economical design.

There are a few DWV terms that need explaining.

Vent stack. A vertical pipe carrying air terminating above the roof to which branch vents and fixtures connect.

Stack vent. The air portion of a vertical waste line above the last fixture connection.

FIGURE 10.8. Continuous venting in an office building.

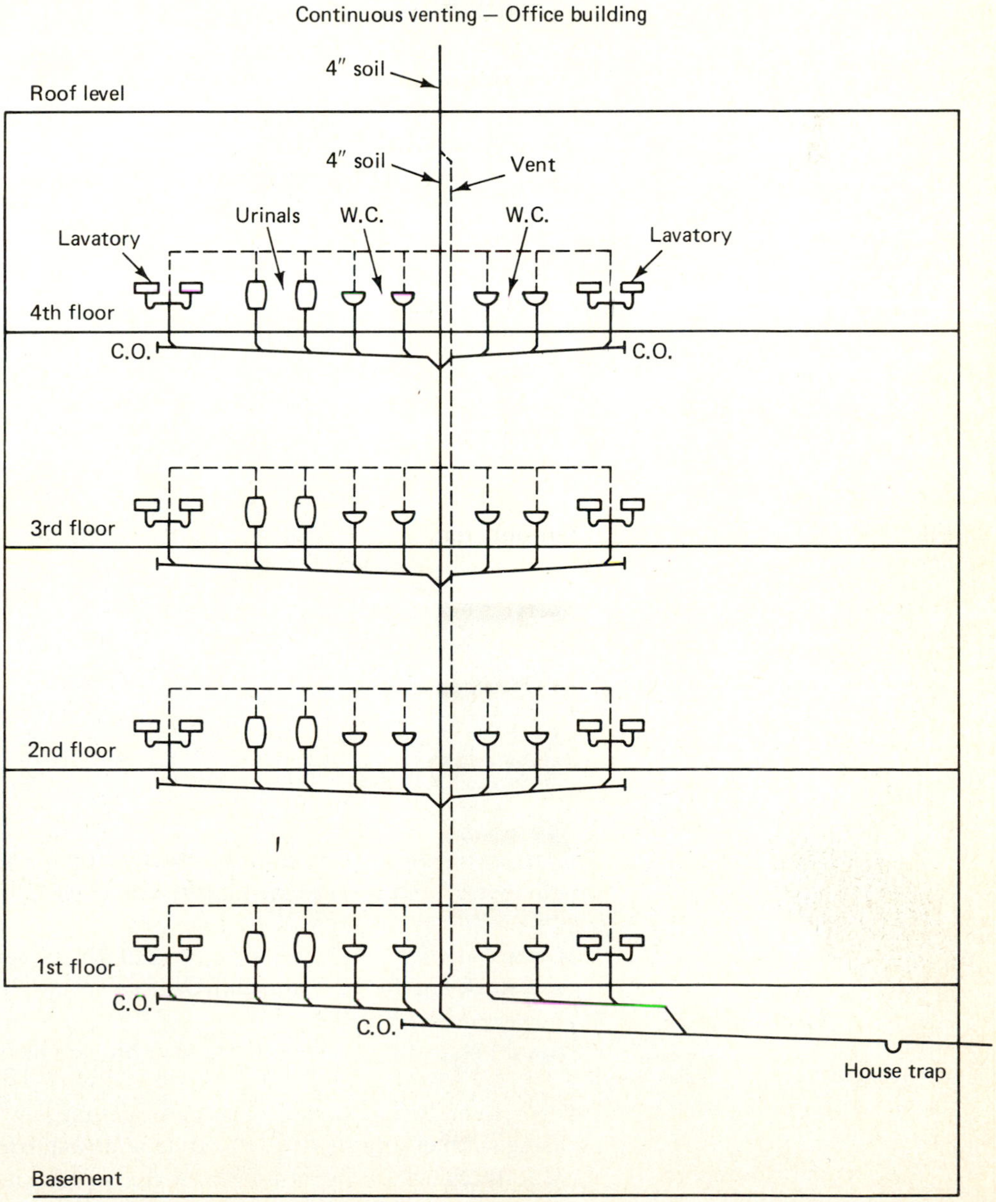

Continuous venting — Office building

FIGURE 10.9.   Vent piping for toilet room fixtures.

FIGURE 10.10.   Vent pipe with lead cylindrical and flat roof flashing.

Battery venting.   Creation of a water and air loop of piping while connecting a number of fixtures into the loop system (Figs. 10.11 and 10.12).

Circuit venting.   A vent loop system in which there is a vent connection between the soil stack and the first branch fixture. It connects to the vent stack (Fig. 10.11).

Loop vent.   Similar to a circuit vent but connecting to a stack vent (Fig. 10.12).

In multistory buildings the battery concept produces economical installations by using simpler venting (Fig. 10.13). In addition, back-to-back connection of a battery of fixtures repeated over numerous floors lends itself to prefabrication.

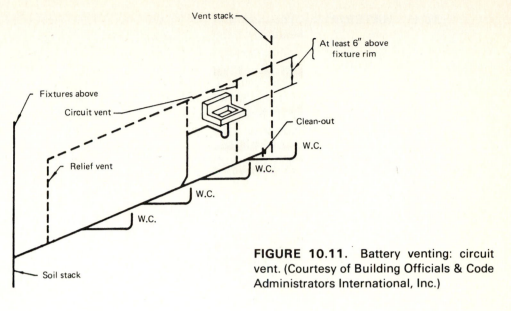

FIGURE 10.11. Battery venting: circuit vent. (Courtesy of Building Officials & Code Administrators International, Inc.)

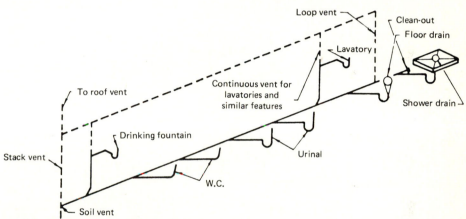

FIGURE 10.12. Battery venting: loop vent. (Courtesy of Building Officials & Code Administrators International, Inc.)

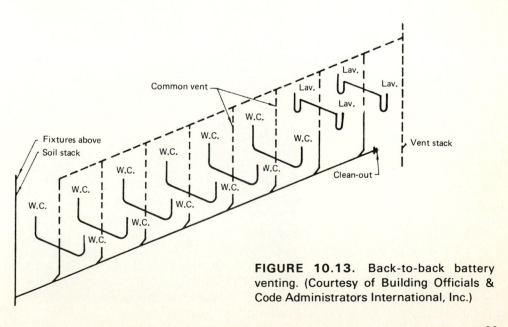

FIGURE 10.13. Back-to-back battery venting. (Courtesy of Building Officials & Code Administrators International, Inc.)

10.5   MATERIALS   The installed price of the material the local code permits usually determines the material designers specify. For two-thirds of the twentieth century, plumbing engineers generally specified bell and spigot cast iron with lead and oakum joints for the drain and waste piping and galvanized steel screw pipe for the vents. In addition to three kinds of cast iron pipe the BOCA code lists brass, copper pipe and tubing, galvanized wrought iron, hubless cast iron, and acrylonitrile-butadiene-styrene (ABS) or polyvinyl chloride (PVC) plastic pipe. These materials can also be used for venting.

Hubless cast iron and plastic pipe have gained wide acceptance in recent years for two important reasons:

1.   They can be assembled quickly and simply.
2.   They lend themselves more readily to prefabrication.

Figures 10.14 and 10.15 show the method of joining no-hub pipe. To protect the pipe and improve ease of shipment and handling by mechanical equipment, cast iron pipe is assembled into secure units (Fig. 10.16). Bell and spigot pipe can also be quickly joined using a gasket and lubricant (Fig. 10.17). Plastic pipe has the advantage of lighter weight, a factor that contributes to lower installation costs. Part of the remaining resistance to its complete adoption by some codes comes from firefighting officials, who claim that it makes their job more hazardous due to the toxic nature of plastics when they burn. Where acids wastes occur, Pyrex glass piping prevents destruction of the DWV system (Fig. 10.18).

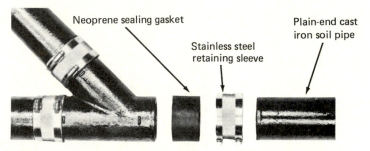

Neoprene sealing gasket

Stainless steel
retaining sleeve

Plain-end cast
iron soil pipe

FIGURE 10.14.   No-hub gasket and sleeve fit over the end of a pipe. (Courtesy of Tyler Corp.)

FIGURE 10.15.   Tightening a retaining sleeve on a no-hub joint. (Courtesy of Tyler Corp.)

FIGURE 10.16. Cast iron pipe is shipped and job stored in multipipe units.

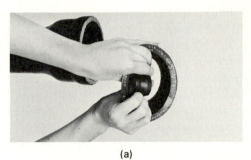

(a)

(b)

(c)

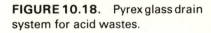

FIGURE 10.17. (a) Installing a gasket; (b) lubricating a gasket; (c) forcing pipe together with a jack. (Courtesy of Tyler Corp.)

FIGURE 10.18. Pyrex glass drain system for acid wastes.

10.6 PIPING AND FIXTURE SUPPORT

The BOCA code in regard to hangers and supports states: "Section P-1302.0 Attachment to Building P1302.1 General: Hangers and anchors shall be securely attached to the building construction at sufficiently close intervals to support piping and its contents."

Most rigid vertical piping is supported at each floor level by a pair of steel straps that clamp around the pipe and bear on the finish floor (Fig. 10.19). Clamping the pipe below a bell or coupling provides superior support. Hangers from the structural frame or slab carry the horizontal piping (Fig. 10.20).

In a structural steel beam or bar joist building, the plumber can readily clamp to any convenient beam as he runs his line (Fig. 10.21). In metal deck or structural concrete slab work, provisions for hanging are usually made prior to pouring the concrete. A dimensional shop drawing showing the layout of the lines locates the hangers. In metal deck construction, hangers are dropped through holes in the metal decking (Fig. 10.22). In concrete form work, the plumber, the pipe fitter, and the electrician follow closely behind the plywood or pan form installation, measuring and fastening hanger inserts (Fig. 10.23). The concrete contractor must allow the mechanical and electrical trades a reasonable time in the forming, pouring, and stripping cycle of concrete work to install this portion of their work. The mechanical and electrical contractors have the obligation to provide the proper number of workers to meet the pace set by the concrete contractor. Job meetings should establish the pace and indicate subsequent crew size adjustments as required.

The plumber must secure the fixtures as well as the piping in accordance with the specifications so that neither one will be dislodged, causing breakage and leaks. Since there is a great variety of each type of fixture, there is a corresponding great variety of fixture carriers.

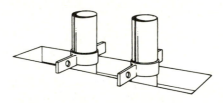

FIGURE 10.19. Riser clamps. (Courtesy of Unistrut Building Systems, GTE Product Corp.)

FIGURE 10.20. Cast iron pipe on hangers.

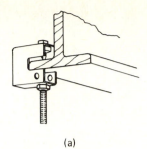

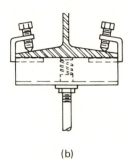

(a)

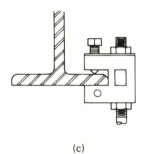

(b)

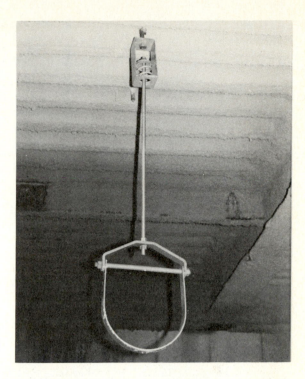

FIGURE 10.22. Isolation hanger hung from a fireproof covered metal deck.

(c)

FIGURE 10.21. Beam clamp mounting methods. (Courtesy of Unistrut Building Systems, GTE Product Corp.)

FIGURE 10.23. Cast iron soil line hung from a concrete slab from inserts cast in concrete.

Water closets, urinals, and lavatories take the most abuse from possible movement and require firm fastening. Carriers for urinals and sinks consist of feet, uprights, horizontal arms, and outlet connections, depending on the design (Fig. 10.24). Firmly fastening the carriers to the floor and wall system and bolting the fixture to the carrier ensures against any movement. In commercial kitchens and other locations where a waterproof membrane is installed, the fixture carriers should be installed before the waterproofing.

Designers of hospitals and high-rise office buildings usually specify wall-hung water closets. The plumbing manufacturers have responded to this demand by developing carrier fittings that perform both support and waste removal functions.

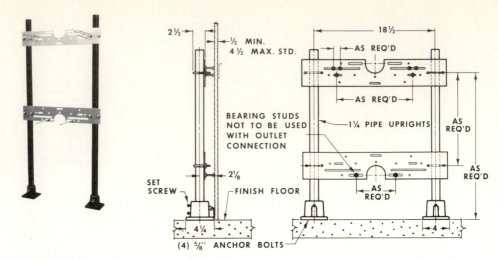

FIGURE 10.24.   Urinal carrier with a fixture hanger support plate and a lower plate with bearing studs. (Courtesy of Tyler Corp.)

These DWV fitting systems are (1) quicker and easier to install, (2) assembled with fewer joints, and (3) come with adjustable supporting legs. There are two general types of water closet: syphon jet and blowout. The syphon jet outlet, being lower than the blowout type, requires a different carrier fitting (see Figs. 10.25 and 10.26).

Carrier fittings accommodate back-to-back water closets, a battery of water closets, and convenient venting fittings (Figs. 10.27 and 10.28). The illustrations used show hubless cast iron carrier fittings, but similar fittings are made for use with lead and oakum joints and screw connections for vent lines.

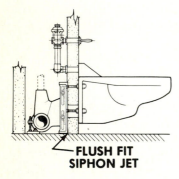

FIGURE 10.25.   Flush-fit siphon jet water closet. (Courtesy of Tyler Corp.)

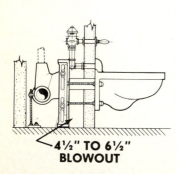

FIGURE 10.26.   Blowout-type water closet. (Courtesy of Tyler Corp.)

FIGURE 10.27.   Installation drawing showing double and right-hand horizontal no-hub carrier fittings connected to a vertical no-hub carrier fitting with double side inlets installed on a stack. A vent is shown on the right-hand side of the stack, but it can be furnished on the opposite side. (Courtesy of Tyler Corp.)

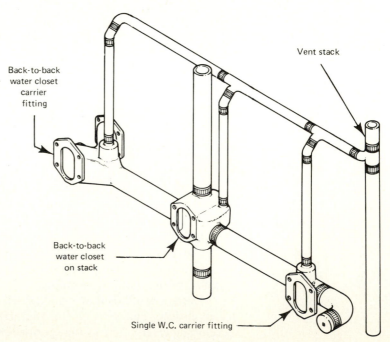

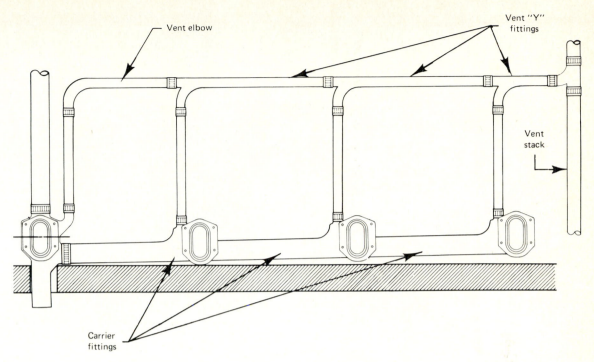

FIGURE 10.28.   Water closet carrier and vent fittings. (Courtesy of Tyler Corp.)

10.7 FIXTURES   In mechanical and electrical construction the roughing phase precedes dry wall, concrete masonry unit, plastering, and tile-setting operations. The plumber returns after the carpenters and wet trades have completed the basic wall and floor finishes to install all the fixtures that have not been built in previously. Bathtubs and plastic bath and shower units generally are installed shortly after the DWV and water supply system have been installed and tested. These are built-in items. Installations of water closets, urinals, lavatories, and service and kitchen sinks follow the architectural finish trades (Fig. 10.29). In the larger installations such as hospitals, apartment houses, and high-rise offices, large-scale deliveries of fixtures arrive by big highway-type tractor-trailer rigs. The plumber makes a concentrated effort to unload, hoist, distribute, and protect these fixtures from damage or breakage. Special crews skilled in plumbing finish work then start setting the fixtures and make the waste and supply piping connections. Items subject to damage or theft, such as water closet seats, are not installed at this time. Corrugated paper shipping containers and polyethylene film fastened with tape provide protection from scratches, plaster, tile cement, paint, and workers using the fixtures as scaffolding (Fig. 10.30). Since all trades at this stage are

FIGURE 10.29.   Setting urinals and installing trim.

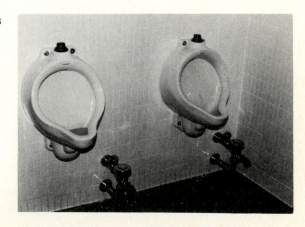

FIGURE 10.30.   Protection for a wall-hung water closet.

installing their finish paint, lights, ceiling, and toilet partitions, one might think such protection would not be necessary, but unfortunately much time and effort in building construction must go toward preventing damage due to carelessness by other trades.

Even if the plumber falls behind schedule, the metal toilet partitions should not be installed until the water closets are connected. Installing the partitions will only make the plumbers' work more difficult and slow them up further.

Types

Types of off-the-floor fixtures supported by carriers include water closets, urinals, lavatories, kitchen and hospital sinks, electric water coolers, and other similar equipment. Wall-type water closets are either syphon jet or blowout. The syphon jet bowl (Fig. 10.31) is the most common because its action is relatively silent and economical in water use. It operates by a jet of water directed through the trapway, which quickly fills and starts the syphonic action immediately upon flushing. The large water surface, among other features, makes it the most desirable type of bowl. The blowout bowl (Fig. 10.32) is particularly adapted for public use, as in airports, stadiums, and plant washrooms. Its action is a driving one of high velocity. It is economical in water use, and has a large water area and an unrestricted trapway. It is also more noisy than the syphon jet bowl because of its direct jet action. The reverse-trap type (Fig. 10.33) is a highly satisfactory floor-mounted water closet.

FIGURE 10.31.   Syphon jet water closet. (Courtesy of Tyler Corp.)

FIGURE 10.32.   Blowout water closet. (Courtesy of Tyler Corp.)

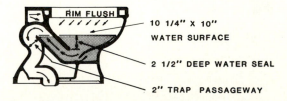

FIGURE 10.33.   Reverse-trap water closet. (From T. Philbin, *Basic Plumbing,* Reston Publishing Company, Inc., Reston, Va., 1977.)

10.8 PREFABRICATION In hospitals, high-rise office buildings, and apartment houses a large portion of the plumbing roughing lends itself to prefabrication. When the toilet room facilities of floor after floor are identical, there is a good opportunity to save labor with prefabrication. Whether performed on the site or at an off-site shop, the repetitious nature of the work encourages prefabrication. Other factors in prefabrication producing labor savings are:

1. Dry, protected, comfortable, well-lighted working space
2. Efficient material stocking and handling
3. Better selection of personnel
4. Production-oriented atmosphere
5. Concentrated supervision
6. Availability of shop tools, machinery, and lifting devices

Offsetting disadvantages are:

1. Bulkier assemblies to handle and store
2. Less efficient transportation
3. Need for some type of crane to lift and swing the assembly to the upper floors

Figure 10.34 shows a prefabricated DWV unit with copper water supply piping for a residential bathroom unit. In Fig. 10.35 the units are laid out on the shop floor ready for assembly. In Fig. 10.36 an electric drill adapted for joining the pipe reduces labor in this operation by 50%. Figure 10.37 shows prefabbed copper water pipe for the unit which has been assembled on a bench using a jig to position the pipe. Bench assembly of copper pipe units is a common practice in plumbing and HVAC work.

FIGURE 10.34. Prefabricated DWV unit. (Courtesy of Tyler Corp.)

FIGURE 10.35. Parts prepared for assembly. (Courtesy of Tyler Corp.)

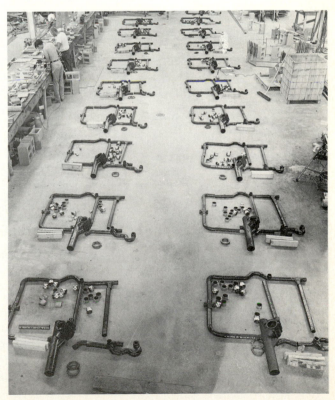

FIGURE 10.36. Use of a jenny wrench speeds assembly. (Courtesy of Tyler Corp.)

FIGURE 10.37. Prefabricated water piping for a DWV unit. (Courtesy of Tyler Corp.)

10.9   THE SOVENT SYSTEM

Most journeymen appreciate the opportunity to work with copper piping. Besides being cleaner and lighter than other metallic piping, workers like the way it cuts and joins. The joining operation consists of cleaning the surfaces to be joined with emery cloth or steel wool and coating them with flux, after which the plumber slips the piping together. Propane torches heat the joint surfaces until the wire solder touching the joint melts and capillary action spreads it over the entire joint area (see Fig. 10.38).

FIGURE 10.38. Joining a sovent fitting to a waste line. (Courtesy of Copper Development Assoc. Inc.)

Sovent plumbing eliminates the usual separate vent pipe and combines drainage and venting in one stack. The system uses an aerator (mixer fitting) at each floor level and a deaerator fitting at the foot of the stack. Since the copper stack vents through the roof, air enters the system at this point. The unusual functioning of the fittings makes this one pipe system operate satisfactorily and accounts for the increasing acceptance of this plumbing concept. This system is now available in cast iron and is currently gaining wider acceptance in this country.

10.10   FLOOR DRAINS

Commercial kitchens, shower rooms, mechanical equipment rooms, and similar areas subject to washing down or flooding require floor drains. These come in a wide variety of designs to accommodate various kinds of construction (Figs. 10.39 and 10.40). Typical of good water disposal design, there are two lines of water defense: (1) the surface grating, carrying off all water from the top of the cement finish; and (2) weep holes to drain off any moisture that penetrates the finish or condenses on top of the

FIGURE 10.39. Floor drain installed in membrane-type floors over insulation where the floor construction is monolithic finished concrete or a cement top coat. (Courtesy of Tyler Corp.)

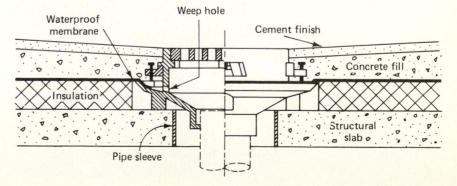

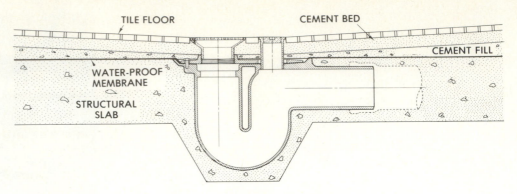

FIGURE 10.40.   Floor drain installed in a ceramic tile floor with a waterproof membrane over the slab. Note the clean-out plug at strainer level, both being adjustable to the required fill. (Courtesy of Tyler Corp.)

waterproof membrane. Pitch, asphalt, or mortar can clog these holes and destroy their function. In exterior applications, unless these weeps drain freely, freezing action can create breaks in the membrane. The plumber should set floor drains to develop maximum allowable floor pitch. Floor drains should be temporarily stuffed with burlap or paper and sealed off to prevent cement paste or plaster from filling up the traps and hardening. The author had the unhappy experience of having traps under an 8-in. structural slab clog up with wastewater and sediment from a masonry saw operation. Holes had to be cut in the structural slab in order to remove and replace the hopelessly filled traps. As previously mentioned, traps connected to floor drains can dry up unless automatically filled by an auxiliary water line (see Section 10.2, Traps).

10.11   TESTING   Plumbing codes require the inspection and testing of the DWV system. When *underground systems* have been properly connected and the piping well supported between exposed joints, the plumber conducts a hydrostatic test. Rubber test plugs (Fig. 10.41) are installed at all openings to seal off the system. Ten feet of vertical piping is temporarily installed at the high point and the system and vertical piping

FIGURE 10.41.   Tightening the wing nut causes the rubber seal to expand, keeping the system sealed and free from mud and trash.

filled with water. Inspection for leaks in the pipe and at the joints begins 15 minutes after filling. Acceptance by the designated authorities must be obtained prior to any backfilling, building in, or covering. A minimum of one day should be allowed in the schedule for each section tested since all productive work ceases during the preparation and conduct of the test.

A second method of testing piping utilizes 5 psi air pressure, equal to the height of 10 in. of mercury. The system must maintain this air pressure for 15 minutes. A small air compressor supplies the required air pressure.

When the plumbing system has been completed, including installing all floor drains, traps, fixtures, and drain and vent lines, the *entire system* is tested for *gas and water tightness*. After all the traps have been filled with water, a pungent thick smoke is introduced into the system. When the smoke appears at the vent openings at the roof, (see Fig. 10.8), they are sealed and a pressure of 1 in. water column created and maintained during a visual inspection of the entire system.

Sometimes the peppermint test takes the place of the smoke test. At each roof vent 2 ounces of oil of peppermint and 10 quarts of hot water are dropped down the vent piping and the vent sealed. Detection of peppermint odor at a trap or other portion of the system indicates an unsatisfactory installation.

10.12 RAINWATER CONDUCTORS

In warehouses and one-story industrial construction much effort and interest centers on the roof drain and rainwater conductor installation. However, on all types of buildings the general contractors and construction managers should insist that the plumber install the roof drains and connect the leader lines (RWCs) to match the progress of the roofing subcontractor. An unconnected roof drain plays havoc with the area below, concentrating large volumes of water. Although specifications generally call for cast iron for underground piping inside the building line and galvanized steel pipe for the vertical aboveground runs, plastic piping is rapidly gaining favor for rainwater installations (Fig. 10.42). Numerous trenches for the rainwater system in industrial buildings run halfway across the building, hampering other construction operations until these lines arc laid, tested, and backfilled. However, the plumber usually has no problem meeting the progress requirements as set by the roofer and the

FIGURE 10.42. Plastic rainwater conductor.

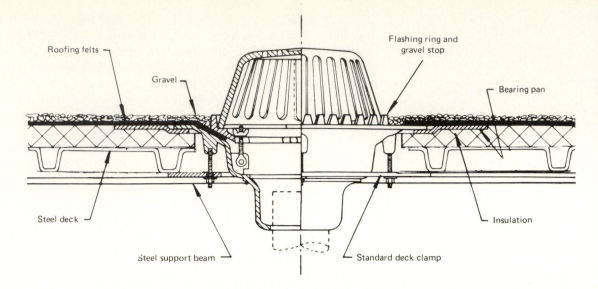

FIGURE 10.43.   Roof drain shown installed in an insulated steel deck with built-up roofing. Metal flashing should be installed under felts and over the bearing pan. (Courtesy of Tyler Corp.)

general contractor. Figure 10.43 shows a typical industrial building installation for a roof drain with metal deck construction. Roofing manufacturers recommend setting the roof drain on lead sheet flashing to improve the connection between the roof drain and the roofing felts. The lead flashing is considered plumbers' work. Figure 10.44 shows another type of roof drain installed in steel and concrete deck. There are very many more types for an equal number of different construction requirements.

It is also important to conduct the water away from the outside of a building. Pitching walks, pavement, and lawns to drain away from the building disposes of much of the surface water. A drainage system located at the base of the foundation wall utilizing perforated plastic piping and well-graded crushed rock or gravel will generally carry off subsurface water (Fig. 10.45).

FIGURE 10.44.   Roof drain installed in a steel and concrete deck. The side outlet permits use in shallow deck construction. (Courtesy of Tyler Corp.)

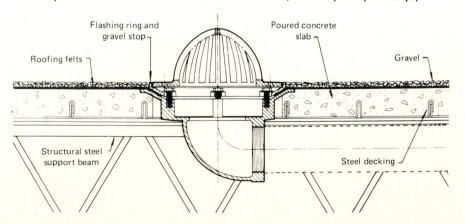

FIGURE 10.45. Perforated plastic pipe and crushed stone drain base of a foundation wall.

10.13 GREASE INTERCEPTORS

Commercial kitchens introduce large quantities of grease into the waste system. This grease, unless caught, eventually fills up drain lines, causing stoppage or sluggish drainage action. Figure 10.46 shows a cross section of a grease interceptor, in which the grease accumulates in the center section. Figure 10.47 shows the sink and vent connections. Periodic removal of this grease is part of a sensible maintenance program. There are many other interceptors, such as the oil type. Most have a special industrial application.

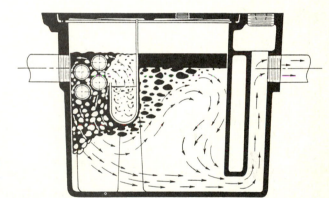

FIGURE 10.46. Sectional view of a grease trap. (Courtesy of Tyler Corp.)

FIGURE 10.47. Installing and venting details of a grease trap. (Courtesy of Tyler Corp.)

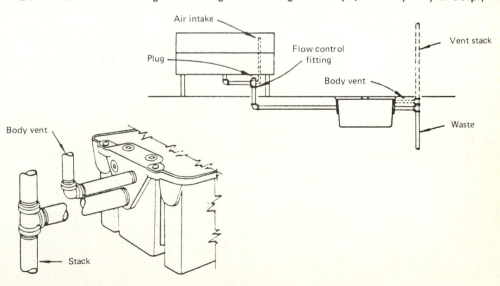

10.14 SEWAGE EJECTORS

Wherever sewage collects below the normal drainage system, sewage ejectors pump the effluent to a higher level so that it will enter the gravity-drain system. Their use extends from one-story residences to mammoth municipal or area authority installations. Figures 10.48 and 10.49 show typical sewage ejector installations.

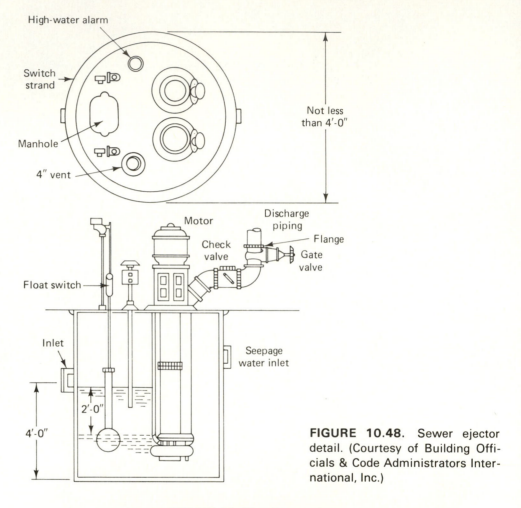

FIGURE 10.48. Sewer ejector detail. (Courtesy of Building Officials & Code Administrators International, Inc.)

FIGURE 10.49.   Sewer ejector.

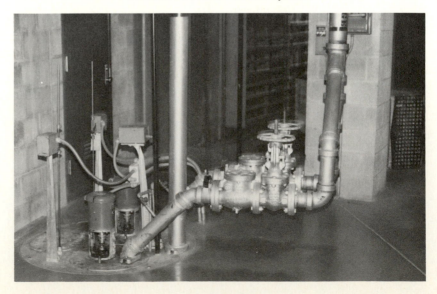

When sewage ejection becomes necessary during the latter stages of a building project, a permanent ejector can be placed in operation. Owners dislike this use of what will become their equipment. The ejector probably will be more severely tested at this stage by plaster, cement, sawdust, and debris than later when the owner puts it to normal use. Frequently, used ejectors are installed temporarily until the building is ready for occupancy. This virtually eliminates possible damage to the new equipment and better satisfies the intended user.

10.15   BACKWATER VALVES

Backwater valves (Fig. 10.50) are recommended for use in localities subject to flood conditions due to heavy rains, inadequate sewer capacity, improper sewer pitch, and in tidewater areas. Backwater valves are normally installed in the sewer line inside the front wall of the building, where access for cleaning and maintenance is readily available.

All mechanical products need attention and a backwater valve is no exception. Good performance depends directly on good maintenance and therefore a backwater valve must be inspected and cleaned regularly. Unless common sewer obstructions are removed, valve performance is impaired.

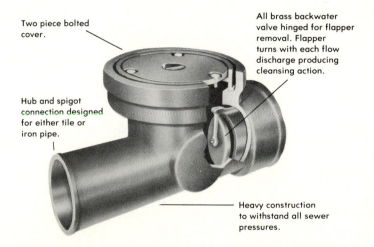

Two piece bolted cover.

All brass backwater valve hinged for flapper removal. Flapper turns with each flow discharge producing cleansing action.

Hub and spigot connection designed for either tile or iron pipe.

Heavy construction to withstand all sewer pressures.

FIGURE 10.50.   Backwater valve. (Courtesy of Tyler Corp.)

11

Fire Protection and Sprinklers

11.1   FIRE IN MODERN BUILDINGS

Despite the publication of some sensational books and the showing of scare-type movies of disastrous fires in high-rise buildings, a modern commercial or public building in the United States represents one of the safest places one can occupy. Thanks to the National Fire Protection Association, insurance companies, fire underwriters, fire departments, protective building codes, and competent architects, engineers, and builders, Americans enjoy the fire safety of well-designed and constructed buildings. The fires and loss of life in nonresidential structures that have occurred have led to thorough investigation and sound recommendations, further improving fire safety in buildings. Two bad killer-fires in our country in the past 40 years took place in nightclubs; the one in Boston involved serious overcrowding, and the more recent one in northern Kentucky involved an unfortunate number of code violations. In the Las Vegas hotel fire, combustion started in a concession area due to short circuiting. The numerous deaths by smoke poisoning resulted from improper fire damper installation.

The relatively good commercial building fire safety record has not lulled the authorities into a state of self-satisfaction. Fire in a high-rise building poses serious problems of getting access to fight the fire and the removal of the occupants to fire-safe areas. Authorities consider complete evacuation of a very large building such as the Sears Tower in Chicago or the World Trade Center in New York during a fire an unrealistic objective. High-rise fire experience to date indicates that part of the occupants will retreat from the fire by going to the upper floors and roof. Fire containment and smoke ventilation present equally difficult situations. Modern air conditioned buildings operate under positive air pressure with large ducts connected to many floors, encouraging the smoke to spread throughout the building. Human habits enter into the fire preventers' thinking as they consider how to program an office worker to take the fire stairway in a fire situation when he or she has ridden the elevator four times a day for the last 10 years. The tragic examples of the high death rate in modern concrete high-rise structures in the Tae Yon Kak Hotel fire in 1971 in Seoul, Korea (163 deaths), and the Joeima Building fire on February 1, 1974 in São Paulo, Brazil (179 deaths), has stimulated fire safety authorities to greater effort toward fire prevention in high-rise construction. We have learned much about our fire vulnerability from high-rise fires that have occurred here in the United States, although the loss of life has been statistically quite low. The National Fire Protection Association, 60 Batterymarch Street, Boston, MA 02110, has published an excellent booklet entitled "High Rise Building Fires and Fire Safety" (reprints from *Fire Journal and Fire Technology,* NFPA, No. SPP-18), which makes fascinating reading on the human and technical side of these very significant fires.

One of the important points of vulnerability brought to light in recent fire experience is the numerous ways in which a fire can spread in a modern fireproof high-rise building. For example, tall buildings lend themselves to the creation of *flue action.* The designers, builders, and individual craftsmen all play an important role in eliminating this type of hazard. Unfortunately, at the exterior of high buildings we have regressed in safety from earlier times when brick and limestone were the prevalent materials. Where formerly the masonry construction built against exterior concrete spandrels and steel beams sealed off each floor, buildings with metal and glass curtain walls and precast concrete exteriors with a generous tolerance for shims have created a ½- to 1½-in. opening between floors. In addition, perimeter ductwork to induction air conditioning units has also created convenient *flues* for passage of fire from one floor to another once the fire breaches the duct system. The increase in *penetrations through the slabs* at the exterior of buildings brought about by perimeter air conditioning is one more regressive factor.

In the interior of most modern high-rise buildings, the area above the hung ceilings has been turned into a return air plenum. Air from any location is drawn up into the space above the hung ceiling and travels horizontally until it enters a return air sheet metal duct and then passes vertically to the return air fan location. This type of construction, during a fire encourages the *horizontal spread* of the fire through the plenum area. Solutions to these problems have been developed but they require conscientious execution and inspection. Specifications now call for sealing off each floor structurally, packing the space between pipes and pipe sleeves with compressible fireproof material, and limiting the size of ceiling plenum areas.

Unfortunately for all of us in the construction industry, our best efforts in designing and building fire safe structures can easily be thwarted by the *highly combustible contents* brought into the occupied space by the tenants. The extensive use of foamed plastic and foamed rubber furniture, wooden desks, drapes, and adhesive-applied carpeting provides marvelous fuel for fires. Certainly improving control over the contents of a building by building owners, insurers, and fire authorities would greatly reduce the potential for fire in high-rise structures.

As we discuss the various facets of mechanical and electrical work we will emphasize the fire protection requirements of each. Fire protection, like construction safety, does not come in a neat package clearly labeled, but must be part of the thought process that goes into all aspects of designing and building process every day.

11.2 HIGH-RISE FIRES AND CONSTRUCTION

A building is most vulnerable to fire during the *construction period* (Fig. 11.1). One or two stories at a time may contain complete sets of concrete forms with many square feet of oil-coated plywood. Form fires are usually quite destructive, ruining both the forms and that portion of the building. Depending on the stage of construction, vertical transportation tends to be slow and rudimentary, fire stairways not enclosed, fire alarm systems incomplete, illumination inadequate, and the whole building subject to high winds. Great quantities of *burnable trash* accumulate daily from shipping cartons and protection padding used for material shipments. A number of trades may be showering the project with *sparks from cutting torches* and *welding arcs* as structural steel gets altered, metal deck installed, curtain wall angle clips fastened in place, and piping joined. The use of temporary heating devices in winter introduces potential sources of fire depending on the equipment used. *Propane heaters* and *oil salamanders* are particularly hazardous. In addition, the trades use highly inflammable sealants, paint, and adhesives. To complicate the fire prevention programming further, a well-run construction project progresses rapidly, changing constantly. A stairway may be usable in the morning and be closed with a scaffold in the afternoon. These constantly changing conditions add to the challenge to the construction supervisor to conduct a sound fire prevention program to keep his project from burning up.

The author would feel remiss if he failed to share his vivid experience with high-rise construction fires. In the first case, a 16-story building under construction was partly enclosed with limestone pilasters and metal and glass curtain wall. Fire broke out at 8:30 A.M. in a wooden shanty on the second floor where most of the temporary offices and shanties were located. When the fire was discovered flames were

FIGURE 11.1. High-rise construction fire. (Courtesy of Factory Mutual.)

shooting out of the shanty, which was completely aflame internally. The fire swept the entire second floor as one propane tank after another, *dictated for use by rulings of the New York City Fire Department* to heat the shanties, popped their relief plugs and fed the fire with mammoth blowtorches. The fire had gotten such a fast and vigorous start that fire extinguishers were completely ineffectual. The 10° F temperature precluded the use of water buckets and temporary hoses.

Living through that hour of panic, fear, effort, frustration, and defeat was an unforgettable experience for all those involved. Despite vigorous effort, the *equivalent of two floors* of construction was lost, with the exception of the structural steel, which the concrete fireproofing protected. The project was pushed back approximately three months.

The second dramatic example came a few years later. One of two competing firms had formed the top floor of a 32-story building, while the other firm was concreting the top floor of a nearby 27-story building. *In the middle of the night* the 32-story building's formwork caught fire and burned in a very dramatic and spectacular fire (see Fig. 11.2). The firefighters were at a great disadvantage fighting the dark, the height, the wind, the shower of sparks, and the fire. The consequences of this fire were many times more disastrous than the earlier one. In this instance, structural steel framing warped by the intense heat had to be removed and replaced. Newly installed elevator motors and generator sets were ruined in the fire, requiring expensive and time-consuming replacements. Work on the exterior facade, which was almost complete, did not resume for nine months. The total resulting delay was so prolonged that certain tenants canceled their leases, to add further to the owner's losses. In each of these cases the job organization underwent a very unhappy and demoralizing period from which it did not recover quickly.

FIGURE 11.2. Night fire in high-rise construction. (Courtesy of Equitable Life Assurance Society.)

The plumber plays a very important role in fire prevention during and after construction. By law and code he must follow the construction progress by keeping the *fire standpipe* (Fig. 11.3) within one or two floors of the top of the construction. At street level he must provide temporary or permanent Siamese fire hose connections clearly marked, easily accessible, and well lit, to which (Fig. 11.4) the fire department pumpers can connect to pump high-pressure water to the top of the building. (Figure 11.5 illustrates an industrial type connection.) As the work progresses, he installs valves and hose connections to the fire standpipe so that hoses can be connected at all levels. Once a fire has broken out, this provides the best means of firefighting in a high-rise building under construction. It may seem redundant to say that all the standpipe valves should be closed at all times except during firefighting and fire tests, but an open valve near the bottom of the standpipe caused serious delay in fighting an upper floor fire in stored sealants during the construction of the Chase Manhattan Bank building in New York City. Some jobs lightly wire and tag the valves in closed position to ensure against this kind of mishap.

Since the spray fireproofing operation consumes large quantities of water, the temporary water supply riser should be equipped with valves and temporary hoses at every second floor for use by job personnel until the fire department arrives. Additional lengths of hose should be carried in the hoist or temporary elevator. The project should provide a key to the temporary elevator or construction hoist in an enclosed glass case for use by the firemen so that the men and their hoses can be sped to the *firefighting floor* during off-hours emergencies.

FIGURE 11.3. Fire standpipe under construction.

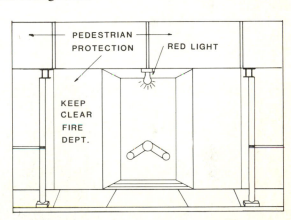

FIGURE 11.4. Construction Siamese connection.

FIGURE 11.5. Siamese connection remote from the factory but adjacent to the street.

We have mentioned the plumbers and elevator constructors in the previous discussion. The electrician also contributes to fire safety by lighting access passageways to stairs, hoists, and elevators. The construction supervisor should work closely with these three trades to provide the best fire protection program possible for high-rise construction. In addition, close liaison with the nearest firehouse should be maintained, with particular effort going toward keeping the firemen advised of access problems and potential hazard areas. Finally, fire protection during construction deserves the same degree of planning, educating, and supervision as all other kinds of construction safety.

11.3   PERMANENT FIRE PROTECTION SYSTEMS

The Fire Standpipe

The fire protection system (Fig. 11.6) for a tall building, which the plumber installs, consists of an incoming water service, pumps, a supply line to the top of the building, a house tank when needed, a Siamese connection (Fig. 11.7), and a fire standpipe line to

FIGURE 11.6.   Fire protection system.

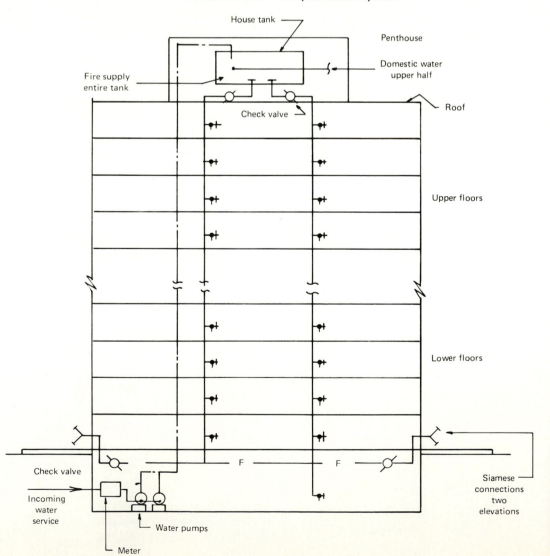

FIGURE 11.7. Siamese connection: Fire Department pumper connects here to supply firefighting water.

the top of the building with valves and fire hose connections at each floor. Most specifications call for threaded steel pipe for the fire standpipe, which makes sense, since it is installed in short sections as the building goes up and has a tee connection at each floor for the fire hoses. The grooved piping system (see Chapter 13) can also be used, for similar reasons. Check valves at the top of the building prevent water from the standpipe from entering the house tank and ensure good pressure reaching the fire hoses from the fire department's pumpers.

During construction the fire standpipe usually remains empty to prevent freezing and for convenience of the plumber while adding to the system. During permanent occupancy it is filled with water for use by building personnel to fight the fire until the fire department arrives. As the permanent work progresses, fire hose cabinets (Fig. 11.8) and fire extinguisher cabinets installed by the plumber are built into the walls. In the areas generally under control of the building operating personnel, simple valves and hose racks suffice. Figure 11.9 shows an interior fire hose connection for fire department use. A dry standpipe does not become useful until water is pumped into it by the fire department.

Since the system in rudimentary form must meet construction progress and be usable at all times, it is gradually upgraded as temporary Siamese connections, valves, and hoses are replaced with the specified permanent installations as the work progresses. Eventually, alarm devices and monitoring signal systems are connected, tests and inspections are conducted, and the entire system is put into operation and turned over to the owner.

FIGURE 11.8. Fire hose and fire extinguisher cabinet.

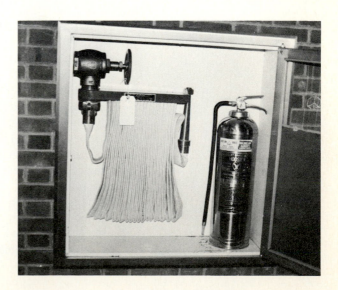

FIGURE 11.9. Internal dry standpipe hose connection for fire department use.

Sprinklers

Industrial use of sprinklers began over a century ago by its inventor, Henry Parmalee, the owner of a piano factory in New Haven, Connecticut. Since that time sprinklers have been used extensively for all kinds of factories, warehouses, department stores, and business enterprises. Owners choose to sprinkle their building primarily because their insurance rates are reduced, and with sprinklers a fire will probably never be severe enough to put them out of business. The automatic, dependable, always available, worry-free nature of sprinkler systems also appeals to them greatly.

The Wet System. The most widely used sprinkler systems consist of an *adequate* water supply feeding a system of mains and branches feeding *properly spaced* heads that are *heat activated* to operate individually when fire breaks out in their area. Figure 11.10 shows this kind of a wet pipe system. The piping is filled with water under pressure that immediately discharges when a sprinkler operates. The water continues to flow, controlling or extinguishing a fire, until it is shut off.

In the diagram a fire has actuated one sprinkler. The main water supply comes from the public water main (1). From the public main, water flows through a series of valves (always kept open) to the yard main (2), to the lead-in (3), through the riser (4), into the feed main (5), to the cross main (6), into the branch line (7), and out through the active sprinkler (8). The fire pump (9) starts automatically when water flow is detected, draws additional water from the stored supply in the suction tank (10), and feeds into the yard main. This additional water from the fire pump and suction tank provides a supplementary supply of sufficient quantity and pressure to meet the demand of the sprinklers.

Any water flowing through the water flow alarm valve (11) sounds a local alarm and may send a signal to a central station, which notifies the local fire department.

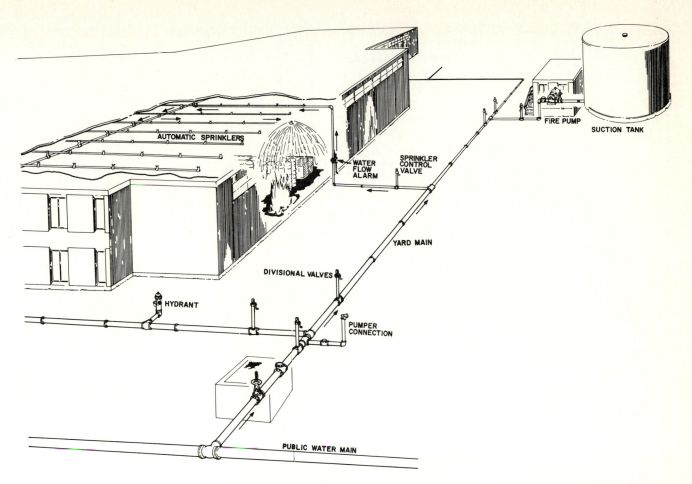

FIGURE 11.10.   Automatic sprinkler system. (Reprinted by permission © 1977 Factory Mutual Engineering Corporation.)

The sprinkler control valve (12) will be used to shut off the water to the system when conditions permit as determined by the person in charge of the firefighting operation.

The Dry Pipe System. In locations subject to freezing, such as loading docks and cold storage warehouses where sprinklers are required, a dry pipe system protects the property. In the cold areas air under pressure fills the piping until one of the heads activates, due to fire. The key device in the system is the *differential-type dry pipe valve* (Fig. 11.11), in which *air* under pressure balances *water pressure.* The activation of a sprinkler head releases the air pressure and causes the valve to open (Fig. 11.12), and forces water to the fire source. A separate heated enclosure, heater unit, and air compressor increase the cost of this special type of installation (Fig. 11.13).

The Deluge System. The deluge system offers a third basic method of utilizing automatically delivered water to fight fire or to prevent fire from starting. The sprinkler heads in a deluge system are already open (Fig. 11.14) and the piping system is dry beyond the operating valve. Heat-activated devices (HADs) strategically spaced trigger the valve to open and *all* heads in the entire system spray water. *Larger piping sizes* enable the system effectively to supply all heads with water. Deluge systems protect high-hazard areas subject to rapid fire spread. They are occasionally installed on the side of a building adjacent to a bad fire hazard, deluging the side of the building when heat triggers the control mechanism.

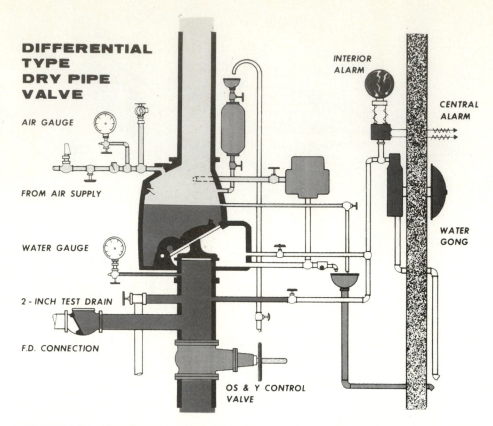

FIGURE 11.11.   Dry pipe system: static state. (Courtesy of Robert J. Brady Co.)

FIGURE 11.12.   Dry pipe system: in operation. (Courtesy of Robert J. Brady Co.)

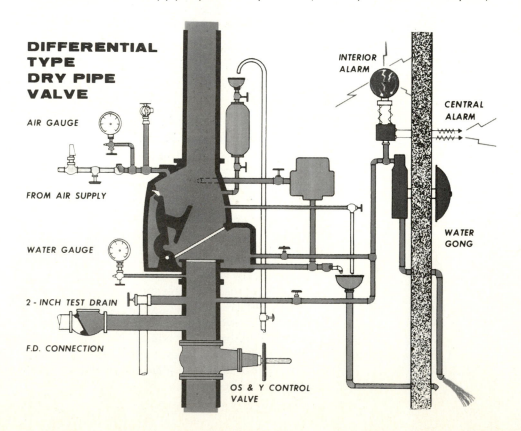

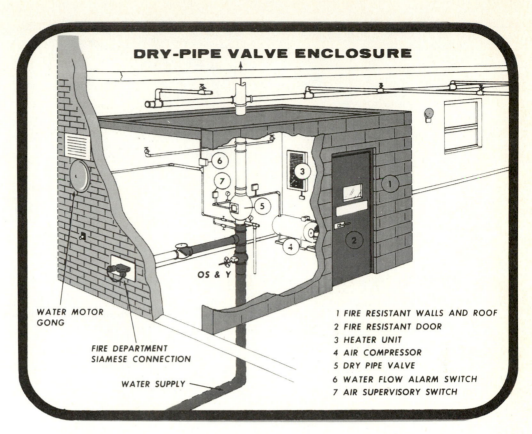

DRY-PIPE VALVE ENCLOSURE

OS & Y

WATER MOTOR GONG

FIRE DEPARTMENT SIAMESE CONNECTION

WATER SUPPLY

1 FIRE RESISTANT WALLS AND ROOF
2 FIRE RESISTANT DOOR
3 HEATER UNIT
4 AIR COMPRESSOR
5 DRY PIPE VALVE
6 WATER FLOW ALARM SWITCH
7 AIR SUPERVISORY SWITCH

FIGURE 11.13. Dry pipe valve enclosure. (Courtesy of Robert J. Brady Co.)

FIGURE 11.14. Deluge system. (Courtesy of Robert J. Brady Co.)

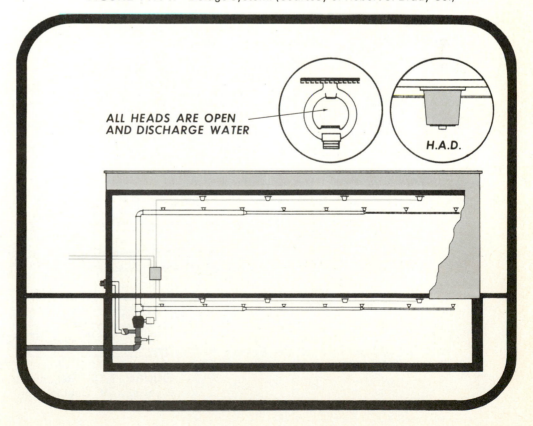

ALL HEADS ARE OPEN AND DISCHARGE WATER

H.A.D.

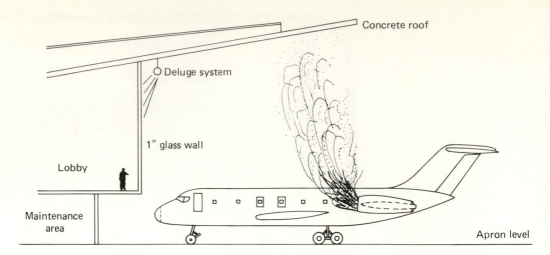

FIGURE 11.15.   Deluge system that protects air terminal passengers.

An interesting use of this method protected the people using the original glass-enclosed Pan American Airways Terminal at JFK airport in New York City. With a rapid rise in temperature from a plane on fire, HADs under the roof were set to activate the deluge system, dousing the 1-in.-thick glass with water, preventing its breakage and entry of the fire (Fig. 11.15).

11.4   WATER SUPPLIES FOR FIRE PROTECTION

Although most water service lines consist of short runs from city mains as mentioned in Chapter 9, industrial systems have many hundreds of feet of piping into and around the factory property. For ordinary conditions, cast iron, asbestos cement, and ductile iron pipe are specified, since they give satisfactory performance while keeping installation costs relatively low. Figures 11.16, 11.17, and 11.18 show three kinds of cast iron joint. Where fire protection pipe may encounter severe shock or when used in tunnels, steel pipe lined and coated may be specified.

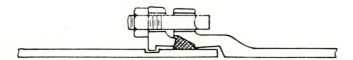

FIGURE 11.16.   Restraining joint. (Courtesy of Ductile Iron Pipe Research Association.)

FIGURE 11.17.   Restraining joint. (Courtesy of Ductile Iron Pipe Research Association.)

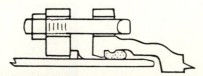

FIGURE 11.18.   Restraining joint. (Courtesy of Ductile Iron Pipe Research Association.)

Each year severe fire losses occur in sprinklered buildings because the water supply has been turned off and left off after repairs and maintenance. To combat this hazard, fire protection agencies recommend locking in the open position the post indicator valve on the line from the main water supply and interior supply valves (Figs. 11.19 and 11.20).

FIGURE 11.19. Interior valve locked in the open position. (Courtesy of Factory Mutual Engineering Corporation.)

FIGURE 11.20. Post indicator valve locked open. (Courtesy of Factory Mutual Engineering Corporation.)

Anchorage

Forces acting on pipe laid in the ground that must be taken into account are: (1) internal static pressure of the water, (2) water hammer, (3) backfill, and (4) live load and impact from trucks and heavy equipment. To prevent joints in cast iron pipes from parting secure bracing and clamping specifications have been developed. Figure 11.21a and b shows method of holding a joint at a T connection with concrete and steel; (Fig. 11.21e), a retainer for a cap at end of pipe. Figure 11.21f shows a method

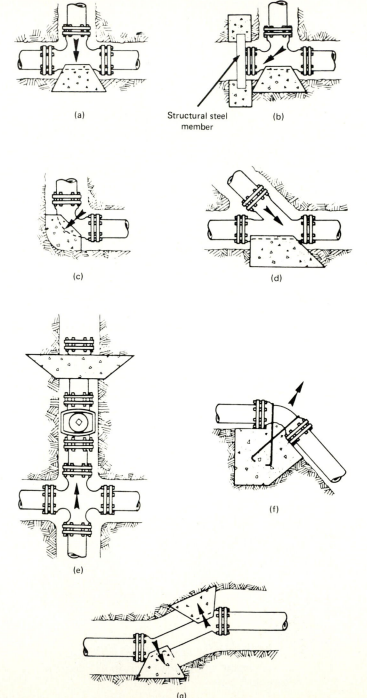

FIGURE 11.21. Thrust block location. (Courtesy of Ductile Iron Pipe Research Association.)

of using a concrete block and ties. Thrust blocks of concrete properly poured against a good bearing soil serve very well to resist forces that would pull the piping apart. Figure 11.21 c, d, and g are three examples.

Testing

Asbestos cement piping is filled with water and allowed to stand 24 hours. Hydrostatic test pressure of 1.5 operating pressure is applied. Allowable leakage by AWWA standards range from 0.71 gallon per hour (gph) per 100 couplings on 4-in. pipe at 50 psi pressure to 13.5 gph per 100 couplings on 36-in. pipe at 225 psi pressure. Cast iron lines are slowly filled with water to expel all air to prepare for test. Lines are hydrostatically tested at 200 psi for 2 hours or at 50 psi greater than operating pressures over 150 psi. Leakage in piping is measured at the specified test pressure by pumping from a calibrated container. Leakage should not exceed 1 ounce liquid measure per hour per inch of pipe diameter per joint.

11.5 SPRINKLER HEADS

Architects and fire protection engineers have available for their designs a wide variety of heads which meet functioning and aesthetic needs. Figure 11.22 shows heads that operate from 165 to 360° F and cover 200 square feet for light hazard and 90 square feet for extra hazard. Whether they stick up or down or are the sidewall variety, their spray will cover the required area. Architects prefer to use the flush type in lobbies, restaurants, and areas, where aesthetic considerations govern. In Fig. 11.22 you can see how the solder links, levers, and valve cap fly off when the solder link melts.

FIGURE 11.22. Automatic sprinkler heads. (Courtesy of Robert J. Brady Co.)

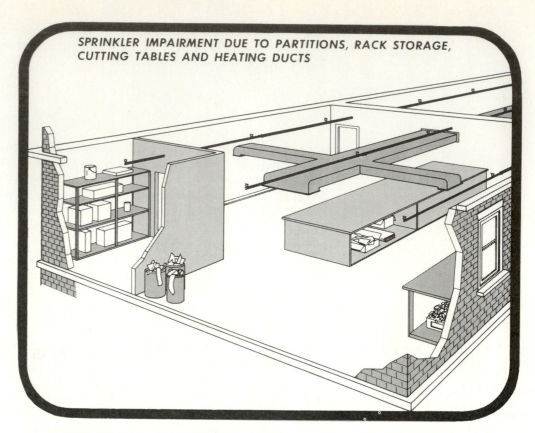

FIGURE 11.23.   Sprinkler action impairment. (Courtesy of Robert J. Brady Co.)

Unless the water can reach the fire, the sprinkler cannot do its job effectively (Fig. 11.23). Impairment shows some problem areas due to partitions, rack storage, cutting tables, and heating ducts. In these cases the system should be added to or altered to make the system more effective.

11.6   TYPES OF HYDRANTS*

"Hydrant bonnets, barrels, and foot pieces are generally made of cast iron with internal working parts of bronze. Valve facings vary and may be leather, rubber, or a composition material. Hydrants are available with various configurations of outlets.

Probably one of the most common types has two 2½ inch outlets and one large pumper outlet. However, hydrants are available with up to six individual 2½ inch outlets and with two or more pumper outlets with or without individual 2½ inch outlets.

There are two types of fire hydrants in general use today. The most common is the base valve (dry barrel) type where the valve controlling the water is located below the frost line between the foot piece and the barrel of the hydrant (see Figs. 11.24 and 11.25). The barrel on this type hydrant is normally dry with water being admitted only when there is a need. A drain valve at the base of the barrel is open when the main valve is closed allowing residual water in the barrel to drain out. This type of hydrant is used whenever there is a chance of the water above the frost line freezing."

* Section 11.6 is reprinted from *OpFlow*, Volume 3, No. 8 (August 1977) by permission. Copyright 1977, the American Water Works Association.

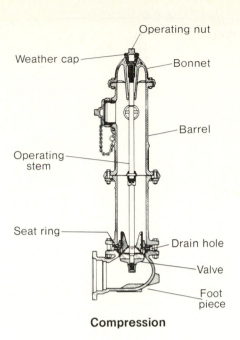

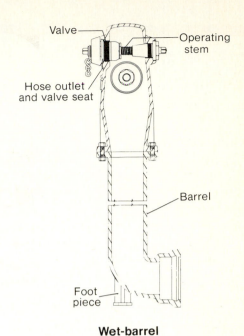

Compression

Wet-barrel

FIGURE 11.24. Compression-type hydrant, used in locations subject to freezing. [Reprinted from *OpFlow*, Volume 3, No. 8 (August 1977) by permission. Copyright 1977, the American Water Works Association.]

FIGURE 11.25. Wet barrel hydrant, used where freezing is not encountered. [Reprinted from *OpFlow*, Volume 3, No. 8 (August 1977) by permission. Copyright 1977, the American Water Works Association.)

11.7 SPRINKLER INSTALLATION

Sprinklers are installed by sprinkler fitters working for firms specializing in the fabrication and installation of sprinkler systems. Sprinkler fitters have their own local unions, which are part of the CIO-AFL "United Association".* Small installations are often tacked on to plumbing contracts, but jobs of reasonable size attract the sprinkler specialists. After the sprinkler contract is awarded, work starts on preparing shop drawings, which locate the sprinkler heads and size the mains and branches in accordance with sprinkler design requirements. In Fig. 11.26 a portion of a sprinkler shop drawing shows a 4-in. cross main feeding branch lines of 2-, 1½-, 1¼-, and 1-in. size. Elevations above finished floor are shown in parentheses, for example, (+8−11). Sprinkler lines pitch from the smallest lines to drain back to the main riser. Notice how for the most part the sprinkler head is located halfway between the wide-flange (WF) beams.

With shop drawings for an area of a building laid out to best suit sprinkler requirements, the sprinkler engineer joins the other mechanical and electrical trades in coordination sessions (see Chapter 6). Since the sprinklers cover such a large area with such precise location for proper distribution and the lines must pitch, other trades usually move their pipes, conduits, and ductwork to accommodate the sprinkler layout. When all the trades agree that the sprinkler shop drawings will accommodate their work, the drawings are sent to the architect and mechanical engineers for their approval.

---

*Short for United Association of Journeymen and Apprentices of the Plumbing and Pipe Fitting Industry of the United States and Canada.

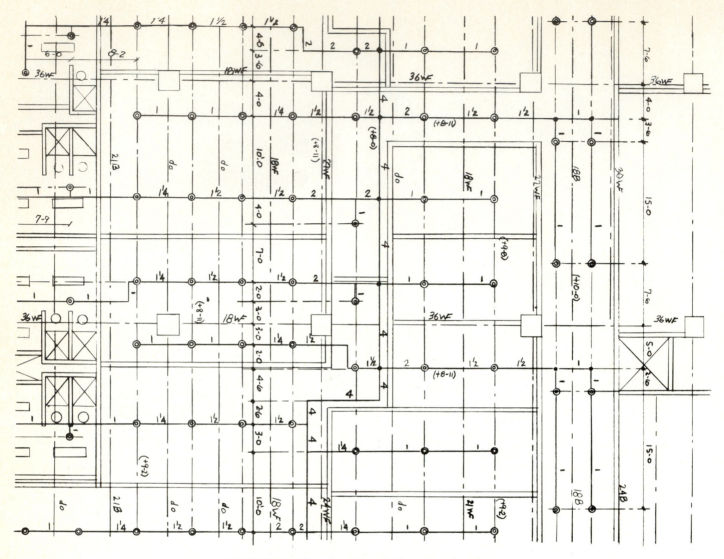

FIGURE 11.26. Partial sprinkler shop drawing.

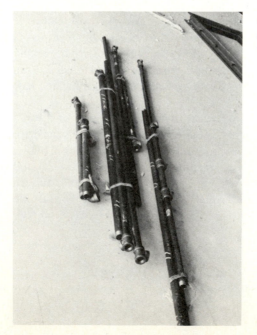

FIGURE 11.27. Sprinkler piping cut, threaded, and assembled ready for installation.

Screw pipe best suits sprinkler design, since a tee fitting interrupts the straight piping at each head.

In the fabrication shop, using the shop drawings for dimensions and sizing, the pipe is cut and threaded, some fittings fastened in place, and system branches bundled and tagged for location (Fig. 11.27). The piping hangs from adjustable pipe hangers, similar to those used to support plumbing. In hung ceiling locations, standard-length drops are installed protruding below the ceiling height line. After the ceiling material has been installed, the drops are marked, cut, threaded, and screwed back in place so that the head meets the required ceiling elevation. The system is then filled with water and tested for leaks.

11.8   USE OF SPRINKLERS IN HIGH-RISE BUILDINGS

The use of sprinklers in high-rise buildings has the enthusiastic support of fire departments and the National Fire Protection Association. New York City, with over 115 high-rise sprinklered buildings, can point to a record of 34 of 41 high-rise fires in sprinklered buildings being extinguished by the operation of one sprinkler head. In the office category of occupancies, automatic sprinklers have demonstrated a 97.4% satisfactory performance according to "Automatic Sprinkler Performance Tables," 1970 edition.

In Japan all buildings over 11 stories must be completely sprinkled, and Sidney, Australia, set the limit at 150 ft.

Fire departments, looking at the negative side of not sprinkling, point to the large numbers of workers and equipment needed to fight a high-rise fire with hoses fed by pumpers and standpipes. Some of the more difficult fires have required from 80 to 150 firefighters.

The Sears Tower in Chicago exemplifies the new concern for high-rise safety. The 110 floors, rising to a height of 1450 ft, contain 4.5 million square feet with an expected population of 16,500 people. The entire building is protected with automatic sprinklers. A description follows:*

"The sprinklers are supplied from two standpipes in the lower levels of the Tower, where two systems are in order and are supplied from one standpipe in the upper levels of the Tower. Control valves are provided on each floor located in the stair towers. Installation of the control valves allows alteration with minimum impairment to protection. When one considers the modular concept of ceiling sprinkler patterns included in the design, it is anticipated that changes will be few and far between.

The basic water supplies to the sprinkler system are from two connections to major circulating city water mains in Jackson Boulevard and Wacker Drive. The normal city pressure is 35 psi.

A 1,500-gpm, 110-pound-net-pressure booster pump that takes suction from those connections supplies Zone 1, which comprises the lower levels through the sixteenth floor (236-foot elevation) [Fig. 11.28]. The Zone 2 pump, which supplies the seventeenth through the twenty-ninth floor (400-foot elevation), is a 1,500-gpm, 105-psi-net-pressure pump that takes suction from the discharge of the Zone 1 pump.

A 10,000-gallon tank located on the thirty-first floor (425 feet) provides suction to the Zone 3 fire pump, a 1,500-gpm, 150-pound-net-pressure pump. That pump serves the twenty-ninth through the forty-sixth floor (634 feet elevation). The Zone 4 pump, which takes suction from the discharge of the Zone 3 pump, is a 1,000-gpm, 90-psi-rated pump that serves the forty-seventh through the sixty-third floor (855-foot elevation). A 5,000-gallon tank located on the sixty-fourth floor provides a

---

* Reprinted by permission from C. W. Schirmer, "Sears Tower: Life Safety and Fire Protection Systems," *Fire Journal*®, September 1972. Copyright © 1972 National Fire Protection Association, Quincy, Massachusetts.

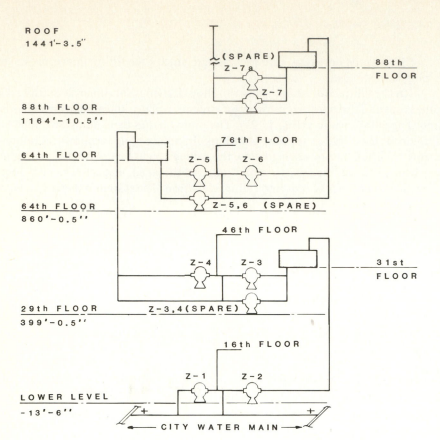

ROOF
1441'-3.5''

88th FLOOR
1164'-10.5''

64th FLOOR

64th FLOOR
860'-0.5''

29th FLOOR
399'-0.5''

LOWER LEVEL
-13'-6''

(SPARE)
Z-7

Z-7

88th
FLOOR

76th FLOOR

Z-5   Z-6

Z-5,6 (SPARE)

46th FLOOR

Z-4   Z-3

Z-3,4(SPARE)

16th FLOOR

Z-1   Z-2

CITY WATER MAIN

31st
FLOOR

FIGURE 11.28. Sears Tower schematic fire protection water supply system. (Reprinted from the September 1972 *Fire Journal*® Copyright © 1972 National Fire Protection Association, Quincy, Massachusetts. Reprinted by permission.)

suction supply to the 1,000-gpm, 131-psi Zone 5 pump, which serves the sixty-fourth through the seventy-sixth floor (1,026-foot elevation). A 1,000-gpm, 65-psi-net-pressure pump that takes suction from the Zone 5 pump serves Zone 6, which comprises the seventy-seventh through the eighty-seventh floor (1,170-foot elevation). An additional 5,000-gallon tank located at the eighty-eighth floor provides suction to a 500-gpm, 180-pound pump that serves Zone 7, the eighty-eighth through the one hundred-tenth floor (1,452-foot elevation).

All the tanks are filled by 2-inch automatic float-valve connections from the domestic pumping system and by 4-inch emergency connections from the domestic supply controlled by manual O.S.&Y. valves. This domestic system has a pumping capacity of 1,400 gpm at the 10,000-gallon tank on the thirty-first floor, 1,200 gpm at the 5,000-gallon tank on the sixty-fourth floor, and 1,000 gpm at the 5,000-gallon tank on the eighty-eighth floor. By means of automatic float-valve control a tank can also be filled automatically by the fire pump supplying the zone immediately below. The tanks therefore are essentially holding tanks, although they have capacities themselves for sprinkler water demand of 66 minutes for Zones 3 and 4 and 33 minutes for Zones 5, 6, and 7, if one assumes no makeup from any of the available fill supplies.

The electrical power to the fire pumps is supplied both from connections to the normal public utility and from the building emergency generator system."

11.9 ANOTHER VIEWPOINT

Most of this chapter deals with firefighting. The plumber and sprinkler fitter provide the means with which we combat fire. In the following chapters we continue to discuss the subject of fire as it pertains to air conditioning, electrical work, and elevator installations. Our attention will go more toward fire prevention and controlling the spread and effects of fires when studying these parts of mechanical and electrical work. Each of these fields create substantial problems which deserve our complete interest where fire is concerned.

12

The Heating, Ventilating, and Air Conditioning Industry

12.1 ENERGY AND COMFORT

In the twentieth century human beings have very nearly reached the ultimate of creating any environment we desire. We have learned to control basic things such as heat, cold, humidity, sound, vibration, and light, not to mention radiation and infectious disease. We can create sterile and absolutely dust-free spaces. We utilize energy from petroleum, coal, hydroelectric installations, wood, the sun, natural gas, manufactured gas, agricultural products, and geothermal sources. We have developed great manufacturing industries devoted to creating and improving a wide range of equipment to heat, cool, dry, clear, and humidify. We have developed a wide range of systems to deliver these means of providing human comfort ranging from the simple and economical to the sophisticated and expensive.

On a worldwide basis we now face the challenge of continuing to design and build for our comfort while strenuously working to conserve our use of energy. Where recently we sold cheap and inefficient devices, we are now designing and manufacturing them with much more emphasis on efficient performance. We are utilizing more control devices to save energy while providing equivalent comfort. We will combine the efforts of architect, mechanical engineer, and manufacturer to get the best energy results from specific building locations and neighboring environments. We will work with building owners to operate and maintain their buildings in the most intelligent manner.

This will still leave us with a vast number of energy-inefficient commercial buildings designed and built after 1945 with minimal concern for energy usage.

Fortunately, these buildings as they continue to age will have to meet modern competitive standards and will be ripe for major recycling efforts to retain their current occupants or attract replacements. Hopefully, the owners will use this period to increase the energy efficiency of their buildings electrically, architecturally, and mechanically. Attractive opportunities for major recycling work for heating, ventilating, and air conditioning companies should prevail during this era.

12.2 THE INDUSTRY STRUCTURE

The heating, ventilating, and air conditioning industry, hereafter referred to as HVAC, consists of some relatively large and financially healthy manufacturers, and numerous moderately-sized and small manufacturers, producing a great variety of fittings, valves, fans, pumps, grilles, and so on; and a highly fragmented group of contractors. In 1980 the two largest mechanical firms secured contracts totaling $343,000,000 and $290,000,000. Of the remaining 15 largest firms throughout the country, contract volume secured for 1980 ranged between $170,000,000 and $42,000,000.* Perhaps the best indication of the extent of fragmentation among HVAC contracting firms is the number of firms engaged in HVAC business in a given locality. Baltimore, Maryland, and vicinity, for instance, lists 55 air conditioning contractors in the "1982 Blue Book Contractors Register."

Mechanical contracting conforms to the typical building construction pattern of a strong single individual starting a firm that grows to a moderate but limited size. The business success stems from the single entrepreneur's ability and good fortune. All too frequently the business does not survive the death of the founder. In the United States the business is generally high risk and domestic in nature.

12.3 THE SIMILARITY TO GENERAL CONTRACTING

Moderate-size to large HVAC firms operate in a manner similar to the larger general contractors. They frequently have the largest single portion of a building under contract. The size and complexity of their work leads to subcontracting work to specialists in other trades. With some exceptions in some parts of the country the HVAC companies basic strength lies in *pipe fitting.* This is readily understandable since a lot of the older and bigger firms started as heating contractors before the real surge in air conditioning in the 1930s and 1940s. As a result, they frequently sublet the *sheet metal ductwork,* a very substantial part of their overall contract. By subletting sheet metal work they eliminate the investment required for a sheet metal fabrication shop and the particular problems of operating such a shop, which include the sales effort required to keep it supplied with a flow of business. In addition, in CIO-AFL building trades union shops they dispose of the necessity of dealing with an additional union, the Sheet Metal Workers International Association, which might otherwise complicate their labor relations somewhat.

They also usually sublet pipe and duct insulation work. *Pipe covering,* as it is commonly called, has become a specialty also involving a separate union, the International Association of Heat and Frost Insulators and Asbestos Workers. HVAC contractors do not have a real interest in pipe covering nor have they been able to install pipe and duct insulation with their own personnel as cheaply as the pipe covering specialist—thus the interest in subletting.

HVAC firms sublet the manufacture, installation, and startup of *large refrigeration machines* and *cooling towers.* Companies engaged in these specialties have the special knowledge to properly and expeditiously install and start up products made by their concern. These companies train their supervisors and workers to become completely familiar with and expert at handling the company's line of

---

* *Engineering News-Record,* August 27, 1981.

products. This is a particularly good basis for subletting, since it taps special expertise while centering responsibility on the product manufacturer.

A third kind of work in this category, HVAC *control work,* differs somewhat from the other two. Where the other two deal with specific items in a complete HVAC system, *control work,* although a separate system of its own, connects with all the operating devices of all the supply air, return air, and ventilating systems in the entire project. The equipment manufactured and supplied by the control specialist and the kind of system installed also vary considerably from the heavy machinery, piping, and ductwork which make up the bulk of HVAC work.

Pipe fabrication is sublet simply because intricate pipe cutting, fitting, and welding can be readily detailed on shop drawings and fabricated better and more cheaply in a pipe fabricating shop. Overall project time and on-site installation time are also substantially reduced.

On high-rise buildings many days are required to hoist heavy and bulky fans, pipe, coils, and tanks to the machine rooms on the upper floors and top of the building. Special booms and engines with which a rigging contractor is well equipped raise the wide variety of items to the appropriate floors. *Rigging specialists* have excellent equipment, trained workers, necessary licenses, and supervisors to take over this dangerous and time-consuming part of the work. Subletting the rigging frees the HVAC contractor's workers to confine themselves to machinery setting and pipe fitting. Employing a rigger also adds the rigger's personnel to the HVAC contractor's overall effort.

Finally, *air balancing work,* measuring the flow of air in the ductwork systems and adjusting the quantities for optimum comfort, can be sublet to firms specializing in this work. These firms employ trained technicians to measure, record, and adjust air quantities. Employing *air balancers* also frees HVAC engineers for other assignments.

12.4 THE CHALLENGE OF SUPERVISION

The most satisfaction in building construction comes from putting together your own organization and doing your own work. Selecting a superintendent and foremen, who in turn assemble a skilled and productive crew of people who take pride in their work, both challenges and rewards those responsible. However, when HVAC company policy decrees subletting, the project manager (the title given to the person primarily responsible for producing a successful project) must operate in a fashion similar to that of a general contractor and produce the best results from the work done by his own forces *and* his subcontractors. He must plan and schedule the work for optimum results. He must sequence and coordinate his trades to have them available when needed with sufficient personnel, equipment, and material to meet project progress. He must monitor their purchasing, fabrication, and deliveries to ensure that necessary equipment and material match schedule requirements. He must conduct labor relations with skill and finesse to avoid jurisdictional trade disputes and costly featherbedding.* He must take bids and award contracts to ensure attractive competitive prices compatible with good subcontractor performance.

12.5 PURCHASING

Subletting reduces the number of individual items the HVAC contractor purchases, but there remains a formidable number of fans, valves, motors, controllers, expansion joints, steam traps, air handlers, and so on, that the contractor orders and has delivered. These he obtains from an equally diverse number of sources. Machinery makes up a large part of what must be ordered, requiring model types, dimensions, metal gauge, horsepower, rpm, and other capacity information. Manufacturers

---

*Featherbedding: A term to describe a work assignment required by union contract but considered unnecessary by the employers.

MODEL R-RD MOUNT DATA

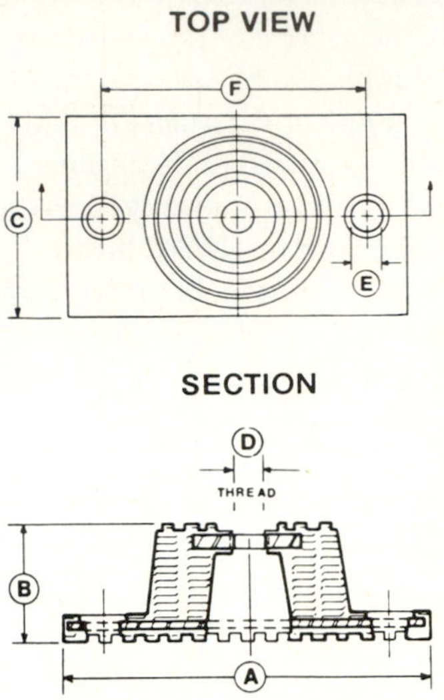

TOP VIEW

SECTION

DATA

MODEL	RATED LOAD POUNDS	COLOR IDENTIFICATION	MAXIMUM DEFLECTION	DIMENSIONS						
				A	B	C	D	E	F	G
RA-40		Black	.11	$3\frac{3}{16}$	$\frac{7}{8}$	$1\frac{11}{16}$	$\frac{5}{16}$-18	$1\frac{1}{32}$	$2\frac{3}{8}$	—
RA-100		Red								
RB-120		Black								
RB-220		Red	.20	$3\frac{7}{8}$	$1\frac{1}{8}$	$2\frac{5}{16}$	$\frac{3}{8}$-16	$1\frac{1}{32}$	3	—
RB-375		White								
RC-250		Green								
RC-600		White	.25	$5\frac{1}{2}$	$1\frac{5}{8}$	$3\frac{5}{16}$	$\frac{1}{2}$-13	$\frac{9}{16}$	$4\frac{1}{8}$	—
RC-1100		Yellow								
RDA-55		Black								
RDA-125		Red	.35	$3\frac{3}{16}$	$1\frac{1}{2}$	$1\frac{11}{16}$	$\frac{5}{16}$-18	$1\frac{1}{32}$	$2\frac{3}{8}$	—
RDB-120		Black								
RDB-220		Red	.40	$3\frac{7}{8}$	$1\frac{7}{8}$	$2\frac{5}{16}$	$\frac{3}{8}$-16	$1\frac{1}{32}$	3	—
RDB-375		White								
RDC-250		Green								
RDC-600		White	.50	$5\frac{1}{2}$	$2\frac{3}{4}$	$3\frac{5}{16}$	$\frac{1}{2}$-13	$\frac{9}{16}$	$4\frac{1}{8}$	—
RDC-1100		Yellow								

FIGURE 12.1. Manufacturer's data or "cut" sheet. (Courtesy of Peabody Noise Control Inc.)

prepare specification data sheets called "cuts" for their product line (Fig. 12.1). The purchaser orders the item he needs from this list, thus simplifying the ordering process. The complexity of HVAC purchasing may be appreciated if a choice must be made from 124 different valves manufactured by one leading valve concern.

12.6 SHOP DRAWINGS, CUTS, AND SAMPLES

Submission of Samples

Specifications usually require the submission of three or more small samples of pipe, welding, insulation, grilles, sheet metal joints, isolation hangers, inserts, and many other items the architect and mechanical engineer may ask to be submitted to them for review of quality, strength, flexibility, finish, color, and appearance. After approval, one of the approved samples should be on hand at the job site for comparison with the work being installed.

Approval of Cuts

For approval of grilles, fans, motors, and many other items, cuts are submitted marked as shown in Fig. 12.2 with an arrow stamp to designate the particular product being ordered. This simplifies the ordering and approval procedure. The mechanical engineer employed by the architect carefully checks the item submitted against the plans and specifications to ensure getting the required performance. For example, the engineer checks gallons per minute for pumps, cubic feet per minute for fans, and so on. He also checks equipment weights since with their liquid content they place heavy loads on the structure, which the structural engineer checks to make sure that the beams, girders, and columns can carry the loads without excessive deflection and vibration. The responsibility for dimensions of inlets and outlets, how the equipment

SPECIFICATIONS

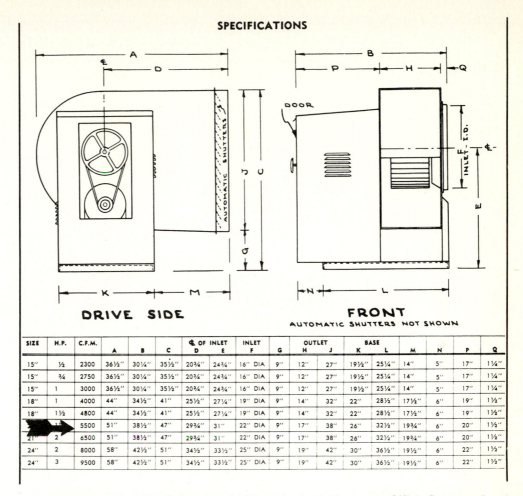

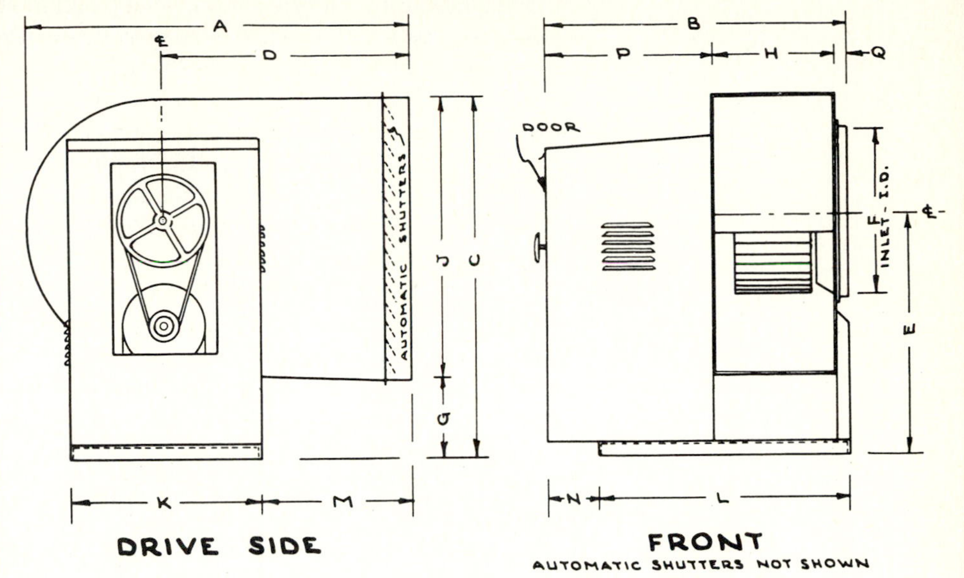
A
D
B
P
H
Q
DOOR
AUTOMATIC SHUTTERS
J
C
G
INLET - I.D.
F
E
K
M
N
L
DRIVE SIDE
FRONT
AUTOMATIC SHUTTERS NOT SHOWN

SIZE	H.P.	C.F.M.	A	B	C	℄ OF INLET D	E	INLET F	G	OUTLET H	J	BASE K	L	M	N	P	Q
15″	½	2300	36½″	30¼″	35½″	20¾″	24¾″	16″ DIA	9″	12″	27″	19½″	25¼″	14″	5″	17″	1¼″
15″	¾	2750	36½″	30¼″	35½″	20¾″	24¾″	16″ DIA	9″	12″	27″	19½″	25¼″	14″	5″	17″	1¼″
15″	1	3000	36½″	30¼″	35½″	20¾″	24¾″	16″ DIA	9″	12″	27″	19½″	25¼″	14″	5″	17″	1¼″
18″	1	4000	44″	34½″	41″	25½″	27¼″	19″ DIA	9″	14″	32″	22″	28½″	17½″	6″	19″	1½″
18″	1½	4800	44″	34½″	41″	25½″	27¼″	19″ DIA	9″	14″	32″	22″	28½″	17½″	6″	19″	1½″
[illegible]	[illegible]	5500	51″	38½″	47″	29¾″	31″	22″ DIA	9″	17″	38″	26″	32½″	19¾″	6″	20″	1½″
21″	2	6500	51″	38½″	47″	29¾″	31″	22″ DIA	9″	17″	38″	26″	32½″	19¾″	6″	20″	1½″
24″	2	8000	58″	42½″	51″	34½″	33½″	25″ DIA	9″	19″	42″	30″	36½″	19½″	6″	22″	1½″
24″	3	9500	58″	42½″	51″	34½″	33½″	25″ DIA	9″	19″	42″	30″	36½″	19½″	6″	22″	1½″

FIGURE 12.2. "Cut" sheet marked for submittal. (Courtesy of King Company.)

will fit into the allowable space, and how piping, ducts, and electric conduit connect, lies with the HVAC contractor. The architect and the mechanical and structural engineers are concerned with conformance with their design, plans, and specifications. This concern extends to the quality of the device and involves low-cost maintenance, freedom from vibration, low power consumption, and long life.

Shop Drawings

When well executed, the architect's and engineer's plans and specifications give the necessary information on appearance, quality, size, capacity, location, dimensions, and method of installation for a complete operational building. As products of the design process, they reflect design computations which establish pipe and duct sizes, fan and pump capacities, number of water closets and roof drains, and information of a similar nature. They also reflect the responsibility the architects and engineers have assumed to carry out the owner's objectives.

One may well wonder why it is necessary to prepare shop drawings when the architects and engineers have done such a fine job of design. To understand the purpose of shop drawings we must realize that the architect's and engineer's drawings contain *key* dimensions and *typical* sectional views. The HVAC contractor, his subcontractors, and his fabricators must hand to their workers the same information, but *with every dimension, every fastening, every connection, and every sectional view clearly indicated.* The objective is to provide the people fabricating material in a shop or installing it in the building with all the information they need to do their part of the work, at the same time clearly indicating the limits of their work and how it must join to the work of other trades and contractors.

One should also realize that the designers, in the interest of the owners, provide for competitive bidding in their documents. In so doing they consider differences in manufacturing design and methods. The HVAC contractors and their suppliers are recognized as the specialists who rightfully should improve their efficiency of fabrication and manufacture. One company may weld where another drills and bolts. Materials, shapes, method of assembly, and other production techniques may vary. The shop drawing, while conforming to design, may substitute a bent plate for a rolled angle, for example. The engineers do not always accept these minor changes from what the plans contain but usually do if the *intent* of the engineers has been satisfied. The basic principle that prevails recognizes that the contractors and manufacturers, in competition with each other, are excellent sources for new ways of keeping down costs.

In like fashion, shop drawings reflect the selection the HVAC contractor has made as to fans, heat exchangers, chillers, and so on. Each manufacturer has his particular model of machinery. The connections of these can vary as to height, plan location, and sometimes size. In a machine room, for example, the HVAC contractor adjusts locations of fans, pumps, and chillers to develop the best arrangement of piping and ductwork, keeping in mind maintenance requirements of owners' operating personnel. In laying out machine rooms it is sometimes even necessary to substitute a heavier beam or shift the location of a beam based on the final layout as developed by the shop drawings.

Most contracts between owners and contractors do not make the HVAC contractor responsible for the adequacy of design. However, as a recognized specialist in his field, he generally has a definite obligation to advise the architect if the design drawings violate the codes that apply. His shop drawings should not contain anything that violates the code in force. Further, as a qualified engineer he would prepare his shop drawings based on good engineering practice. Should the designer fail to take into proper account the resistance required for incoming steam lines at high pressures or forces created by steam and hot water lines expanding and contracting, the HVAC engineer would correct this condition by adding necessary thrust blocks and hangers while preparing his shop drawings.

Figure 12.3 shows part of a sheet metal shop drawing. Section A-A shows a 28-in.-wide by 25-in.-high duct with a minimum resistance elbow. The joint in the upper left-hand corner is dimensioned 12½ in. above the second floor. The drop through the second-floor opening is 46½ in. long. Note that in *their own* work the sheet metal detailers only use inches for measurements whereas ceiling heights and so on are given in the standard feet and inches. Note the pipe under the elbow and SLR "A," a stream line register, also called a linear diffuser. In section B-B a large 10-gauge black iron kitchen exhaust duct penetrates the second floor. On the horizontal run it might seem preferable to show the dimension as 28×80 since it is 80 in. wide over the ceiling. However, in sheet metal practice the 28 represents the dimension of the duct section which is shown in the sketch. Note the different graphical method of showing the air duct joint and the flange joints in the black iron. Note the work of the duct insulator, which is shown but noted "by others." Note the dimensions 12, 7 and 7, 33 that create the two 45° elbows. Note in section C-C how the 30×16 duct splits into 23×16 and 7×16 and that the elbow is a 90° rectangular fitting.

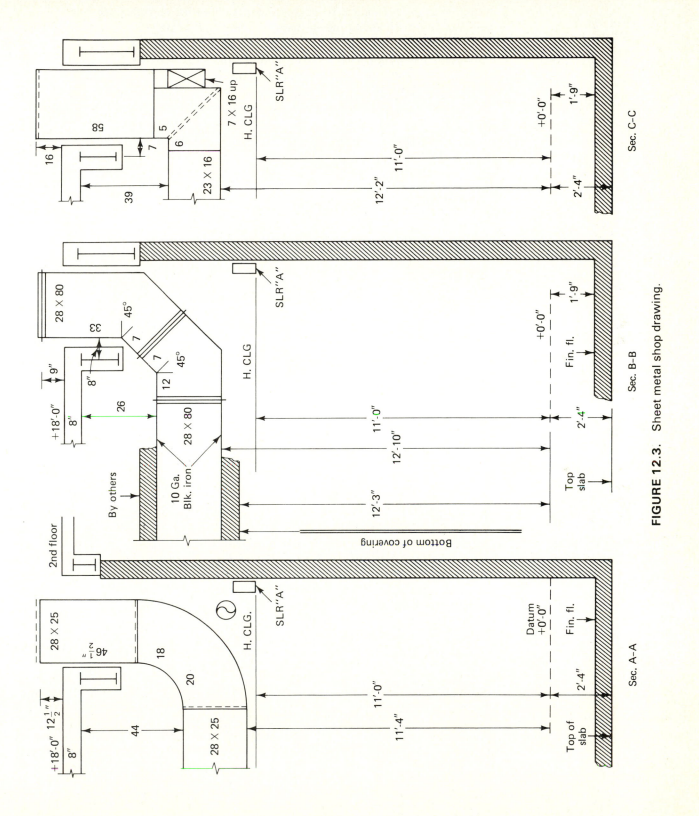

FIGURE 12.3. Sheet metal shop drawing.

12.7   COORDINATION   Throughout most of the United States, current building construction contracts require mechanical and electrical contractors to coordinate mechanical and electrical installations. Mechanical and electrical engineers develop their design drawings only to the extent that they believe all the pipes, mains, drains, lights, sprinklers, ducts, and conduits can be detailed to fit into the architectural space provided. For the most part, the greatest amount of time and effort goes toward solving the problems in the space between the bottom of the structural slab and the top of the hung ceiling beneath it and in the entire mechanical equipment rooms. Architects and engineers generally provide adequate openings for ducts, pipes, and conduits in vertical systems.

Coordination meetings are often conducted by the general contractor as each mechanical and electrical engineer brings his fairly complete shop drawings to scheduled meetings. A room with enough plan table space is essential to let the trades spread out their drawings. The objective of the meetings is to make the best use of the available space and to resolve all interferences between piping, ducts, and so on. Each trade has valid reasons for not relocating. For design reasons, the architect and illumination engineer do not want electric light fixtures moved. Sprinklers and air diffusers also must drop through the ceiling at fixed dimensions. Sprinkler lines, plumbing drain lines, and steam supply and return lines all must maintain proper pitch. Many problems are solved by rerouting the HVAC contractor's ductwork, but this is the bulkiest and most cumbersome of the trades.

Through cooperative effort, problems are reasoned out and resolved. Trades relocate their work horizontally and vertically. Ductwork often changes shape from say a 30×30 equaling 900 square inches to a 20×45. Electrical conduits and ductwork frequently raise or drop to miss another trade's work, utilizing flattened S-shaped fittings or conduit bends. Occasionally, relief is sought from the designers in the form of a request for piping or hangers to pass through ductwork. When all else fails, the plumber or HVAC contractor and the general contractor may suggest a beam or girder cut for piping or ductwork. Cutting through a structural member requires the approval of the structural engineer, with the mechanical and electrical designers concurring. The surrounding part of the beam must be reinforced with plate welded around the perimeter of the opening. Considerable expense results, which generally the owner pays to the steel contractor as a structural change. As for the costs of the coordination process, each contractor must cover this expense in his original bid. Perhaps this method of resolving coordination works best. Certainly, the mechanical and electrical contractors, looking after their own interests, will seek the most economical coordination solutions.

When coordination has been completed for a floor or space, every trade involved signs off, freeing the others to put the final touches on their shop drawings before submitting them for approval. Prompt approval by the design engineers ensures a smooth flow of work into the various shops and out into the field. Controlling the submission and approval of shop drawings is a major responsibility of the general contractor or construction manager. In a well-run project shop drawings are returned to the firm that submitted them no later than 14 days after submission. Messenger service minimizes the time the drawings are unavailable because they are in transit.

The failure to approve shop drawings within a reasonable period of time seriously slows a project. Poorly prepared work by the subcontractor will mean disapproval. Unacceptable substitution of materials, joining methods, failure to properly detail the meeting with the structure or work of other trades, and anything that will not perform as specified will cause rejection. Reworking and resubmission of the drawing then requires valuable project time.

Occasionally, an architect or engineer will utilize the shop drawing period to improve an initially poor design. Certainly, the shop drawing period should produce the most practical solution consistent with the original design intent. However, shop drawings are intended to further detail the basic design and develop installation information, not to start designing all over again. Redesigning on shop drawings is an unfair use of the contractors' allotted time to build. However, in the interest of safety and the client's needs, when absolutely necessary, redesign in the shop drawing stage can be excused. As the saying goes: "Better late than never."

12.8 INDUSTRY TRENDS

The 1970s have had a pronounced effect on the HVAC industry. Some of the changes that have occurred are attributable to the energy situation facing countries such as ours which are dependent in whole or part on imported oil. Since 1973 the amount of energy used to heat and cool our buildings has been the subject of much interest and debate on the part of designers, manufacturers, building owners, energy producers, and governmental agencies. The nation has been asked in some cases and directed in others to accept 65°F for the heating of buildings and 78°F for their cooling as a substantial effort to conserve energy. Standards for designing buildings within limits of energy use have been established while still being tested, evaluated, questioned, and in some cases criticized. Tax credits have been offered by the U.S. government to some of those willing to upgrade their homes and commercial buildings with energy-saving materials and devices.

The combination of increased construction costs and higher interest rates during most of the 1970s, which have been tied to and caused by the general inflationary trend, have also encouraged designers to seek more economical HVAC solutions. As a result, there has been a return on the part of some to simpler HVAC designs. At the same time, the impact of better construction management is being felt as projects are being completed in shorter time spans. Quicker job startups and faster construction help keep the inflation increment as low as possible. Reducing the length of the construction loan time also has a definite economic advantage.

Industry spokesmen notice a trend toward improvement in the skill, training, and attitude of workers. The young people entering the trades now receive very good apprentice training. Their companies also conduct more training sessions for improving skills and working safely. A surprising number of workers now have college degrees or have attended college for a few years.

Whether a trend has started or not, it is also worth noting that in recent years two of the largest mechanical contractors have been bought out by bigger organizations. It will be interesting to observe the effect this has on these companies that have been among the industry's leaders.

13

Heating

13.1 INTRODUCTION

The separate operations of heating, ventilating, and air conditioning in most up-to-date commercial buildings are performed by a combination of steam, hot and chilled water, and air equipment assembled in fairly complex systems. We use the term and think of HVAC as a single entity that all goes together. However, since heating systems in many cases are separate systems of their own, we can examine this part of HVAC separately. In our study of air conditioning in Chapter 14 we will learn the part that heating plays in the air conditioning of buildings.

The consumption of energy to produce the necessary heat for a building can occur at a distant location such as a boiler plant for a university, industrial complex, or at a city steam plant. Most heat comes from boilers and furnaces located in basements. Finally, the heat can be generated in the room itself or the factory by using gas burners or electric resistance heaters.

13.2 BOILERS

High-pressure steam delivered by mains to the basement of a building from the street or utility tunnel (Fig. 13.1) naturally simplifies the design and construction to some degree. At a convenient point a steam station contains a pressure-reducing valve, making low-pressure steam available as the heat source (Fig. 13.2). A large proportion of nonresidential heating systems are designed as hot water or steam systems. The smaller space required for the piping is an advantage over hot air systems. Most systems operate with low-pressure steam supplied by boilers having maximum ratings

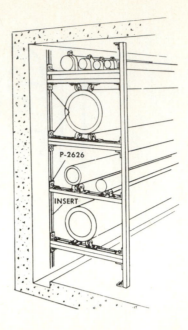

FIGURE 13.1. Utility tunnel piping. (Courtesy of Unistrut Building Systems, GTE Product Corp.)

FIGURE 13.2. Steam pressure-reducing station with low pressure on the right side.

of 15 psi gauge steam pressure, 30 psi gauge water working pressure, and 250F° temperature. Most building boilers are package units that only have to be properly set and connected. In some designs the hot gases pass through the boiler tubes, which is called fire tube, while in others they pass around them, called water tube (Fig. 13.3). Some are modular packages that can be assembled to create the desired capacity (Fig. 13.4). The following is one company's description of systems using special boilers:*

"*High* Temperature and *Medium* Temperature Water Systems (Special Boilers): Although conventional water systems have maximum temperatures below 250F°, much higher temperatures may be obtained by maintaining correspondingly higher pressures. The lowest pressures in the system (either at the highest elevation or at the pump suction) must be kept above the boiling point of water at that pressure. As pressures increase, the allowable temperature increases do not advance equally. For example, an increase in pressure from 100 to 200 psig allows a temperature increase of 50F°, but the 100 psi increase from 300 to 400 psig allows a temperature increase of only 26F°. Because of this relationship, present practice is to stop at about 425 psig which permits temperatures to 455F°.

*Johnson Controls, Inc., *Johnson Controls Training Manual.*

(a)

(b)

(c)

FIGURE 13.3. (a) Fire tube boiler; (b) water tube boiler; (c) cutaway of a fire tube boiler. (Courtesy of Cleaver-Brooks.)

FIGURE 13.4. Cutaway section of modular boiler units. (Courtesy of Burnham Corporation.)

"Under such conditions, it is feasible to use temperature drops in heating units up to 200F° instead of the 20F° or 30F° common in low temperature water systems. To appreciate the advantages of high temperature water, especially where heat must be transmitted over long distances, consider that one-tenth as much water is required to release a given amount of heat if the temperature drop is 200F° as would be required when the temperature drop is 20F°. This means smaller mains, smaller pumps, and lower pumping costs.

"Water systems from 250F° to 350F° are classified as 'Medium Temperature'; water systems above 350F° are classified as 'High Temperature.'"

13.3   CONVERTERS   With steam available and a boiler unnecessary, the preference could still be for a hot water heating system. A converter produces hot water in this case in a steel shell containing copper tubes through which the steam passes (Fig. 13.5a and b). An exchange of heat takes place, raising the supply water temperature to the desired level. A converter changes (*converts*) steam to hot water. A very similar device called a heat exchanger utilizes high-temperature water to produce low-temperature water, and lower-temperature chilled water to produce higher-temperature chilled water.

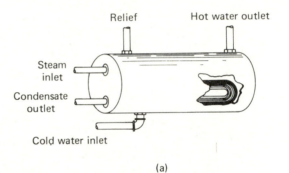

(a)

(b)

FIGURE 13.5.   (a) Converter. (b) A converter produces hot water using low-pressure steam.

13.4 WARM AIR FURNACES

Warm air furnaces, although used mainly for residential heating, are also installed in low-cost commercial buildings, schools, and churches. Capacities run from around 50,000 Btu/hr to 400,000 Btu/hr. Figure 13.6 shows a ceiling-suspended model which is gas or oil fired. Rooftop models are also available. In industrial applications a manufacturing process might require that high volumes of air be exhausted to the atmosphere. For these situations forced air furnace *makeup* units pull in fresh air, heat it with gas or LP (liquefied petroleum) gas, and discharge it into a supply system. Fresh air can be combined with recirculated air in the interest of economy. These units range from 11,000 to 100,000 cubic feet per minute (CFM) with inputs of 1,089,000 to 10,900,000 Btu/hr. Figure 13.7 shows a floor-mounted unit.

FIGURE 13.6. Oil-fired ceiling-hung warm air furnace. (Courtesy of York-Shipley Inc.)

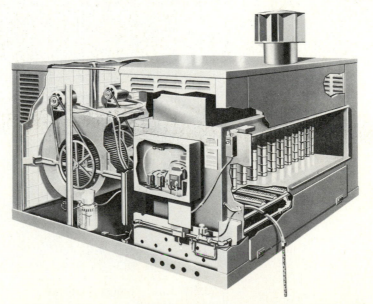

FIGURE 13.7. Gas-fired makeup unit. [Courtesy of ARES (Air Replacement Engineering Systems), a member of Bastian Industries.]

13.5   A GAS HEATING DEVICE   The Roberts Gordon Appliance Corporation has patented an infra-red vacuum gas heating system called Co-Ray-Vac, consisting of interconnected gas burners under a reflective metal cover. "Each burner depends upon vacuum to introduce a filtered air-gas mixture to complete the combustion process (Fig. 13.8). The products of combustion are safely exhausted to the outdoors by means of a vacuum pump-exhauster."* Thus, the system fires and operates under a unique vacuum principle claimed to be highly efficient. Combustion chambers are spaced 15 to 21 ft apart at a height of 10 to 15 ft above the floor (Fig. 13.9).

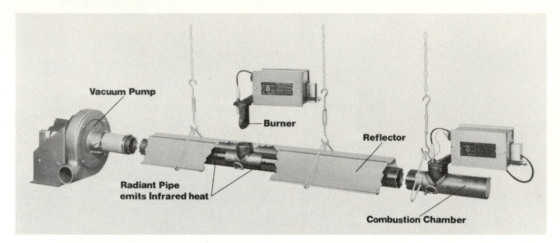

FIGURE 13.8.   Parts of an infrared vacuum gas heating system. (Courtesy of Co-Ray-Vac Div. of Roberts Gordon.)

FIGURE 13.9.   Industrial application of gas heating. (Courtesy of Co-Ray-Vac Div. of Roberts Gordon.)

*Quotation taken from undesignated advertising brochure of the Roberts Gordon Appliance Corp.

13.6   DISTRIBUTION NETWORKS†

Steam Distribution

"One of the earliest arrangements was called a one-pipe system [Fig. 13.10] and consisted of a single pipe leaving the boiler, making a circuit to radiators throughout the building and returning as condensed steam, or "condensate," to the lower part of the boiler. Since these risers were connected to only one end of the radiator, the condensate had to return through the same pipe to the boiler.

Air always gets into steam distribution systems and must be eliminated by air vents, or eventually the steam would be entirely displaced, because air cannot easily follow the path of the condensate.

Today steam systems are two-pipe systems, using separate piping for supplying the steam and returning the condensate."

Supply Mains. "The most common systems in the heating field have a maximum pressure of 15 psig and are called low pressure systems. The steam leaves the top of the boiler and is carried through a horizontal main running below all, or the majority of, the terminal heating units in a so-called up-feed system [Fig. 13.11] or above the

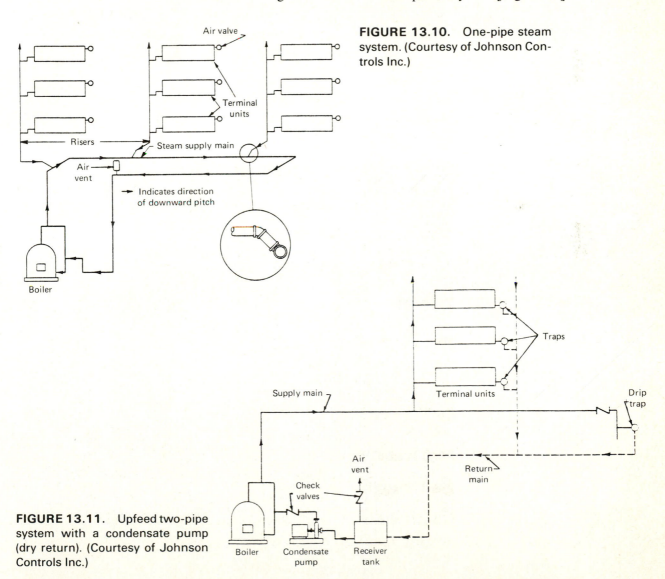

FIGURE 13.10.   One-pipe steam system. (Courtesy of Johnson Controls Inc.)

FIGURE 13.11.   Upfeed two-pipe system with a condensate pump (dry return). (Courtesy of Johnson Controls Inc.)

†Section 13.6 is taken from Johnson Controls, Inc., *Johnson Controls Training Manual.*

terminal units in a down-feed system [Fig. 13.12]. In either case, the horizontal runs are pitched downward in the direction of steam flow so that any condensate forming in the pipes will flow in the same direction as the steam."

Classification of Systems. "The foregoing remarks about supply lines will apply to any type of steam system, but condensate or return lines are run in various manners depending on the type of system. These are classified by the ASHRAE Guide as high pressure, low pressure, vapor or vacuum systems, depending on the pressures under which they operate; as dry-return or wet-return, depending on whether the return lines are generally above or below the water line of the boiler; and as gravity or mechanical return systems, depending on whether the condensate returns by gravity or is forced into the boiler by a condensate pump, a vacuum pump, or a boiler return trap."

Vacuum Systems. "Vacuum return systems are most commonly used in large buildings. Mechanically operated pumps have their suction connected so as to produce a pressure below atmospheric in the condensate return line. This pressure is usually between 4 and 8 inches Hg. vacuum. The vacuum helps draw condensate and air through the traps and pipe lines. When water accumulates in a line, air flow is restricted. A vacuum in the return line will help to relieve this situation."

High- Vacuum Systems. "Some systems are designed for use with vacuums as high as 20 to 25 inches Hg. The function is not to increase the difference in pressure between the steam and return lines, but to vary the pressure in the entire system from 2 psi or more to as low as 25 inches Hg. A vacuum as large as this may be carried back to the source of steam. Under some operating conditions, it is the difference in vacuum rather than steam pressure which forces the steam, air and condensate through the circuits.

A difficulty with high vacuum systems is in keeping the piping and equipment tight enough to prevent air from entering."

Vapor Systems. "Vapor systems are those which may operate at a vacuum without the use of a pump. They may have a dry return or a combination dry and wet return. The vacuum is produced by the condensation of steam. As each cubic foot of steam becomes roughly a cubic inch of water when condensed (provided all pipes are tight) there will not be enough steam left to fill the system and a natural vacuum is created."

FIGURE 13.12.   Downfeed two pipe system with dry return. (Courtesy of Johnson Controls Inc.)

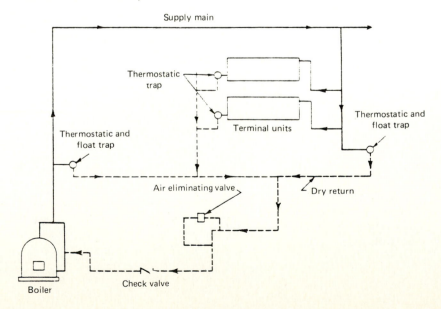

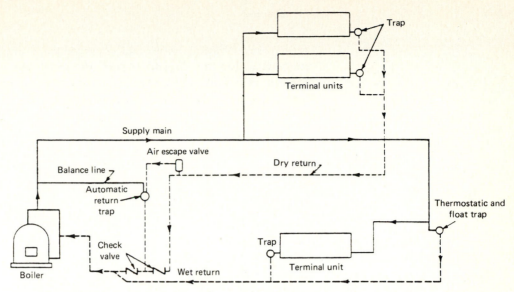

FIGURE 13.13. Two-pipe system with wet and dry returns. (Courtesy of Johnson Controls Inc.)

Return Line Systems. "Return lines are run to the equipment room in any of the three following methods. A dry return system is one in which all lines are carried back above the water level of the boiler so that they handle condensate flowing at the bottom and air at the top [see Fig. 13.10].

A wet return system is one in which the horizontal return main is below the boiler water level. With a vacuum pump, the air can easily be pulled along with the condensate and eliminated.

The third method uses both dry and wet returns. [Figure 13.13] is an example of the combination."

Return Components. "Vacuum pumps have been mentioned for the production of mechanical vacuum. They eliminate the air which accumulates in storage tanks on which they are mounted, and also force the condensate into the boilers.

Condensate pumps, like vacuum pumps, have tanks for the accumulation of air and liquid. The air is eliminated through an air valve of the non-return type, and the motor is operated from the level of condensate in the tanks. These pumps are sometimes used for only one or more parts of the return line system which are below the boiler water level. In such cases, they may simply lift the condensate up to a point where it can flow to the boiler by gravity or to another pump as shown in [Fig. 13.14]."

FIGURE 13.14. Condensate pump as a mechanical lift in a vacuum system. (Courtesy of Johnson Controls Inc.)

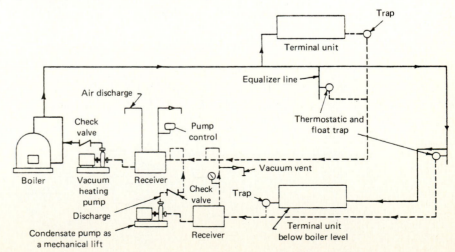

Traps. "Various types and styles of traps are used to regulate the flow of condensate and to eliminate air from the system. [Figures 13.15 to 13.18] illustrate some of the various types and their application. [Figure 13.19] shows a strainer used to trap dirt and scale. [Figure 13.20 illustrates a terminal unit with cooling leg and dirt pocket.]

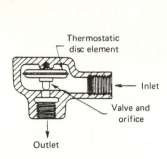

FIGURE 13.15. Disc-type thermostatic trap. (Courtesy of Johnson Controls Inc.)

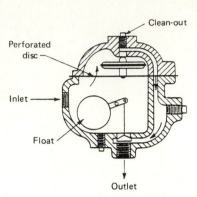

FIGURE 13.16. Float and thermostatic trap. (Courtesy of Johnson Controls Inc.)

FIGURE 13.17. Steam trap.

FIGURE 13.18. Automatic drain trap. (Courtesy of Johnson Controls Inc.)

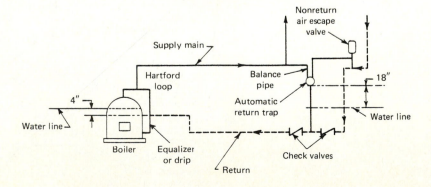

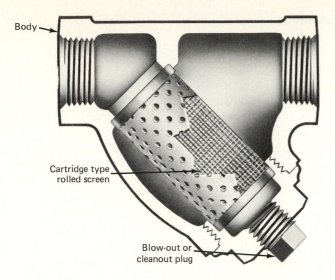

FIGURE 13.19.   Cross section through a Y-pattern strainer. (Courtesy of Crane Co.)

FIGURE 13.20.   Terminal unit with a cooling leg and dirt pocket. (Courtesy of Johnson Controls Inc.)

High-Pressure Systems. "High pressure systems require special attention because they handle saturated or superheated steam. In the case of superheated steam, which is less common for heating buildings, there may be no condensation in the main, and if so, the mains can be run up and down without traps. With saturated steam, however, thermostatic traps, or float and thermostatic traps, are used the same as with low pressure systems. The thermostatic trap must be selected for the steam temperature. For example, at 150 psig steam, the trap will work at 360F. The great difference between high and low pressure systems is at the discharge side of the trap. As the high temperature condensate is discharged, all or much of it flashes into steam at the lower pressure.

Instead of thermostatic traps, float traps and check valves may be used as shown in [Fig. 13.21]. The pressure in the return line depends on the amount of flash steam, the extent of cooling in the lines, and the speed of emptying the return.

Steam and hot condensate may be used in heat exchangers to heat domestic hot water or make-up water for boilers. Heat exchangers such as these are called "economizers" and are so named because they provide an economical way of heating water."

FIGURE 13.21.   High-pressure system with a float trap and check valves. (Courtesy of Johnson Controls.)

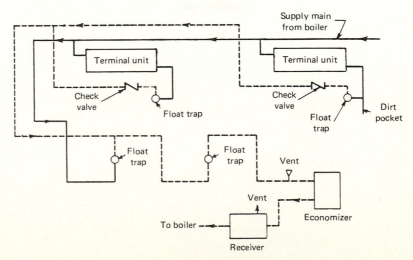

Water Distribution

"Hot water has one prime advantage over steam in piping systems: return lines and supply lines may be pitched in any direction. They are always filled with water so that any flow from the boiler forces the return back to it.

Most water systems have motor-driven circulating pumps which force the water to circulate regardless of temperature. Water systems, like steam systems, must have provisions for air elimination. Thus, an air vent is usually located at the top of each heating unit and at all high points where air might accumulate.

A disadvantage of water is its weight. In very tall buildings, it is usual to divide the system into two or more parts for the upper and lower sections. This avoids abnormally high pressure in the basement. Every 100 feet of elevation adds 43 psi pressure at the bottom. The use of steam imposes no such problem.

This section deals with forced hot water systems only. Gravity systems which obtain circulation from the difference in temperature between the supply and return have been used extensively in the past, mainly in residences, but are no longer of commercial importance."

Hot Water Systems. "Hot water piping systems are less complex than steam systems. There are five general types.

1. *A one-pipe system* which consists of a single circuit around the building, or in each section of a building, with connections to both ends of the heating unit so that a part of the water in the main is passed through each unit. A fitting in one of these connections, usually the return, is installed to permit adjustment of the amount of water flowing through the unit [see Fig. 13.22].
2. *A two-pipe, direct return system* [Fig. 13.23] is difficult to balance because the unit with the shortest supply line also has the shortest return.
3. *A two-pipe, reverse return system* [Fig. 13.24] provides uniform pressure drop through all of the heating units. The first unit on the supply main has little drop from the boiler, but the maximum drop is in the return line back to the boiler. The last unit to obtain hot water has the shortest return main.
4. *A series loop system* [Fig. 13.25] is suitable for small systems only because the water loses heat in each unit through which it passes. Hence, the last unit must be designed to give the required amount of heat from a considerably lower water temperature.

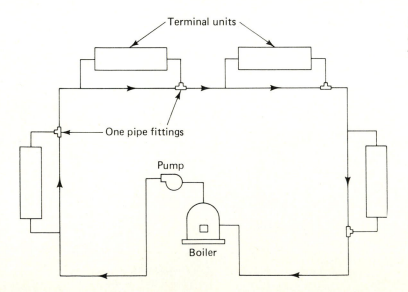

FIGURE 13.22. One-pipe water circuit. (Courtesy of Johnson Controls Inc.)

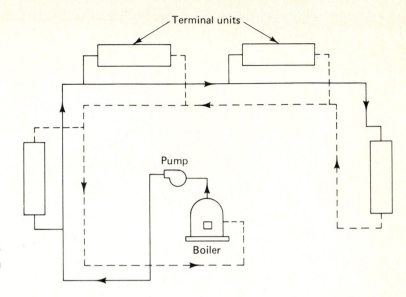

FIGURE 13.23. Two-pipe direct-return circuit. (Courtesy of Johnson Controls Inc.)

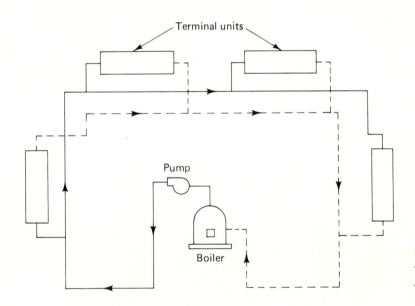

FIGURE 13.24. Two-pipe reverse-return circuit. (Courtesy of Johnson Controls Inc.)

FIGURE 13.25. Series loop circuit. (Courtesy of Johnson Controls Inc.)

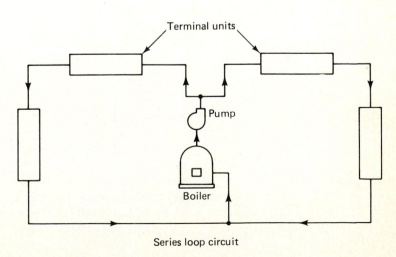

Series loop circuit

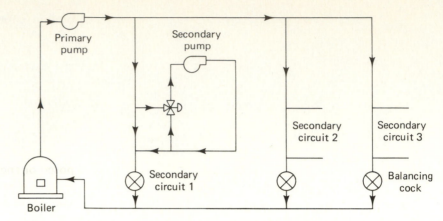

FIGURE 13.26. Primary–secondary circuit. (Courtesy of Johnson Controls Inc.)

5. *A primary-secondary system* [Fig. 13.26] has one primary pump and secondary pumps in each secondary circuit. The basic piping arrangement can be either direct or reverse return. Flow in the secondary circuit occurs only when the secondary pump is running.

Of the piping systems discussed, the two-pipe reverse return system is most extensively used because it does not require careful pipe sizing and elaborate manual balancing as does the direct return system. The one-pipe system and the series loop system offer piping economies. Many of the larger systems employ a combination of the various types of circuits, each of which offers a particular advantage or economy for a portion of the system."

13.7 HEATING APPLIANCES*

Radiators

"Cast iron radiators and fan-coils were the earliest terminal heating units. Usually these units were installed exposed in rooms, and either stood on legs or were fastened to walls. In either case, the heat was given off to the room partially by radiation and partly by convection, the relative amount depending on the shape of the radiator. Convection is the transfer of heat secured by speeding up the moving air. In the case of a radiator, the air motion is caused by the tendency of the lighter, warmer air to rise.

The capacities of cast iron radiators are given as "sq. ft. EDR," defined as square feet of equivalent direct radiation. Originally, this referred to the square feet of metal exposed to the air, but design changes have affected this relationship. Now it is only an indication of capacity; one square foot EDR equals 240 Btu per hour."

Convectors

"Convector" is a word applied to a rather general replacement for radiators. A convector has its heat transfer surface recessed in the wall or enclosed in a cabinet so that most of the heat is given off by convection. Air enters the enclosure through grilles or openings below the heating element and leaves by similar openings above the element. The "draft" of a convector, like that of a chimney, is increased by the length of travel through a heated space. The greater the height of the discharge grille above the

*Section 13.7 is taken from Johnson Controls, Inc., *Johnson Controls Training Manual.*

heating element, the stronger the convective force, the more rapid the air travel, and the greater the heat transfer.

Convectors are rated in square feet EDR or in Mbh (thousands of Btu per hour). Four square feet EDR is equal to approximately 1 Mbh."

Exterior Wall Radiation

"To produce maximum room comfort, radiation is preferably located along cold walls defined as those exposed to the weather or to an adjoining cold space [Fig. 13.27]. Baseboard radiation consists of exposed or enclosed pipe along the entire cold wall. Usually, the pipes are copper and have fins attached [Fig. 13.28]. These fins add to the amount of heat transfer surface in contact with the air."

FIGURE 13.27. Convector with fin tube radiation along the exposed glass curtain wall.

FIGURE 13.28. Fin tube radiation. (Courtesy of Slant/Fin Hydronic Heating Division.)

Radiant Panels

"Radiant heating describes a method of space heating in which water piping, air ducts, or electrical resistance elements are embedded in or located behind ceilings, walls or floor surfaces. Surface temperatures are usually maintained in the range of +80F to 125F.

Radiant panels in concrete have large heat storage capacity and continue to emit heat long after the heating medium supply has stopped.

FIGURE 13.29.   Snow melting coils ready for concrete.

It is very important to take precautions against the introduction of excessively hot water to radiant panels which could cause extensive damage to the concrete slab or plaster.

[Radiant panels are also used externally for snow melting. An antifreeze solution prevents freezing of solution and breaking of piping and concrete. Snow melting devices should be turned on well in advance of the actual snowfall (Fig. 13.29).]"

13.8   PIPING MATERIALS

The common materials for heating systems for steam mains and return lines are steel and wrought iron. Black iron can also be used for hot water system mains along with steel. Some designers require wrought iron- and copper-bearing steel for return lines. Most piping is schedule 40,* meeting basic standard internal pressure requirements.

For branch piping threaded steel and wrought iron continue in common use, particularly in additions and industrial applications. The use of copper tubing has gained tremendous favor, particularly in large commercial installations, during the last 20 years. Copper tubing and copper fin tube radiation are modern-day favorites for economical heating installations of excellent quality.

13.9   HEATING PIPE INSTALLATION

The three basic means of joining the larger sizes of steam and hot water heating pipe are:

1.   Use of threaded connections
2.   Welding
3.   Use of grooved pipe method

At valves and other key points, flanged connections are utilized for all three methods (Fig. 13.30).

---

*The pipe schedule number is the ratio of the internal pressure in psi divided by the allowable fiber stress multiplied by 1000. James J. O'Brien, *Construction Inspection Handbook,* Van Nostrand Reinhold, New York, p. 368.

FIGURE 13.30.   Flange pipe connection.

Threaded Pipe

Pipe to be threaded on the job site is cut to length with an electrically powered band saw, power or hand cutter (Fig. 13.31). A power threading machine cuts the proper thread. The pipe is tightened with large pipe wrenches or long-handled chain wrenches (Fig. 13.32). Pipe dope for steam and hot water lines applied to the pipe threads helps ensure a leak-free joint. Fittings basically are plant manufactured with internal threads. Internal threads are not cut on the job site.

FIGURE 13.31.   Hydraulic power pipe cutter. (Courtesy of E.H. Wachs Company.)

FIGURE 13.32.   Chain wrench for tightening large pipe.

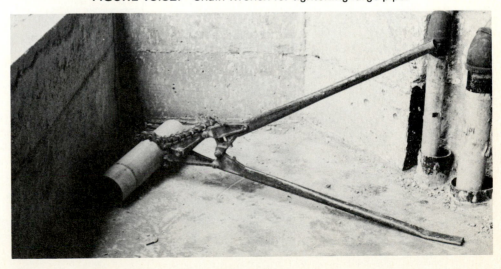

Welded Pipe

Threaded pipe has given way to welding to a very substantial degree. This is due in part to improvements in welding technology, an increase in the availability of certified welders, and the acceptance of and skill in the use of welding by designers. Welders can readily be submitted to test, and work in place can be x-rayed or tested with magnoflux or ultrasonic testing devices. Once completed and subjected to pressure test, the welded system in a building should be trouble-free.

   Pipe for welded work comes from the manufacturer with beveled ends prepared to receive weld material (Fig. 13.33). Preassembly of lengths of pipe or pipe and fittings takes place on the floor or workbench. The piping sections are laid in roller holders (Fig. 13.34), which permit the rotation of the work as the welding takes place so that the welder always can weld from a position above the joint. This is the *easiest, fastest,* and *best position* for welding. After the pipe sections are aligned and clamped into position, the welder fixes their position permanently by tack welding the joint (Fig. 13.35), that is, placing ¼ to ½ in. of weld material in three or four equally spaced

45° bevel

FIGURE 13.33. Pipe beveled to facilitate welding.

FIGURE 13.34. Roller holder horses for pipe welding.

FIGURE 13.35. Pipe risers tack-welded only. Short angles align pipe; note lifting rings for rigging.

locations around the joint. The entire weld can then be completed immediately or the *production* welding completed after many sections of pipe and fittings have been assembled by tack welding (Fig. 13.36). Pipe guides keep the riser piping properly aligned during expansion and contraction (Fig. 13.37).

Elbow, tee, cap, and reducing fittings to meet most piping needs are readily available (Fig. 13.38). Pipe fabricating firms can make any unusual piping fitting not

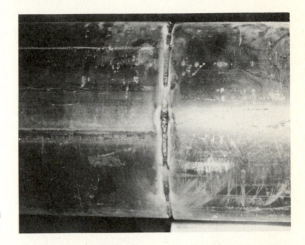

FIGURE 13.36. Early stage in production welding of piping.

FIGURE 13.37. Pipe guides.

FIGURE 13.38. Welding fittings. (Courtesy of Tube-Line Corporation.)

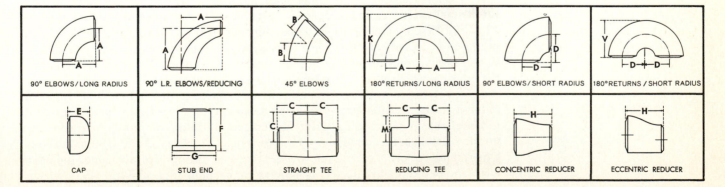

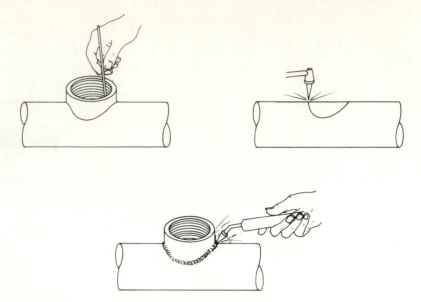

FIGURE 13.39. Threaded tee-fitting installation. (Courtesy of Tube-Line Corporation.)

FIGURE 13.40. Tee welding fitting. (Courtesy of Tube-Line Corporation.)

available in the supplier's regular line of products. For the numerous taps into the larger lines for threaded branches, welding outlets are cut and welded into the mains (Figs. 13.39 and 13.40).

Grooved Pipe

The Victaulic Company of America manufactures fittings, valves, and cutting tools for their method of joining steel pipe or cast or ductile iron piping. In this system (Figs. 13.41 and 13.42), a malleable or ductile iron housing fits into grooves near the ends of piping to be joined. The housing encloses a resilient elastomeric gasket which seals tighter as water pressure increases. Oval neck, track head bolts fit into oval openings in the housing, where they cannot turn as the nuts are tightened (Fig. 13.43).

The grooved method provides some expansion, flexibility, and vibration reduction. The manufacturer claims it to be up to three times faster than welding and more reliable than threading. Grooving replaces threading for joint preparation and no special skills are required to assemble the housing and gasket and tighten the bolts. Check and butterfly valves and a wide range of fittings are available, including adapters for joining to threaded or flanged connections (Fig. 13.44).

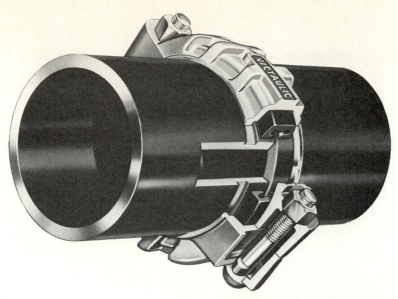

FIGURE 13.41. Cutaway section of a grooved joint assembly. (Courtesy of Victaulic Co. of America.)

FIGURE 13.42. Grooved pipe. The pipe is purchased grooved or is field prepared.

FIGURE 13.43. Assembly procedure following the grooved pipe method. (Courtesy of Victaulic Co. of America.)

Components – Ready to assemble
Victaulic Couplings are supplied with housings, gasket, nuts and bolts.

Check Pipe Ends
Groove pipe to Victaulic specifications. Check to be sure pipe, from end to groove, is clean.

Lubricate Gasket
Apply thin coat of Victaulic Lubricant to gasket lips and complete exterior of gasket.

Position Gasket
Place gasket on one pipe end, being sure gasket lip does not overhang pipe end.

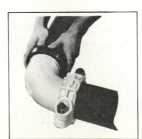

Center Gasket
Butt fitting or mating pipe ends and slide gasket to center between grooves.

Apply Housing
Assemble housings over gasket being sure keys engage into the grooves on both sides.

Start Nuts and Bolts
Insert bolts with track head fully into housing. Apply nuts and tighten finger tight.

Tighten Nuts
Tighten nuts *uniformly* until bolt pads are together firmly– *metal-to-metal.* No special torque is required.

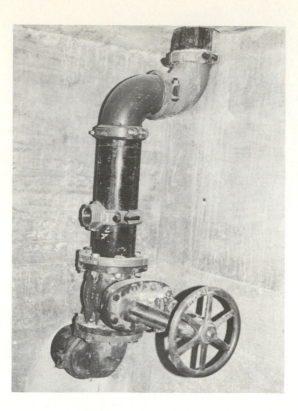

FIGURE 13.44. Grooved pipe elbows and tee connection. The valve is connected with a bolted flange connection.

13.10   HEATING PIPE SUPPORTS

The choice of pipe supports depends on whether the pipe will remain static, as, for example, a sprinkler line, or will expand and contract depending on the water temperature or flow of steam. Pipe supports also vary with the covering used.

Figure 13.45 shows a static location on a steam line and Figs. 13.46 and 13.48 show a roller support within 15 ft which permits pipe movement. The pipe load does not crush the insulation against the roller support since the metal sheet distributes the concentrated load. In Fig. 13.47 the support is anchored into the concrete by drilling a hole into which an expansion anchoring device is inserted.

A number of pipes can be supported on a trapeze arrangement (Fig. 13.49). Rods, angle irons, channels, and even small beams can be designed to hold the required loads. A versatile material for this purpose is Unistrut, which has innumerable applications in all fields of mechanical and electrical work.

FIGURE 13.45.   Static pipe support.

FIGURE 13.46.   Roller pipe support.

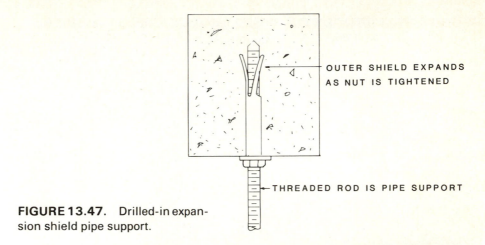

OUTER SHIELD EXPANDS
AS NUT IS TIGHTENED

THREADED ROD IS PIPE SUPPORT

FIGURE 13.47. Drilled-in expansion shield pipe support.

FIGURE 13.48. Trapeze pipe support.

FIGURE 13.49. Versatile trapeze material. (Courtesy of Unistrut Building Systems, GTE Product Corp.)

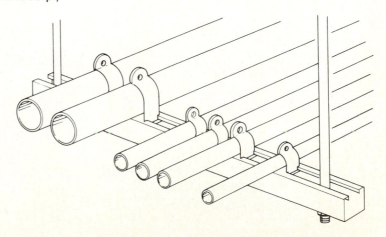

13.11   METHODS OF HANDLING PIPE MOVEMENT

Piping movement occurs in the ways shown in Fig. 13.50. Although piping can be fabricated or installed to compensate for movement in the form of an expansion loop, this older way of doing things requires quite a bit of pipe, pipe fabrication, supporting members, and space. A more modern and better solution uses rubber or metal expansion devices which fit right into the line being installed. Figure 13.51 shows a small-diameter expansion unit for operating pressures to 150 psi. A stainless steel bellows flexes in a steel housing. Pipe ends can be steel with threaded, flanged, or welded connections, or copper with sweat ends. Other types of metal bellows devices for larger pipe sizes which absorb movement in various combinations of direction are shown in Figs. 13.52 and 13.53. Again the bellows are constructed of stainless steel.

One superior solution for vibration and sound control utilizes rubber or rubber compounds. Figure 13.54 shows construction details of this type. Another frequently used expansion device, made of braided bronze or stainless steel, is the flexible metal connector.

Axial compression.

Reduction of face-to-face dimension measured along the axis.

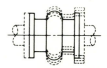

Axial elongation.

Increase of face-to-face dimension measured along the axis.

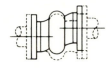

Transverse or lateral movement.

The movement of the joint perpendicular to the axis.

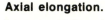

Vibration absorption.

The movement of the joint due to vibrations which are effectively intercepted and insulated against transmission to remainder of system.

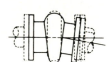

Angular movement.

The displacement of the longitudinal axis of the joint from its initial straight line position (a combination of axial elongation and axial compression).

FIGURE 13.50.   Definition of movement. (Courtesy of Unaflex Rubber Corp.)

FIGURE 13.51.   Expansion unit for small pipe. (Courtesy of Unaflex Rubber Corp.)

FIGURE 13.52. Bellows-type expansion joint with fixed flanges. (Courtesy of Unaflex Rubber Corp.)

FIGURE 13.53. Bellows-type expansion joint with welding nipples configuration. (Courtesy of Unaflex Rubber Corp.)

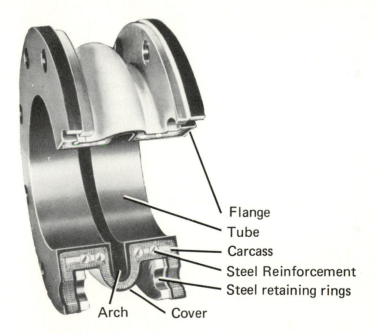

Flange
Tube
Carcass
Steel Reinforcement
Steel retaining rings

Arch   Cover

FIGURE 13.54. Rubber expansion joint. (Courtesy of Unaflex Rubber Corp.)

13.12 TEMPORARY HEAT

The building construction industry engaged in commercial work in the United States operates basically 12 months of every year. In the colder areas, roofing, caulking, glazing, and masonry are the types of trades where work is discontinued or slowed during cold, windy, and snowy periods. The rate of inflation, the cost of borrowed money, taxes, loss of income on property while it is under construction, and the expense of stopping and restarting a project all preclude shutting a project down during the winter months. Instead, general contractors and construction managers plan and schedule their work recognizing the impact of the seasons on the intended progress. Building excavation begun before appreciable frost makes good sense since it leads to pile driving, caisson, and foundation work, which all are less affected by poor weather, particularly snow, wind, and cold, than are steel erection, formwork, or masonry. Starting an 18- to 24-month project in the fall is an intelligent decision, with excavation and preparation of the foundations proceeding in the winter months allowing superstructure work to start in the spring and the building to be enclosed by the following winter. The second fall season should be a period of intense activity when roofing, brick work, precast concrete panels, and metal portions of curtain walls are pushed to completion and windows and glazing installed before winter weather takes control. With the building enclosed, reasonably lighted with a temporary system, and partially heated, work proceeds in an orderly, predictable fashion unaffected by the season.

Most modern specifications require a building under construction, once it has been enclosed, to be heated sufficiently to maintain a minimum temperature of 50° F. This prevents excessive contraction and stabilizes the structure so that marble, plaster, drywall, cabinet work, mill work, and other finishes will not be subject to detrimental forces caused by temperature changes.

Temporary heat should be included as a *separate item* in the bidding documents. A specification should be prepared for this item, indicating temperatures to be maintained, a method of providing temporary heat, maintenance, and many other items to make a complete contractual arrangement which the general contractor can readily administer. In Appendix C the temporary heat specification covers all the interested parties involved in enclosing and temporarily heating a building.

The requirements can state the number of heat units required, their type, size, and capacity. Steam or electric unit heaters or gas burning units without fans can be used. The systems should be simple. Where steam or hot water lines are used, they can be left uncovered to help heat areas through which they pass. If the *heating contractor* can prepare it in time, the whole or part of the permanent system can provide the heating required. The method of heating selected usually depends on economic factors which combined produce the required result at the lowest overall cost. Equipment, salvage, cost of installation and removal, fuel, workers required by union contracts, hours of operation, and any rehabilitation of permanent equipment or the structure itself must enter into the computation of the entire cost. Use of the permanent system for temporary heating may overlap with the owners' partial occupancy of the building, in which case fuel costs and maintenance labor charges should be prorated between construction costs and permanent building operational costs.

14

Air Conditioning

14.1 INTRODUCTION

Personal comfort in an enclosed space is governed largely by the activity of the occupants, their physical makeup, their mode of dress, and of course the condition of the air. The engineer designs to meet standards of quiet, temperature, humidity, flow of air, and air cleanliness. He often gets simultaneous complaints from someone who has overdressed, overeaten, and been very active, and a nearby occupant who is lightly dressed, physically thin, and has been quietly sitting at a desk. The former says that it is too hot, the latter that it is too cold.

Owners and designers have a wide range of systems and equipment from which to choose to meet comfort standards. An owner may decide that initial minimum cost best serves his or her purposes. The result may be a building that costs more to operate and never achieves the comfort of a moderately more expensive design. The construction manager can casually install these systems by faithfully following the design drawings and never understand the type of system and its functioning. This *can* be left to the design experts and the HVAC contractor and his subs, but to control the construction process professionally, the construction supervisor will serve his client better if he has a general understanding of the HVAC system the designer chooses and its functioning.

14.2 THERMAL LOADS

A good insight into air conditioning design is acquired by studying HVAC in a typical rectangular-shaped high-rise office building of moderate height. The designer must supply heating and cooling based on the thermal loads imposed on the building.

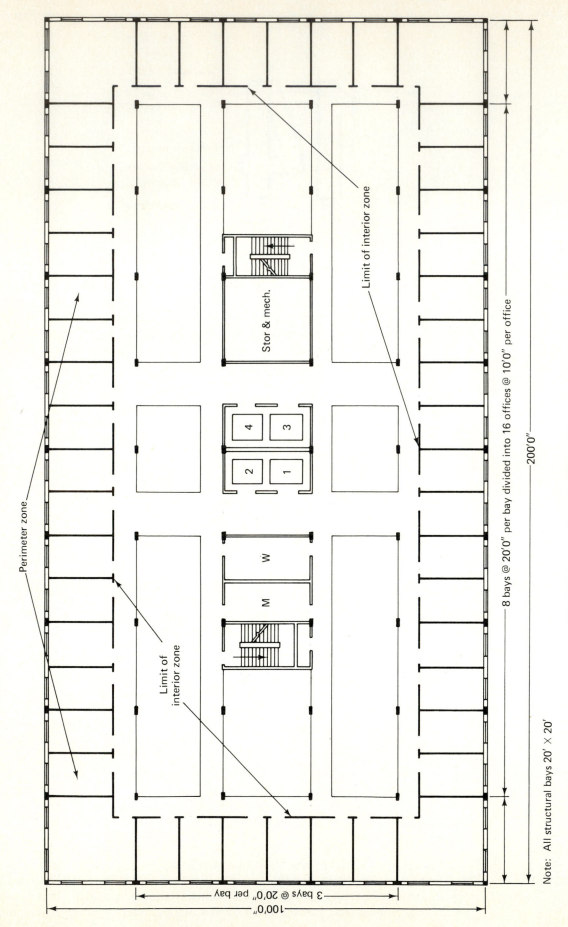

FIGURE 14.1. Typical office building floor plan.

Note: All structural bays 20' × 20'

8 bays @ 20'0" per bay divided into 16 offices @ 10'0" per office

200'0"

3 bays @ 20'0" per bay

100'0"

Perimeter zone

Limit of interior zone

Stor & mech.

M W

People, light, and equipment produce internal loads. The sun, wind, and outside air temperature produce a wide spectrum of external thermal loads that infiltrate or are transmitted into the building through walls, roofs, doors, and windows. The intensity and movement of the sun through the sky and the change of the force of the wind and its direction typify the wide swing of conditions the designer must accommodate during any one day, season, or year. To satisfy this wide variance in loads the designer utilizes multiple systems. He generally designs for an interior zone where conditions fluctuate less, and separate exterior areas that will be quite hot in the sun while northern areas in the shade will be very cold (Fig. 14.1).

14.3   FOUR BASIC SYSTEMS

The four basic systems are:

1. All *air,* in which conditioned air enters the room through diffusers
2. All *water,* in which chilled water circulates through a coil over which the room air passes
3. *Air* and *water,* where each is supplied to a unit containing a coil
4. *Unitary,* which are package units either through the wall, typical of motel design, or rooftop units, used in supermarkets, schools, and industrial plants, which often heat only or cool only

14.4   ALL-AIR SYSTEMS

All air systems fall into two basic categories: (1) *constant volume* and (2) *variable air volume* (VAV). In the latter case the *amount* of air entering an internal space together with its quality satisfies the design criteria.

Constant Volume

Constant volume systems can be:

1. Single zone
2. Single zone with reheat
3. Blow-through reheat
4. Dual duct
5. Multizone

Single Zone. In this system a centrally located fan outside the space operates continually at a constant speed supplying, for example, 10,000 CFM. The supply air passes over a cooling coil supplied with chilled water from a central source which cools the air to approximately 58° F. The thermostat in the zone turns the flow of chilled water going to the coil on and off. Because of its simplicity, the system has the disadvantage of supplying air that is either too cool or not cool enough. Air from the zone can be returned to the fan area to be filtered and used again (Fig. 14.2).

Single Zone with Reheat. To overcome the wide fluctuations in air entering the zone supplied by the fan, reheat coils are installed close to the location where the air enters the occupied area. In this manner the air can be raised to a more comfortable 68° F. In addition, the single zone can be broken down into smaller zones each with its own thermostatic control (Fig. 14.3). The reheat consists of a coil installed in the ductwork which heats the now-cooled air to a higher temperature.

Reheat coils function using hot water, steam, or an electric resistance device (Fig. 14.4). The reheat process is energy wasteful since energy is expended to cool the air and more energy is then used to raise its temperature again. The exception would be a system designed to utilize a recoverable heat source that would otherwise be wasted.

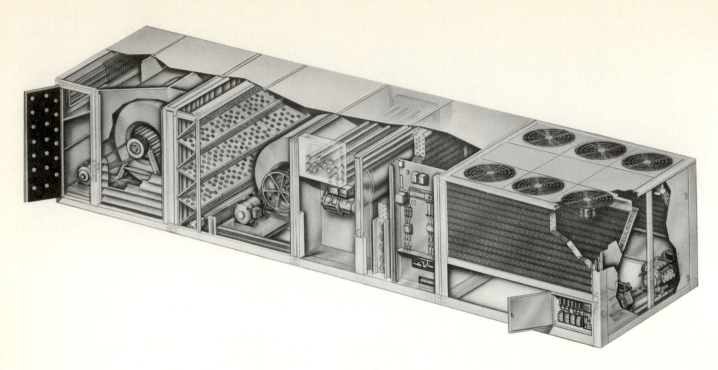

FIGURE 14.2. Single-zone roof-mounted unit. (Courtesy of McQuay-Perfex Inc.)

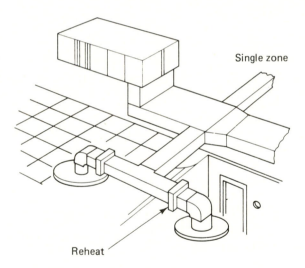

Single zone

Reheat

FIGURE 14.3. Single-zone unit with reheat.

FIGURE 14.4. Reheat coil. (Courtesy of McQuay-Perfex Inc.)

Unfortunately, our cities contain many buildings with this type of reheat design which are energy liabilities in the 1980s.

Blow-through Reheat. In this system *conditioned air* from a central source is brought to a room terminal located below a window (Fig. 14.5). The heating takes place in the terminal enclosure just prior to entering the air conditioned space (Fig. 14.6). Depending on the season and the thermal loads in any elevation of the building, the system supplies quiet draft-free heated or cooled air. Because of its versatility this system gained widespread use in office buildings prior to the petroleum crises of the 1970s and became the virtual standard for medium-priced rental space. Circular high-velocity ductwork of perimeter risers and branch lines supply air to the units, and a similar system using steel pipe risers and copper branch lines provides the heat. Architects have devised innumerable attractive means of incorporating the terminal enclosures into the exterior walls of buildings to create handsome interior spaces.

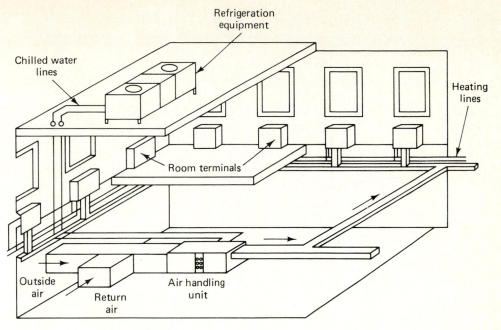

FIGURE 14.5. Blow-through reheat system.

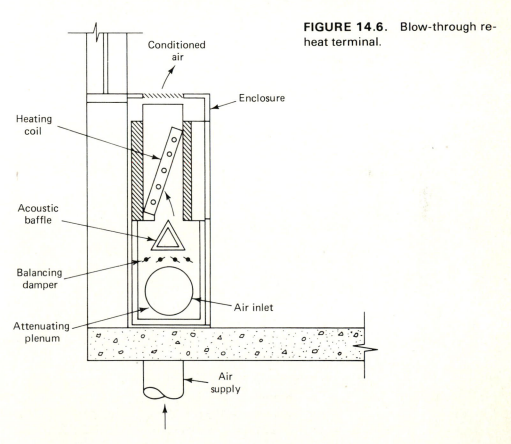

FIGURE 14.6. Blow-through reheat terminal.

Dual Duct. As the name implies, in the dual duct system two high-pressure ducts supply hot and cold air to a mixing box which reduces air velocity, eliminates airborne noise, and blends the hot and cold air to meet comfort requirements (Fig. 14.7). Running two small high-pressure ducts instead of one larger low-pressure one generally makes this system, with its mixing boxes and higher-pressure fans, less competitive than other systems (Fig. 14.8). Some designers find it particularly desirable for the many individual rooms of a hospital.

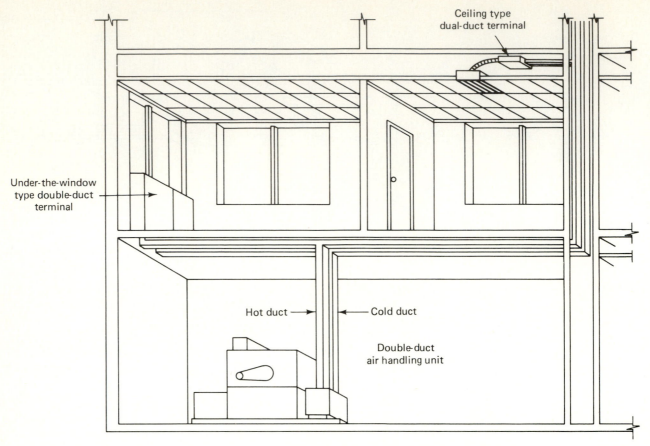

FIGURE 14.7. Dual duct: high pressure.

FIGURE 14.8. Dual-duct supply unit. (Reprinted by permission of The Trane Company, La Crosse, WI.)

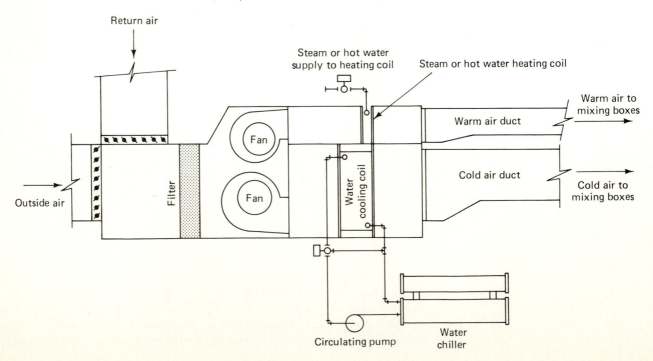

Multizone. The multizone system solves the problem of two ducts by mixing hot and cold air as it leaves the multizone unit (Fig. 14.9). This unit in effect is a medium-high-pressure reheat system. The key to this equipment lies with the *mixing dampers,* which blend 120° F air with 55° F air to discharge into a number of supply ducts, each serving a separate zone (Fig. 14.10). The thermostatic controls in a particular zone regulate the mixing dampers for that zone to supply the conditioned air desired. Figure 14.11 shows the various sections of a blow-through unit.

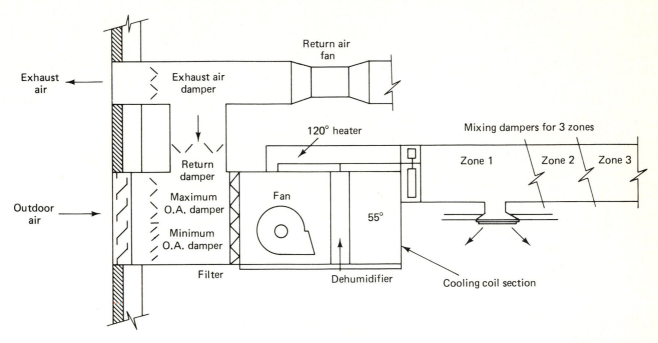

FIGURE 14.9. Multizone unit.

FIGURE 14.10. Five pistons operate dampers to regulate the multizone airflow.

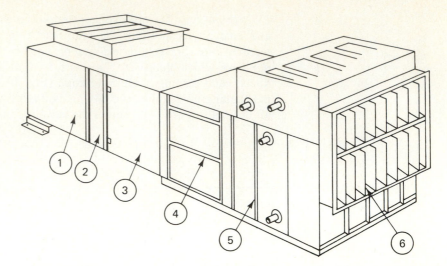

FIGURE 14.11. Typical multizone blow-through unit. 1, mixing box section; 2, high-velocity filter section; 3, access section; 4, fan section; 5, coil section; 6, zoning dampers. (Reproduced by permission of Carrier Corporation © Copyright 1982 Carrier Corporation.)

Variable Air Volume

The systems discussed previously are all *constant volume* types in which the fans operate at a constant speed. In a VAV system the air temperature remains constant but the *volume of air* delivered varies. In the simplest installation a lesser volume of air passes into the duct system from its source as part of the air is either bypassed or exhausted by utilizing discharge dampers. This reduces the required horsepower somewhat but does not compare in efficiency with the terminal-type VAV device, in which the vanes controlling the amount of air discharging into a space increase or decrease the discharge slot. In Fig. 14.12 the two types of VAV are compared. Figure 14.12 is a typical comparison of horsepower reduction as airflow is modulated. Assuming that you have a 40-horsepower motor delivering 40,000 CFM against 4-in. static pressure, and recognizing that a VAV system will typically have an average load of 70% of design, you can quickly determine the savings resulting from the use of the inlet vane type.

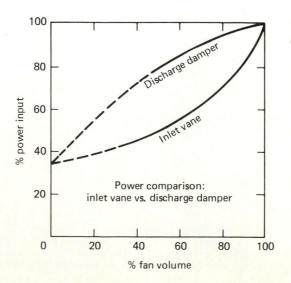

FIGURE 14.12. VAV comparison. (Courtesy of McQuay-Perfex Inc.)

Discharge dampers at 70% load   90% = 36 hp
Inlet vanes at 70% load   60% = 24 hp
    Savings   12 hp

$$\$ \text{ Savings} = \frac{12 \text{ hp} \times 0.75 \text{ kW/hp} \times 5\text{¢/kW-hr} \times 3000 \text{ hr}}{0.85 \text{ motor efficiency}}$$

Thus, the use of inlet vanes produces $1588 annual potential operating savings.* Figure 14.13 shows an example of the vane type, which becomes part of the ceiling construction. Figure 14.14 shows a VAV multiple self-contained terminal which connects to the supply ductwork and can service up to five diffusers (Fig. 14.15) through low-velocity flexible ducts. An air-inflated bladder actuates a damper to control air volume. The thermostat can be wall mounted or an integral part of the VAV.

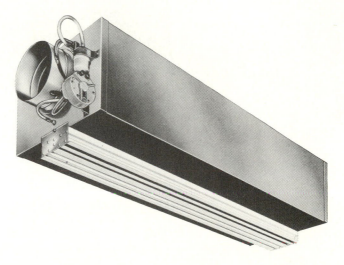

FIGURE 14.13. Vane-type VAV for ceiling installation. (Courtesy of York Division of Borg Warner Corporation.)

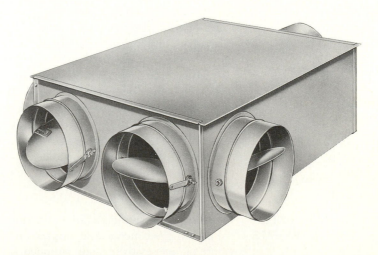

FIGURE 14.14. VAV multiple terminal. (Courtesy of York Division of Borg Warner Corporation.)

* Bulletin A/Sp-160C, McQuay-Perfex Inc., Minneapolis, Minn.

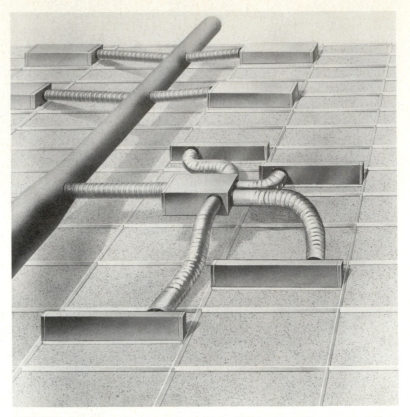

FIGURE 14.15. Single and multiple VAV units. (Courtesy of York Division of Borg Warner Corporation.)

Return Systems

In the constant volume systems we have discussed, the conditioned air loses its desirable qualities and must be returned to be partly exhausted, freshened with outside air, filtered and further dried, dampened, and heated or cooled as conditions require. Where practical, this air collects in an enclosing space such as a hung-ceiling area and is drawn off through horizontal and vertical ductwork by a return air fan frequently located not far from the original supply point. The architect accommodates this return air by providing means for the air to exhaust from a space through door grilles, shortened doors, return wall and ceiling grilles, ceiling perforations, and by using ceiling lighting fixtures with return air slots. The return air generally enters the ductwork system at some advantageous point on every floor in high-rise work and is brought vertically to the air conditioning unit.

The hot and chilled water and steam supplied to terminals and reheats also returns to the supply point for reuse. This principle is so basic to HVAC designs that the absence of a return line is quite exceptional.

14.5 ALL-WATER SYSTEMS

In the *all-water systems* a fan, filter, and coil are housed in a metal cabinet installed on the building perimeter. The simplest and cheapest is the *two-pipe system* (Fig. 14.16), where air passes over a coil supplied with hot water in winter and chilled water in summer. The two-pipe system has the disadvantages of noise, no introduction of fresh air, and the system must be changed over from winter to summer operation. This causes many comfort complaints during intermediate weather during the spring and fall.

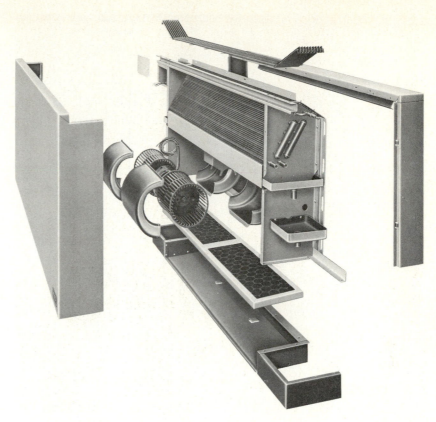

FIGURE 14.16. Two-pipe all-water terminal. (Courtesy of McQuay-Perfex Inc.)

The *four-pipe system* has both a hot and cold coil, which makes it more versatile. However, *condensation* forms with this arrangement and a condensate drain must be installed from each terminal. The four supply and return pipes, plus a small piping system for the condensate, make an expensive installation that still does not provide fresh air.

14.6 AIR–WATER SYSTEMS

Until the petroleum crisis grew acute in the 1970s, the air and water system utilizing induction units was frequently used by designers for high-quality high-rise office buildings. In cross-sectional view (Fig. 14.17), high-pressure *conditioned air* from a *central source* ejects into the space, *inducing* room air to flow over the water coil. This mode of operation provides great flexibility obtained with these operating temperatures.

Season	Load	Air temp.	Water temp.
Summer	Full	55° F	58° F
Fall/spring	Variable	55° F	110° F

Although previously very popular, this system has lost favor due to higher intial cost and higher operating costs. There have also been some problems with high-frequency noise in particularly quiet buildings. Figure 14.18 shows how the conditioned air and water are produced in a mechanical equipment room and the distribution systems for the water and air.

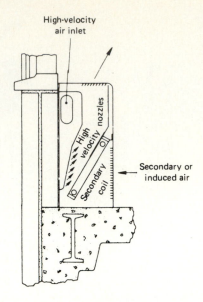

FIGURE 14.17. Cross section of an induction unit. (Reprinted by permission of The Trane Company, La Crosse, WI.)

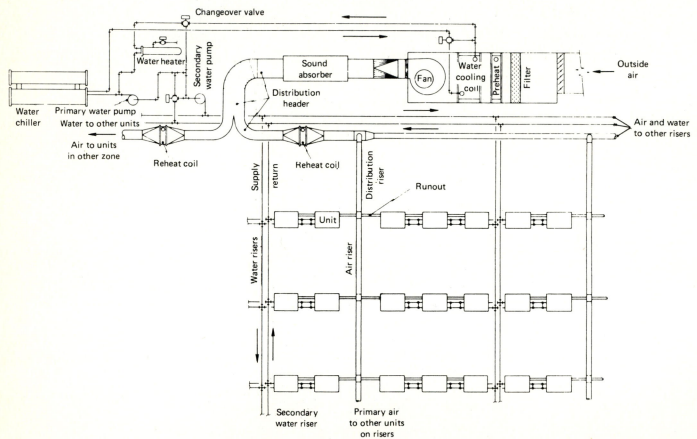

FIGURE 14.18. Sectional view of an induction system. (Reprinted by permission of The Trane Company, La Crosse, WI.)

14.7 THE CENTRAL STATION

Human comfort can be satisfied to a degree by some rather simple installations. A *unit heater* operating on hot water, steam, or electricity can heat an industrial working area. The unitary system mentioned earlier requires only a source of electricity to heat or cool apartments and motels. These simple systems should offer no appreciable challenge to the construction manager or HVAC contractor. The structural and

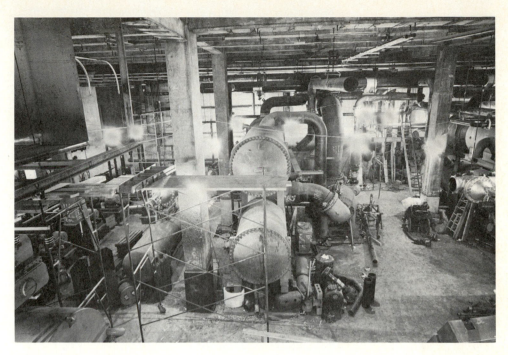

FIGURE 14.19.   Central station under construction. (Photo by Gil Amiaga, N.Y.C.)

architectural construction of a *central station* and its mechanical and electrical installation is quite another matter. Most of our large buildings have a large primary area called a central station containing boilers or an incoming source of steam, refrigeration machines to produce chilled water, chilled water pumps, condenser water pumps, an emergency electric generator, fresh air and exhaust air shafts, fire and domestic pumps and tanks, heat exchangers, alarm systems, and air conditioning units serving systems supplied from this part of the building.

The construction of a central station (Fig. 14.19) challenges the management skills of the general contractor or the construction manager and their team of architectural, structural, mechanical, and electrical subcontractors. The following section looks at some of the systems in a central station.

14.8   THE CHILLED WATER SYSTEM

The chilled water system consists of three principal parts:

1.   The *refrigeration machines,* which produce the chilled water
2.   The *condenser water system,* which draws off and dissipates the heat created by the condensing unit of the refrigeration machine
3.   The *chilled water distribution system,* which pumps and delivers the chilled water to whatever central units or terminals it supplies

The Compressor-Type Chiller

Refrigeration machines are commonly called *chillers.* Like our household refrigerators, the smaller units operate on the principle of compressing a gas such as Freon with a piston-type compressor. The larger units that are generally used in central stations compress the gas using a turbine. Figure 14.20 shows how the low-pressure gas leaving the evaporator is *compressed* into high-pressure vapor. In the condenser it gives off heat as it condenses to high-pressure liquid. As it passes into the evaporator the Freon *expands* into gaseous form and in this state absorbs heat from the returning chilled water circulating through the evaporator section.

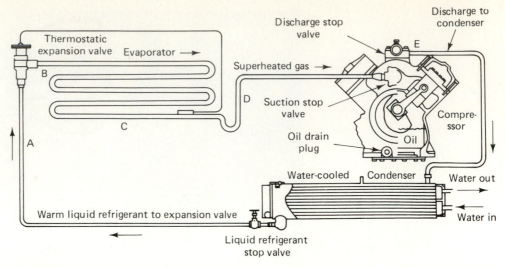

FIGURE 14.20. Reciprocating compression refrigeration cycle. (Reproduced by permission of Carrier Corporation © Copyright 1982 Carrier Corporation.)

The Absorption Machine

The development of the *absorption machine* (Fig. 14.21) grew out of interest in finding a more economical substitute for electric power used to operate compressor-type chillers. The absorption machine operates on a supply of steam or hot water to the concentrator (see Fig. 14.22). Particularly in large central station installations, the refrigerant is water and the absorbent is lithium bromide. The following explains the operation of the absorption water chiller:*

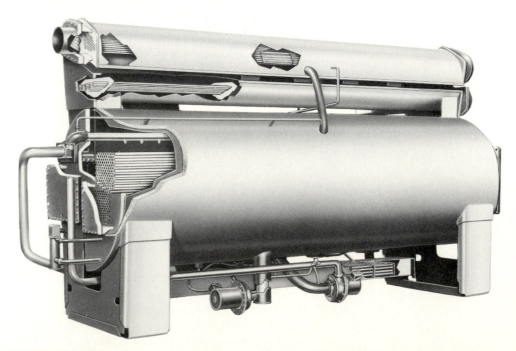

FIGURE 14.21. Absorption chiller. (Reproduced by permission of Carrier Corporation © Copyright 1982 Carrier Corporation.)

*Trane AC manual, June 1973 printing.

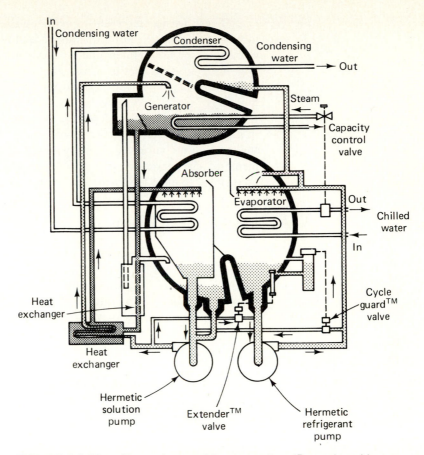

FIGURE 14.22. Absorption machine operation. (Reproduced by permission of Carrier Corporation © Copyright 1982 Carrier Corporation.)

"Basically, the absorption water chiller is not too different in operation from the more familiar, mechanical compression water chiller. Both machines accept heat to evaporate a refrigerant at low pressure in the evaporator, and thereby create a cooling effect. Both also condense the vaporous refrigerant, at a higher pressure and temperature in the condenser, in order that the refrigerant can be re-used in the cycle.

In both cases, the capacity of the machine depends upon the pressure that exists in the evaporator since this determines the evaporator temperature.

In mechanical compression systems, the vapor formed when the liquid refrigerant absorbs heat, to provide the refrigerant effect, is drawn to a lower pressure area created by the mechanical movement of the pistons. In an absorption machine this vapor is also removed to a lower pressure area. However, the low pressure area in the absorption machine is created by controlling the temperature and concentration of a water–lithium bromide solution.

In a compression system the refrigerant vapor is mechanically compressed and moved from the low pressure to the high pressure side of the system. In an absorption system the vapor is first condensed and mixed into a solution of lithium bromide. This solution is then pumped to a higher pressure area and heat applied. Heat causes the solution to boil, driving off the refrigerant vapor at the higher pressure.

It is therefore evident that exactly the same function—that of taking low pressure refrigerant vapor from the evaporator and delivering high pressure refrigerant vapor to the condenser—has been performed in both the compression and absorption cycles. The only difference here has been in the method of transporting the vapor from the low to the high pressure side.

There are no reciprocating parts on the absorption machine; in fact, the only moving parts are the solution pumps and a vacuum pump. Thus, the machine is vibrationless and operates quietly. The absorption machine is lightweight and has a low floor loading factor. This, combined with its quiet operation, makes the absorption machine ideal for hospitals, hotels, apartment buildings, and office buildings."

In large installations designers use both absorption and centrifugal compressors, in series. High-pressure steam drives the centrifugal compressor and exhausts this steam at low pressure, say 12 psig, where it is used as the heat source for the absorption machine. The operating expense of the combination of machines is less than either one used singly to meet the load.

The Condenser System

Dissipating the heat from the condenser portion of a large chiller requires husky pumps capable of forcing large quantities of water to the top of the building, a pair of large pipes to and from the cooling tower in a mechanical riser shaft, and a cooling tower itself. Figure 14.23 shows a rooftop cooling tower installation. In recent years, the trend has been away from custom-designed towers assembled in place, to completely factory manufactured units that are used singly or in multiples. Figure 14.24 shows condenser water main and branch lines to cooling tower units.

The rigging of these large bulky units from ground to rooftop requires planning, special high-capacity lifting equipment, and a highly competent rigging crew.

Structurally, the designers must provide for a substantial structure with a significant wind surface. The structural cost of placing the cooling towers on the roof and the operating expense of pumping large quantities of water to rooftop level has caused some designers to solve this problem by developing ingenious ground-level solutions. The Trans America building in San Francisco utilizes a ground-level cooling tower to good advantage.

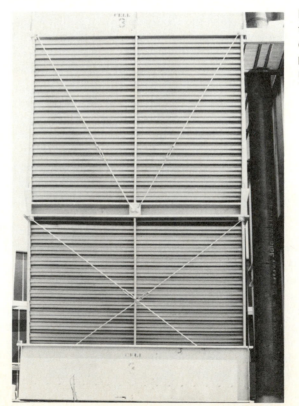

FIGURE 14.23. One cell of a three-cell cooling tower. Each cell comes from the manufacturer complete and ready for piping connections.

FIGURE 14.24. Condenser water piping to cooling tower units supported by a threaded trapeze hanger.

The Chilled Water Distribution System

The chilled water distribution system would naturally depend on the designer's approach to air conditioning. Usually, *chilled water risers and returns* run the full height of the building and occupy a shaft with the *condenser water lines* to and from the cooling tower (Fig. 14.25). Frequently, a central station located in the basement

FIGURE 14.25. High-rise pipe riser installation.

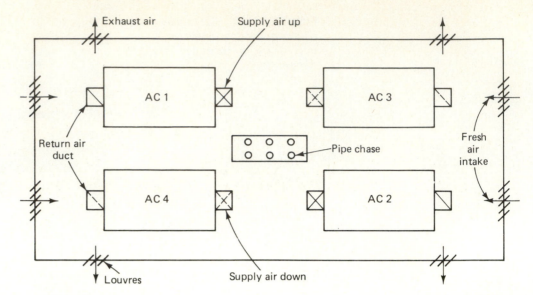

FIGURE 14.26. Midbuilding mechanical equipment floor. Ac units 1 and 2 supply conditioned air to 6 to 10 upper floors, while ac units 3 and 4 supply an equal number of lower floors. The pipe chase contains (1) chilled water supply and return, (2) hot water supply and return, and (3) condenser water lines to the cooling tower on the roof. All piping would come from the central station located in the basement.

supplies a mechanical equipment room at the top of the building. Both areas contain multiple air conditioning units supplying various zones in their parts of the building. Some designers of our taller buildings (over 30 stories) locate an entire mechanical equipment floor halfway up the building (Fig. 14.26). This makes it possible to split the building into three parts with the middle mechanical equipment floor feeding risers to upper floors and drops (risers that feed down) to lower floors. It also permits smaller chilled water and reheat piping while reducing the pumping head, thereby conserving energy. This holds true particularly for the sizing of the much bulkier ductwork.

14.9 THE MECHANICAL EQUIPMENT ROOMS: CONSIDERATIONS

Designing and constructing above-grade central stations and mechanical equipment rooms challenges the skills of designers and builders. Conformance with the building's appearance requires tasteful use of *louvers,* both active and purely decorative, since large quantities of air are both brought in and exhausted. This dramatic change in the facade of the building from windows to louvers would require the owners' approval (Fig. 14.27). Sometimes to gain the necessary height for the equipment and layers of large ducts in very tall buildings, the mechanical equipment floor is two stories in height.

Structural engineers provide a floor system consisting of a structural concrete slab which is waterproofed. Much of the electrical control conduit is installed over this waterproofing and a concrete fill completes the floor construction. The engineers must also design to eliminate vibration and noise emanating from these areas. Sometimes these solutions are worked out only after the tenant or user complains.

The construction manager should develop a separate schedule for each of the mechanical equipment rooms in a big building. He should allow four to six months for the mechanical and electrical installation in each one. He must also realize that these areas must be made workable for the HVAC and electrical contractors as soon as possible. Since these areas have differing construction requirements from standard office space, special attention must be paid to the architectural shop drawings, particularly the large louver areas. He may find it expedient to do some of his concrete slab and masonry work out of normal sequence to enable an early mechanical and

FIGURE 14.27.   Architectural louvers for air conditioning intake or exhaust. Dampers are behind louvers.

electrical start. Particularly where mechanical systems feed downward, preparing the area that will supply these lower floors requires careful planning and scheduling to have the necessary air conditioning operable when the office space has been completed architecturally.

Coordination of the ductwork and piping in a machine room area rests almost entirely with the HVAC contractor. The owners' operating personnel may have some very firm opinions about the space between pumps and chillers and proper access for equipment replacement or maintenance. The structural engineer may lay down some rules on the location of heavy chillers and fans. From a basic schematic layout the HVAC contractor then details the layers of piping and high- and low-pressure ductwork. In a central station containing large, intricate chilled water and condenser water pumps and piping systems, the *pipe fabrication* is sublet to pipe fabrication specialty firms. These firms shop-fabricate the piping into convenient lengths for rapid assembly on the job, which reduces overall construction time. The individual assemblies may have a strange appearance, but it is readily recognized that many elbows, special bends, straight runs, and flanges have been joined to make a single unit easily joined in proper position (Fig. 14.28).

FIGURE 14.28.   Prefabricated piping for machine room.

Contractors have differing approaches to sequencing air conditioning unit assemblies. Some install overhead piping and ductwork and then install the coils, filters, fans, and sheet metal enclosure, whereas others prefer to install the fans, coils, and rest of the unit first and then follow with piping and ductwork. Perhaps a lot depends on the equipment, material, and crews that are available and common local practice. Flexible fabric connections isolate the ductwork from the fans, which makes either approach less dependent on the other (Fig. 14.29).

FIGURE 14.29.   Flexible connections at a fan. Consult the SMACNA *Duct Design Manual* and AMCA Publication 201 regarding performance of various inlet and outlet conditions. (Courtesy of Sheet Metal & Air Conditioning Contractors' Assoc. Inc.)

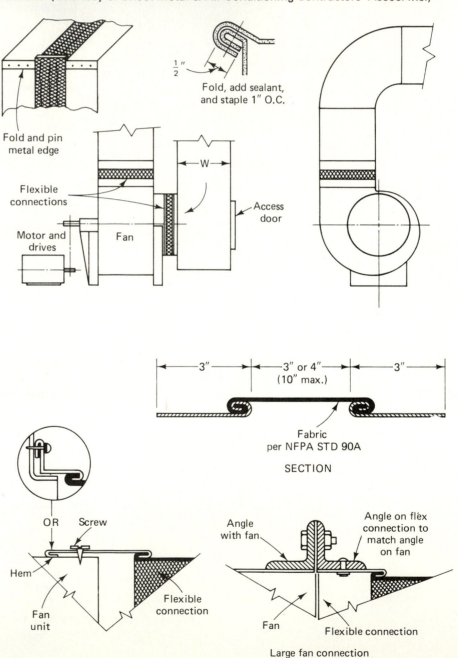

FIGURE 14-30. Inertia block for a large fan. Block will be raised off the floor and supported on springs under the side brackets. (Photo by Gil Amiaga, N.Y.C.)

To prevent transmission of undesirable sound and vibration to the building, the rather large fans are mounted on heavy concrete *inertia slabs* (Fig. 14.30), which in turn are supported by large spring devices called *vibration eliminators*. The general contractor usually coordinates the pouring of these slabs with raised slabs called *housekeeping pads,* on which the coils, filters, and air conditioning unit enclosure are mounted.

14.10 RIGGING   The HVAC contractor has the problem of hoisting many large cumbersome items to the mechanical equipment room of a large building. The general contractor's hoist can handle small items such as ductwork, motors, valves, pipe hangers, and so on. For larger and longer things such as tanks, fans, coils, and pipe, he usually sublets this rigging to a rigging contractor. With the cooperation of the general contractor or construction manager, an opening is left in the side of the building and the rigging contractor installs a Chicago boom (Fig. 14.31) at that floor level. Rigging contractors are specially trained and licensed, making a safe, efficient organization available to the HVAC contractor. In addition, it frees the HVAC contractor's personnel to proceed with the installation of the various systems, leaving this logistical problem to others. This is particularly beneficial during periods of shortages of skilled pipe fitters and sheet metal workers.

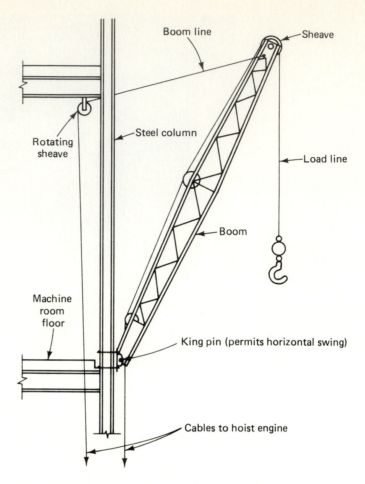

Boom line

Sheave

Steel column

Rotating
sheave

Load line

Boom

Machine
room
floor

King pin (permits horizontal swing)

Cables to hoist engine

FIGURE 14.31.   Chicago boom at a machine room floor.

14.11 THERMAL INSULATION

Both plumbers and HVAC contractors sublet pipe and duct insulation work to insulation contractors. In union insulation shops, contracting firms employing AFL-CIO construction workers, the craftsmen hired are members of the International Association of Heat and Frost Insulators and Asbestos Workers. Since asbestos carries the stigma of being a carcinogenic substance, other materials have taken its place for mechanical insulation work. Pipe coverers, as insulators are commonly called, install coverings for hot water, steam, and chilled water systems, including piping, ducts, devices, and equipment. Together with the prevention of heat loss or gain, coverings control undesirable condensation. The specified materials include numerous products made of fibrous glass, cellular glass, foamed plastic that resembles sponge rubber, urethane, polystyrene, and magnesia (magnesium oxide). Some materials come with jacketing fabrics and foils. They may be rigid or flexible, flat board or blanket type, molded or plain.

For pipe installations *molded types* are widely used in single (Fig. 14.32) or multiple (Fig. 14.33a) layers, where possible joints are staggered and end joints wrapped (Fig. 14.33c). A wide variety of adhesives selected for the insulations application join the longitudinal and end joints. Bands or wires also hold the insulation tightly in place against the pipe and fittings. Factory-fabricated fitting and valve covers (Fig. 14.33b) facilitate this portion of the work, as well as flanges. At hangers the clevis type with support plates, coupled with more rigid, less compressive insulation, distributes the load adequately without deforming the insulation material. Where specified, the insulator enhances the appearance of the insulation with brushed-on sealers and finishes.

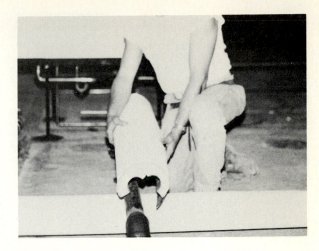

FIGURE 14.32. Placing pre-formed fabric-covered pipe insulation around a pipe.

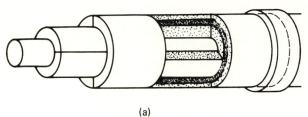

(a)

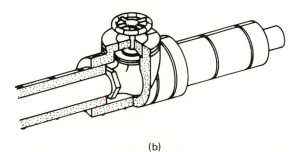

(b)

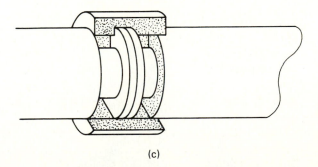

(c)

FIGURE 14.33. (a) Multilayer pipe insulation; (b) insulation detail at the valve; (c) insulation detail at the pipe flange. (Courtesy of Dow Chemical Company.)

The exterior of round ductwork usually gets covered with *blanket-type* fibrous glass with a foil or glass fiber-reinforced aluminum foil. Careful cutting and fitting, proper use of adhesives, and final taping reflect the skill of the insulator. Fine wire provides mechanical fastening.

Rectangular ductwork, fans, and plenums can be covered with either *blanket* or *board-type* material (Fig. 14.34). Closely spaced thin pins are glued or spot-welded to areas to be covered. The board or blanket material is then forced over the pins and nonslip round or rectangular *"washers"* are then clipped to the pins, securely fastening the insulation in place.

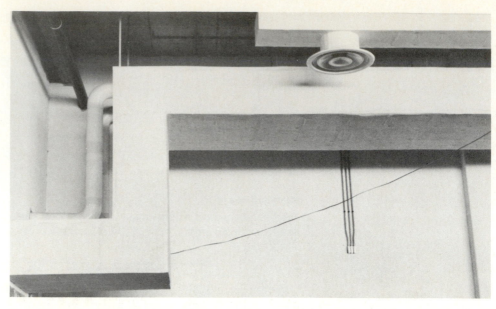

FIGURE 14.34.   Ductwork covered with insulation with a fabric outer layer.

For kitchen exhaust systems and other high-temperature flues, the *10-gauge black iron* used requires an insulation better suited to high-temperature requirements. Layers of *magnesia blocks* cemented and wired in place meet these specifications.

The insulation contractor's work follows and conforms to the progress of the piping and ductwork contractors. Many of these systems, particularly ones carrying hot, chilled, and condenser water, must be tested by the HVAC contractor. Frequently, the insulator covers the greater portion of the piping but leaves all pipe and fitting joints exposed while pressure water tests are being conducted (Fig. 14.35). Immediately following a successful test, the construction manager insists on the completion of covering at the exposed joints so that shafts can be built and piping "covered up." Because his work follows piping and ductwork, he frequently is among the last contractors on a project, particularly in central stations and mechanical equipment rooms (Fig. 14.36).

FIGURE 14.35.   Piping covered with exposed joints and fittings ready for pipe testing.

FIGURE 14.36.   Machine room pipe covering.

15

Air Distribution

15.1 INTRODUCTION   Much like the general contractor, the HVAC contractor works with many crafts. His strength usually lies with the pipe fitting trade in the installation of steam, hot, and chilled water systems. Some companies have a strong sheet metal department, but this is probably the exception rather than the rule. Even the largest mechanical contractors at times sublet their sheet metal work to subcontracting specialists. In this decision these factors are involved:

1. Because of competition in the subcontractor's field, it is cheaper to sublet than to do the work with one's own shop and workers.
2. Almost all of the ductwork is fabricated in a sheet metal shop. A company's own shop might be busy with other work or be too small.
3. Sheet metal work is quite different from piping. The detailing, shop operation, sheet metal workers, and trade associations are alien to piping specialists—it is a different business.

15.2 SHOP DRAWINGS   An air conditioning duct system follows the dictates of the building it serves. Essentially of a custom characteristic, it twists and bends, fattens and thins, to fit in the spaces provided. It must avoid pitched pipes, sprinkler heads, and lighting fixtures. In some buildings it passes through openings provided in girders or through bar joists. Maintaining ceiling heights challenges the sheet metal coordinator continually.

Sandwiching layers of ductwork, particularly in and adjacent to mechanical equipment rooms, requires considerable shop drawing skill and drafting manpower (Fig. 15.1). The detailing must conform to good practice despite numerous changes in cross section and alignment. Deviation from acceptable standards reduces air flow due to excessive friction or produces undesirable turbulence resulting in comfort problems when the air distribution system is placed in operation. The work must be detailed to provide the fabricating workers in the shop with information sufficient to cut, bend, and preassemble each piece to the desired condition. Designers and detailers use a standard set of symbols to simplify the graphic process (Fig. 15.2).

FIGURE 15.1.   Challenge of coordination. (Photo by Gil Amiaga, N.Y.C.)

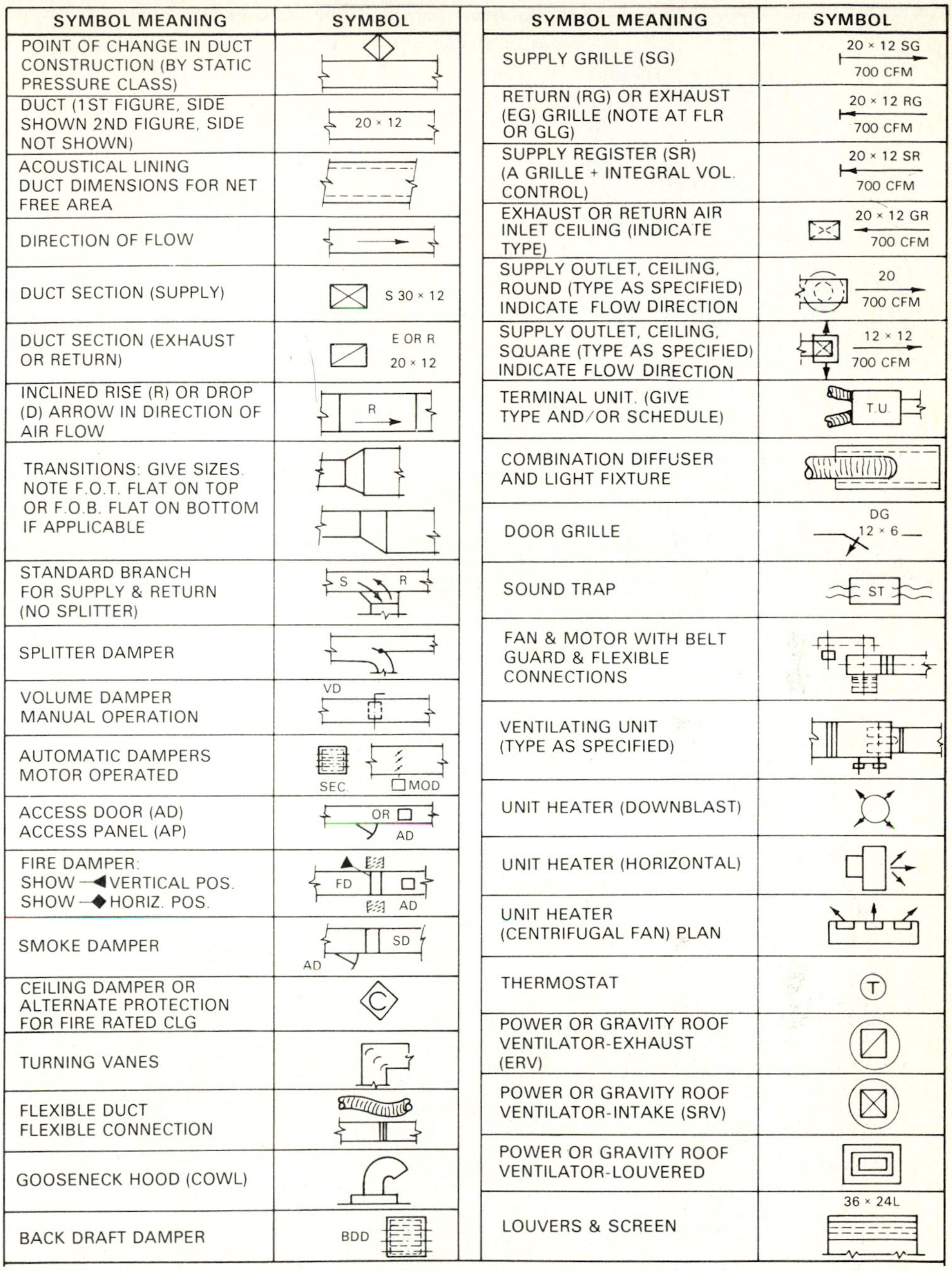

SYMBOL MEANING	SYMBOL	SYMBOL MEANING	SYMBOL
POINT OF CHANGE IN DUCT CONSTRUCTION (BY STATIC PRESSURE CLASS)		SUPPLY GRILLE (SG)	20 × 12 SG 700 CFM
DUCT (1ST FIGURE, SIDE SHOWN 2ND FIGURE, SIDE NOT SHOWN)	20 × 12	RETURN (RG) OR EXHAUST (EG) GRILLE (NOTE AT FLR OR GLG)	20 × 12 RG 700 CFM
ACOUSTICAL LINING DUCT DIMENSIONS FOR NET FREE AREA		SUPPLY REGISTER (SR) (A GRILLE + INTEGRAL VOL. CONTROL)	20 × 12 SR 700 CFM
DIRECTION OF FLOW		EXHAUST OR RETURN AIR INLET CEILING (INDICATE TYPE)	20 × 12 GR 700 CFM
DUCT SECTION (SUPPLY)	S 30 × 12	SUPPLY OUTLET, CEILING, ROUND (TYPE AS SPECIFIED) INDICATE FLOW DIRECTION	20 700 CFM
DUCT SECTION (EXHAUST OR RETURN)	E OR R 20 × 12	SUPPLY OUTLET, CEILING, SQUARE (TYPE AS SPECIFIED) INDICATE FLOW DIRECTION	12 × 12 700 CFM
INCLINED RISE (R) OR DROP (D) ARROW IN DIRECTION OF AIR FLOW	R	TERMINAL UNIT. (GIVE TYPE AND/OR SCHEDULE)	T.U.
TRANSITIONS: GIVE SIZES. NOTE F.O.T. FLAT ON TOP OR F.O.B. FLAT ON BOTTOM IF APPLICABLE		COMBINATION DIFFUSER AND LIGHT FIXTURE	
STANDARD BRANCH FOR SUPPLY & RETURN (NO SPLITTER)	S R	DOOR GRILLE	DG 12 × 6
SPLITTER DAMPER		SOUND TRAP	ST
VOLUME DAMPER MANUAL OPERATION	VD	FAN & MOTOR WITH BELT GUARD & FLEXIBLE CONNECTIONS	
AUTOMATIC DAMPERS MOTOR OPERATED	SEC. □ MOD	VENTILATING UNIT (TYPE AS SPECIFIED)	
ACCESS DOOR (AD) ACCESS PANEL (AP)	OR □ AD	UNIT HEATER (DOWNBLAST)	
FIRE DAMPER: SHOW ◄ VERTICAL POS. SHOW ◆ HORIZ. POS.	FD □ AD	UNIT HEATER (HORIZONTAL)	
SMOKE DAMPER	SD AD	UNIT HEATER (CENTRIFUGAL FAN) PLAN	
CEILING DAMPER OR ALTERNATE PROTECTION FOR FIRE RATED CLG	C	THERMOSTAT	T
TURNING VANES		POWER OR GRAVITY ROOF VENTILATOR-EXHAUST (ERV)	
FLEXIBLE DUCT FLEXIBLE CONNECTION		POWER OR GRAVITY ROOF VENTILATOR-INTAKE (SRV)	
GOOSENECK HOOD (COWL)		POWER OR GRAVITY ROOF VENTILATOR-LOUVERED	
BACK DRAFT DAMPER	BDD	LOUVERS & SCREEN	36 × 24L

FIGURE 15.2. Symbols for ventilation and air conditioning. (Courtesy of Sheet Metal & Air Conditioning Contractors' Assoc. Inc.)

15.3 SHOP FABRICATION

Most ductwork is fabricated from *galvanized sheet steel* 26 gauge down to 20 gauge, with some running to heavy 10 gauge. Where corrosion may occur due to snow or rain in fresh air intake systems, designers specify 32- to 14-gauge *aluminum* and 16- to 32-ounce *copper*. For kitchen exhaust systems and flues, heavy (10-gauge) black iron may be specified (Fig. 15.3). The shop process consists primarily of layout, cutting, bending, and joining. The many ways of joining the metal have been established as standards by SMACNA, the Sheet Metal and Air Conditioning Contractors National Association, Inc. (Figs. 15.4 and 15.5).

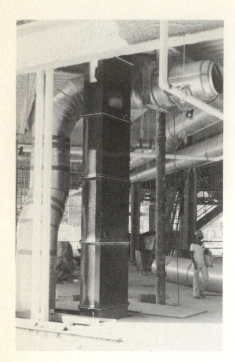

FIGURE 15.3. Black iron exhaust duct riser. Note the flange joints.

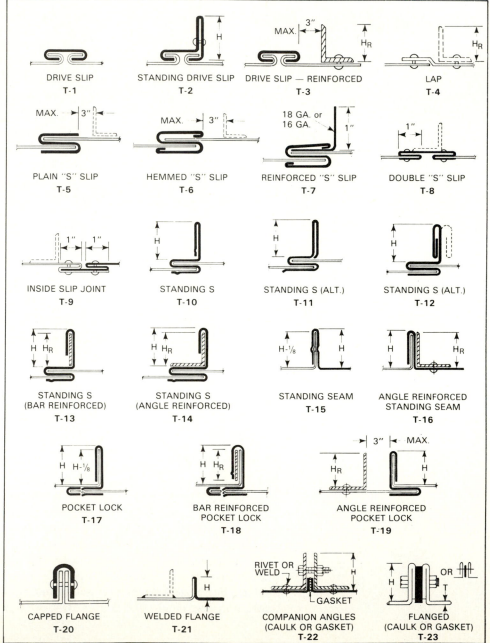

FIGURE 15.4. Transverse (girth) joints. (Courtesy of Sheet Metal & Air Conditioning Contractors' Assoc. Inc.)

194

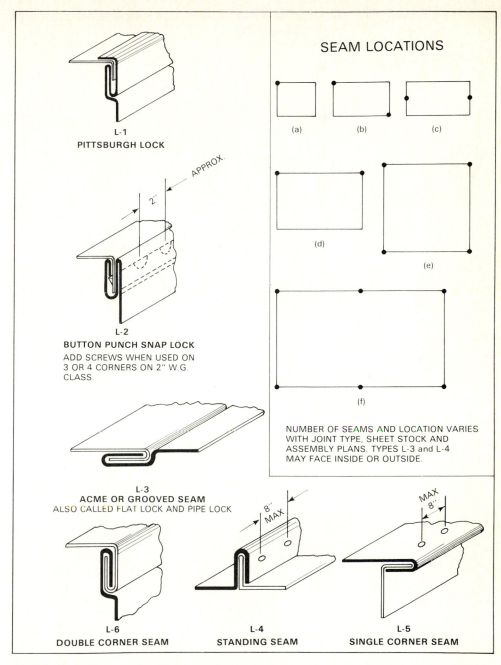

FIGURE 15.5. Longitudinal seams. (Courtesy of Sheet Metal & Air Conditioning Contractors' Assoc. Inc.)

Figure 15.6 shows the recommended type of *longitudinal* and *cross joints* for ducts 18 in. and less and the limits for using the different metal gauges. Figure 15.7 shows the standards for ducts 19 through 30 in. Note the use of stiffener angles, different cross joints, and the limit of 10 ft. 0 in. between joints. The elbow shown in Fig. 15.8 is one example of the many fittings used. In sheet metal work the bulky nature of sheet metal makes shipping and storage an important consideration in the shop, in transit, and on the project. In one extreme the parts are cut to shape and joining edges prepared to permit shipment as flat sheets. Some sections are fabricated and shipped L-shaped. Many L-shaped pieces can be placed on end one against another. Fully fabricated ducts are nested one inside another for shipping and storage

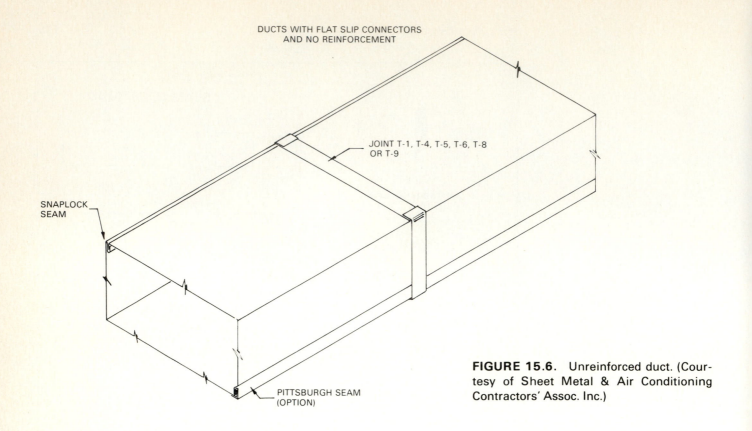

DUCTS WITH FLAT SLIP CONNECTORS
AND NO REINFORCEMENT

JOINT T-1, T-4, T-5, T-6, T-8
OR T-9

SNAPLOCK
SEAM

PITTSBURGH SEAM
(OPTION)

FIGURE 15.6. Unreinforced duct. (Courtesy of Sheet Metal & Air Conditioning Contractors' Assoc. Inc.)

FIGURE 15.7. Duct reinforced on two sides. (Courtesy of Sheet Metal & Air Conditioning Contractors' Assoc. Inc.)

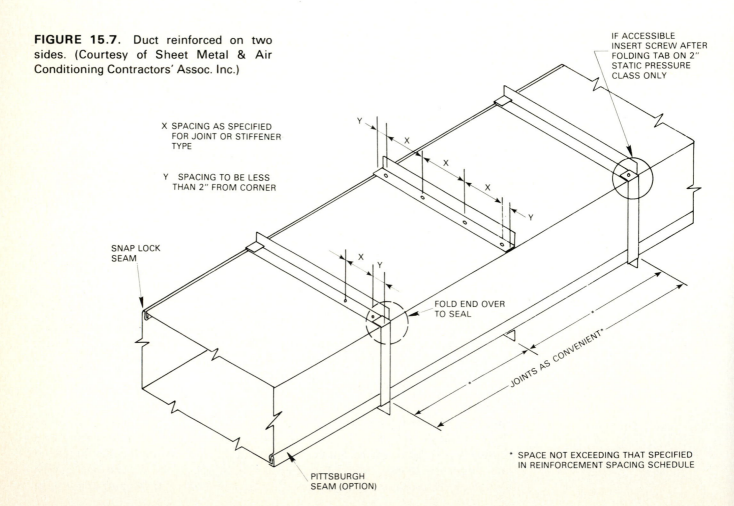

IF ACCESSIBLE
INSERT SCREW AFTER
FOLDING TAB ON 2"
STATIC PRESSURE
CLASS ONLY

X SPACING AS SPECIFIED
FOR JOINT OR STIFFENER
TYPE

Y SPACING TO BE LESS
THAN 2" FROM CORNER

SNAP LOCK
SEAM

FOLD END OVER
TO SEAL

JOINTS AS CONVENIENT*

PITTSBURGH
SEAM (OPTION)

* SPACE NOT EXCEEDING THAT SPECIFIED
IN REINFORCEMENT SPACING SCHEDULE

FIGURE 15.8. Shop-fabricated ductwork elbow showing turning vanes which improve elbow efficiency and ensure a smooth, even flow of air.

efficiency. Each separate piece of ductwork carries the contract number, the system designation (such as AC3), the dimensions, and the piece number (for example, AC3-5). On the job this facilitates segregating and delivery of shipments to their location of erection (Fig. 15.9). Shop assembly and field preassembly help reduce erection time (Figs. 15.10 and 15.11). Light lifting devices raise and position the ductwork for hanging and joining (Fig. 15.12).

FIGURE 15.9. Sheet metal ductwork storage occupies construction areas on the job site.

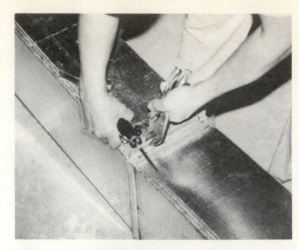

FIGURE 15.10. Ductwork is assembled into conveniently handled sections by bolting and riveting prior to connection into the system.

FIGURE 15.11. Duct sections assembled in the shop reduce field labor.

FIGURE 15.12. Ductwork is elevated and held in place by rolling hand-operated machinery.

15.4 THE SHAPE OF DUCTS*

"For air ducts a *round shape* is the most efficient and economical. Round duct is used extensively in all systems because it excels in strength, rigidity, and ease of making airtight connections. Very light material can contain even the highest positive pressures found in air-conditioning systems. The dominant factor in round duct construction is the material's ability to withstand the physical abuse of installation and the negative pressure requirements.

High pressure metal round duct is normally manufactured with *spiral seams* and with fittings of continuous welded construction [Fig. 15.13]. Low pressure metal round ducts are generally constructed of light gauge spiral aluminum or steel, or use longitudinal snaplock seams. Fittings are made with mechanically locked or riveted connections [see Fig. 15.14].

Flat oval steel duct combines some advantages of round duct and rectangular duct, in that it will fit in spaces where there is not enough room for round duct, and can also be joined using the techniques of round duct assembly [Fig. 15.15]."

FIGURE 15.13. High-pressure spiral ductwork.

FIGURE 15.14. Oval ductwork being assembled prior to hanging.

*Section 15.4 is reprinted with permission from the 1979 Equipment Volume, *ASHRAE Handbook & Product Directory*, pp. 1.2, 1.3.

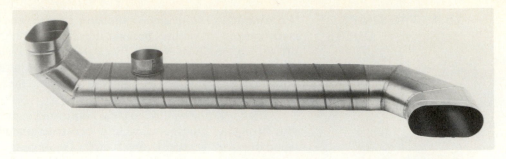

FIGURE 15.15.   Oval ductwork. (Courtesy of United McGill Corporation.)

15.5   THE SEQUENCE OF OPERATIONS

In high-rise work the HVAC contractor will be directed by the construction manager to put his greatest early effort into *mechanical equipment rooms* and *vertical risers* (Fig. 15.16). The latter work permits the general construction of the core areas containing toilet rooms, mechanical, electrical, and telephone shafts, stairways, and elevators. On the perimeter of the building the riser work must proceed in advance of numerous architectural items, such as column enclosures, venetian blind pockets, and air conditioning units and enclosures. These vertical systems, both supply and return, are the mains to which the numerous branch lines connect. Naturally, any testing and insulation must precede their being enclosed (Fig. 15.17). Horizontal and vertical runs of ductwork increase or decrease in size depending on their function. Figure 15.18 shows this characteristic changing of ductwork size.

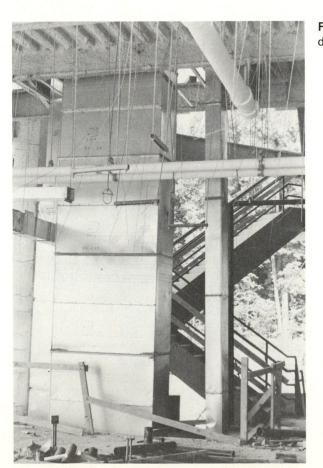

FIGURE 15.16.   Sheet metal duct riser shaft.

FIGURE 15.17. Ductwork sealed prior to testing.

FIGURE 15.18. Typical ductwork transition from rectangular to round.

15.6 HANGERS, FASTENERS, ANCHORS, AND CLAMPS

The means of supporting horizontal ductwork depends on the size gauge and dimensions of the run to be supported. SMACNA has established standards for this purpose (Fig. 15.19). Smaller sizes of ductwork have sufficient rigidity to be held in place directly by hanger straps or rods (Fig. 15.20). Angles are used extensively to support the ductwork from the bottom and for very large ductwork small channels and I-beams. Rods using either threaded or friction fasteners are most popular for the heavier installations. Vertical runs are supported at the floors with angles or channels spanning the opening through the slab.

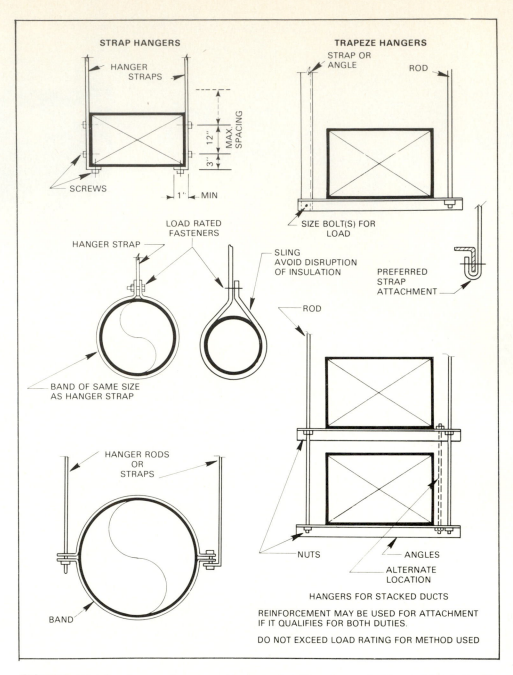

FIGURE 15.19. Lower hanger attachments. (Courtesy of Sheet Metal & Air Conditioning Contractors' Assoc. Inc.)

FIGURE 15.20. Simple ductwork supports clipped to a beam and fastened to ductwork.

There are a great variety of ways of attaching hanger rods and straps to the structure (Figs. 15.21 and 15.22). For structural concrete using metal decks, flat bars and rods are dropped through holes in the metal deck (1). Where wood forms are used for structural concrete slabs, concrete inserts of bent or cast metal are lightly nailed to the formwork (2a and 2b). Parts 3a and 3b show the use of powder-actuated fasteners shot into the concrete by explosive force. In parts 7a, 7b, and 7c, three types of drilled in anchors provide the required holding force. Drilling the proper-size hole is essential to get the desired results. Parts 4a, 4b, and 6 depict popular means of fastening to steel beams and bar joists. Parts 5a and 5b show the fastening made with a welded stud.

FIGURE 15.21. Upper attachments. (Courtesy of Sheet Metal & Air Conditioning Contractors' Assoc. Inc.)

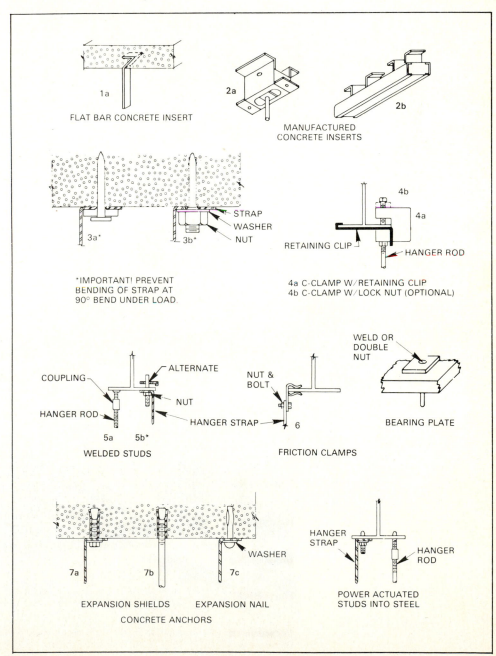

FIGURE 15.22. Flat bar type insert used with a metal deck form.

15.7   TYPES OF DUCTS

Fibrous Glass Duct

"*Fibrous glass duct* and duct-board are a composite of rigid fiber glass and a factory applied jacket. This material is available in molded round sections or in board form for fabrication into rectangular shapes. Fibrous glass duct has come into common use for low velocity and low pressure applications where thermal insulation or acoustical treatment is required. The approved joining systems result in a substantial leakfree duct system."*

Qualified contractors using semiautomatic equipment fabricate the duct board into rectangular and other straight-line shapes (Fig. 15.23). Folds and longitudinal joints are made with shiplap joints. The longitudinal joint is stapled and taped. The outward clinch staples and SMACNA-approved tapes make strong, airtight joints. Fittings are constructed of the same shiplap stapling and joining techniques. On the construction site the duct is supported by trapeze hangers.

FIGURE 15.23. Fibrous duct showing a linear diffuser in an acoustical ceiling and lighting fixture layout. Note the sprinkler piping drop in the foreground.

*Reprinted with permission from the 1979 Equipment Volume, *ASHRAE Handbook & Product Directory*, p. 1.4.

Flexible Duct

"Flexible duct is usually used to connect air distribution equipment such as mixing boxes, light troffers induction units, air distribution boxes, and diffusers. It is also used where air terminal devices are subject to relocation.

The length of flexible duct used should be long enough to provide a sweeping configuration without undue installation restrictions."*

Flexible air ducts are manufactured using a steel wire core laminated within layers of polyester film. Ducts may be purchased with or without insulation and with or without end fittings. To obtain an airtight connection to air devices, the mechanical fastening is reinforced by taping the joining of the material (Figs. 15.24 to 15.26).

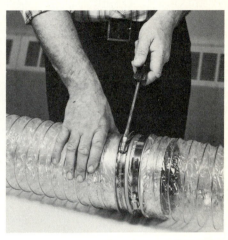

FIGURE 15.24. Taping and clamping flexible ducts. (Courtesy of Wiremold Company.)

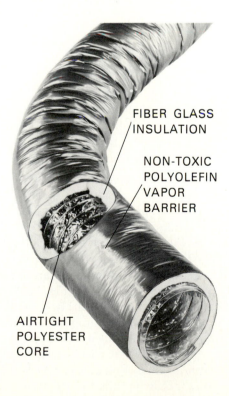

FIBER GLASS
INSULATION

NON-TOXIC
POLYOLEFIN
VAPOR
BARRIER

AIRTIGHT
POLYESTER
CORE

FIGURE 15.25. Insulated flexible duct. (Courtesy of Wiremold Company.)

*Reprinted with permission from the 1979 Equipment Volume, *ASHRAE Handbook & Product Directory,* p. 1.4.

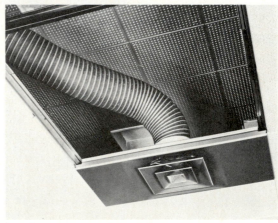

FIGURE 15.26. Flexible duct connects ductwork with the diffuser. (Courtesy of Wiremold Company.)

15.8 AIR DIFFUSION

Conditioned air leaves the ductwork and enters a space through many different means. Grilles, diffusers, and slot diffuser outlets are widely used. Air is also discharged through ceiling lighting fixtures, air distributing ceilings, and luminous-type ceilings. Stamped grilles are primarily decorative and inexpensive (Fig. 15.27), while fixed bar grilles have fixed vanes that are not adjustable (Fig. 15.28). Adjustable bar grilles can have a single set of vanes or both horizontal and vertical vanes that control the airstream in both planes. Round, square, and rectangular ceiling diffusers set against or flush with the ceiling material (Fig. 15.29) are aerodynamically designed with curved surfaces to diffuse the air in draft-free fashion. They usually consist of an outer shell with duct collar and a readily removable inner assembly. Linear diffusers or slot diffuser outlets are installed in ceilings high up on walls and at floor level. Ceiling and floor types generally are also located at the perimeter of the space. These types give a very even air distribution at ceiling level. Vanes behind the slots permit uniform discharge throughout the full length of the diffuser. The supply ductwork and the diffusers themselves make this a more expensive installation; however, architects prefer them, particularly in lobbies and special rooms.

FIGURE 15.27.   Stamped grille.

FIGURE 15.28.   Fixed grille and movable vanes.

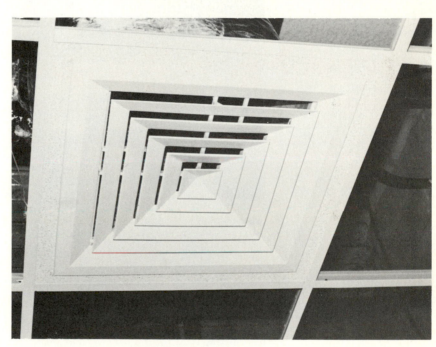

FIGURE 15.29.   Ceiling diffuser.

15.9   DUCT LINING   Certain portions of air distribution systems require *acoustical lining* of ductwork. The dimensions of the metal ductwork are increased to accommodate the thickness of the lining for the areas to be lined so as to maintain the required airflow size. Most of this internally fastened material is secured to the metal with mechanical fasteners and the entire metal surface is covered with fire-retardant adhesives.

The material used for duct lining must resist *erosion* and *fire* and have other properties to ensure ease and quality of fabrication and erection. Flexible and semirigid fibrous glass blankets made with special surface treatment meet these requirements. Some specifiers prefer rigid-coated fibrous glass board for masonry air shafts, large plenums (chambers), and air conditioning unit casings. Shop and field workmanship of high quality ensures against flaking and delamination of the lining material, which could cause malfunction of volume controls, terminals, and coils due to fouling.

15.10 FANS   In heating, ventilating, and air conditioning the discharge of spent air, the introduction of fresh air, the supply and return of conditioned air, and the circulation of interior air all depend on *fan operation*. Some of our important ventilation systems do not depend on conditioned air but simply remove undesirable odors, fumes, smoke, or heat. Separate fan-operated systems remove kitchen, toilet room, and laboratory impure air. Industrial buildings utilize large numbers of high-volume exhaust fans for similar air problems (Fig. 15.30).

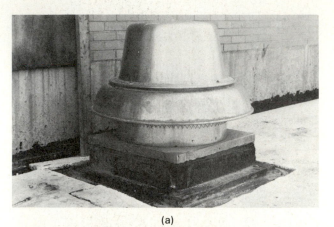

(a)

(b)

FIGURE 15.30. (a) Roof ventilator exhaust; (b) roof-mounted centrifugal fan used for exhaust purposes.

Centrifugal Fans

In air conditioning systems the two basic types of fans are *centrifugal* and *axial*. Figure 15.31 shows the common terminology for a centrifugal fan. Centrifugal fans for HVAC use come with three types of impeller design: air foil, backward-inclined backward curved, and forward curved. Delivered air velocity, volume, pressure, and operating efficiency vary with impeller design, offering the designer the opportunity to choose the most efficient fan for his purpose.

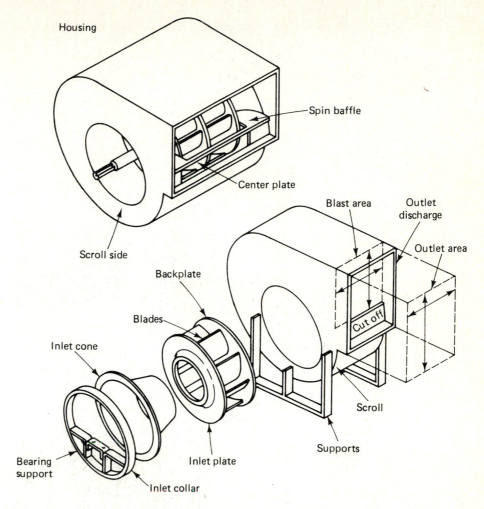

FIGURE 15.31. Terminology for common fan components. (Courtesy of Aerovent Inc.)

Rotation and *angle of discharge* must be specified when ordering fans. Figure 15.32 gives 4 of 16 possible configurations for centrifugal fans. Figure 15.33 shows the proper method of designating the position of the belt-driven motor. Centrifugal fans for large installations are bulky and awkward to hoist to upper fan rooms. Improper rigging can damage the vulnerable impeller and casing. Along with many other mechanical items, fans should be protected from the elements until they have been properly installed.

FIGURE 15.32. Four of 16 examples of designations for the rotation and discharge of centrifugal fans: (a) clockwise, top horizontal; (b) clockwise, top angular up; (c) counterclockwise, up-blast; (d) counterclockwise, bottom horizontal. (Courtesy of Aerovent Inc.)

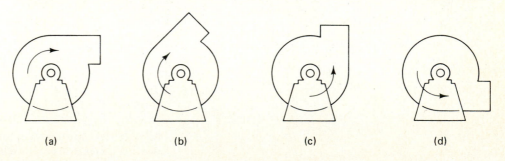

(a)   (b)   (c)   (d)

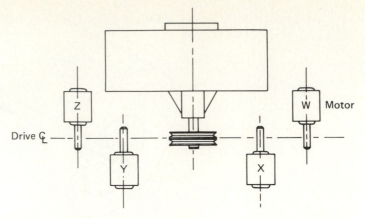

FIGURE 15.33. Motor positions for belt-driven centrifugal fans; plan view of a centrifugal fan. The location of the motor is determined by facing the drive side of the fan and designating the motor position by the letters W, X, Y, and Z. (Courtesy of Aerovent Inc.)

To achieve rated fan performance, air should enter the inlet area in a smooth, even flow. Reduced fan performance can usually be traced to poor inlet conditions. Figure 15.34 shows examples of poor and good inlet connections. Fan discharge also affects performance. A straight run of duct the size of the outlet opening should continue for several duct diameters for best results. Figure 15.35 shows poor and better outlet connections.

FIGURE 15.34.   Fan inlet connections. (Courtesy of ASHRAE.)

Poor: restrictive duct and transformation

Good: open inlet or full size straight approach duct

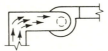

Poor: spin with wheel — reduced volume and pressure; spin against wheel — power increase required

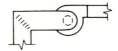

Good: turning vanes for uniform distribution, no swirl at inlet

Poor: high duct entry loss

Good: low duct entry loss

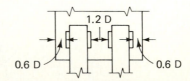

Poor: D = less than one wheel diameter
Good: D = more than one wheel diameter

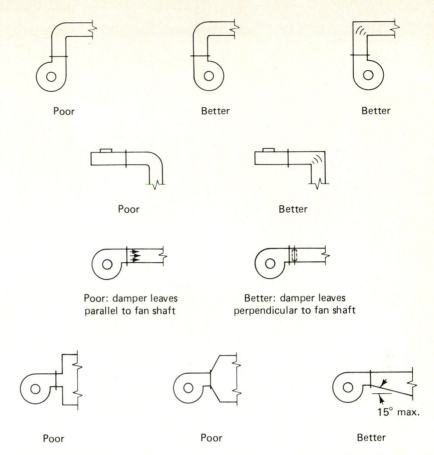

FIGURE 15.35. Fan outlet connections. Poor, without perforated baffle plate; fair, with perforated baffle plate. (Courtesy of ASHRAE.)

Axial Fans

Axial fans (Fig. 15.36) are manufactured with *propeller, tubeaxial,* and *vaneaxial* design. The propeller type delivers low-pressure, high-volume air for circulation or ventilation without ductwork. The tubeaxial and vaneaxial types connect directly into the duct system.

FIGURE 15.36. Common names associated with axial fan components. (Courtesy of ASHRAE.)

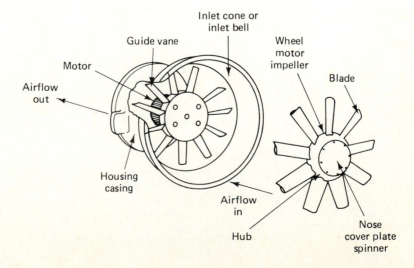

Tubeaxial Fans

"Direct drive and belt driven tubeaxial fans, or duct fans as they are called are built to be installed in the system so as to become a part of the duct. The fan may be located at the entrance, in the middle of, or at the discharge end of the duct.

The construction of the belt driven fan places the motor outside the airstream. The belt drive and bearings are enclosed in a tube and cover situated on the suction side of the propeller. For normal air handling and high temperature operation, the bearing cover is open at the propeller end. This design induces airflow through the belt tube and bearing enclosure, creating an air insulation barrier which constantly purges contaminant and, in the case of high temperature applications, provides drive and bearing cooling. In situations where the air is wet and very contaminated with corrosive chemicals, the fan may be built with the bearing cover completely closed and with a seal around the propeller shaft."*

Figures 15.37, 15.38, and 15.39 show various installations that affect pressures and efficiency. Figures 15.40, 15.41, and 15.42 show suggested connections for fans in duct systems. Installation of a tubeaxial fan is illustrated in Fig. 15.43.

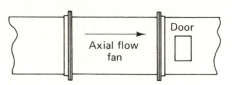

FIGURE 15.37. Tubeaxial fan—same size duct normal fan installation. Rating tables are based on this arrangement. Variations in installation indicated in Figs. 15.38 and 15.39 will provide more economical fan performance and fan selection but may require a higher installation cost. (Courtesy of Aerovent Inc.)

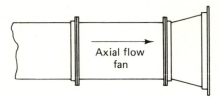

FIGURE 15.38. Tubeaxial fan. Tapered discharge allows regain of a portion of the velocity pressure to static pressure. The fan can then be chosen at a lower pressure for more economical operation. (Courtesy of Aerovent Inc.)

FIGURE 15.39. Tubeaxial fan. Converging inlet and discharge provides lower duct velocities and less friction to the airflow. This allows choosing a fan at lower pressure for more economical operation. (Courtesy of Aerovent Inc.)

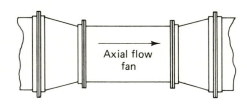

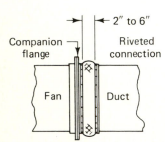

FIGURE 15.40. Suggested connections for fans in duct systems. Flexible connection: minimum vibration using canvas-type material. (Courtesy of Aerovent Inc.)

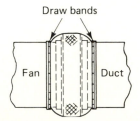

FIGURE 15.41. Suggested connections for fans in duct systems. Flexible connection: minimum vibration using canvas-type material. (Courtesy of Aerovent Inc.)

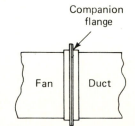

FIGURE 15.42. Suggested connections for fans in duct systems. Rigid connection. (Courtesy of Aerovent Inc.)

* *Fan Engineering and Application Data,* Aerovent Inc., Piqua, Ohio.

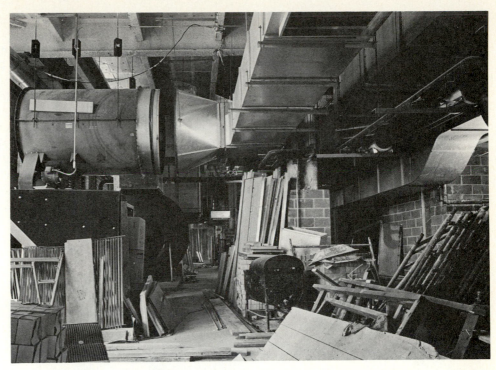

FIGURE 15.43. Tubeaxial fan installation in progress. (Photo by Gil Amiaga, N.Y.C.)

15.11 SOUND DAMPENING Lined Ducts*

"Sound intensity from fans can be greatly reduced by using acoustically lined ducts [Fig. 15.44]. A minimum duct length of three fan diameters should be used. Longer ducts and elbows or turns will increase the sound reduction. Ductwork can be made of acoustic type construction materials if application or local codes permit.

The following material is recommended for lining ducts: fibrous glass duct liner 1″ thick, neoprene coated. This duct liner should be cemented in place. The coated side must be exposed to the airstream. This prevents eroding of the fiberglass by the airstream. Covering joints and raw edges is necessary.

Oversized ducts will aid in keeping the sound level down by allowing reduced velocity and a greater area of duct lining.

Flexible connections between the fan and duct section will reduce vibration to the duct and decrease noise. Mounting the fan on vibration isolators is also necessary for noise dampening."

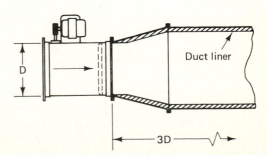

FIGURE 15.44. Sound dampening with lined ducts at a tubeaxial fan. (Courtesy of Aerovent Inc.)

*This section is taken from *Fan Engineering and Application Data*, Aerovent Inc., Piqua, Ohio.

Duct Silencers

Noise from outside the building and noise generated inside the building by fans, motors, and the rush of air can be effectively and economically reduced with manufactured duct silencer units (Fig. 15.45). Duct silencers decrease unwanted noise, whether the sound travels with or against the airflow (Fig. 15.46). The air passes through a narrowed passage of perforated metal behind which is acoustical fill. The silencers can be installed in the ductwork itself (Fig. 15.47c–e), in the intake structure (Fig. 15.47a), or in the fan and coil casing (Fig. 15.47b).

A companion to the duct silencer is the acoustical air transfer unit. These units permit passage of air from one room to another without disconcerting or irritating noise or conversation. Figure 21.30 shows examples of wall and ceiling installations.

Sound carried by the metal material and vibration are prevented from traveling from fans and air conditioning units to the rest of the ductwork by the use of fabric closures between duct sections.

FIGURE 15.45.   Duct silencer. (Courtesy of United McGill Corporation.)

FIGURE 15.46.   (a) Forward flow: the noise field propagates in the same direction as the airflow. (b) Reverse flow: the noise field propagates opposite to the airflow. (Courtesy of Industrial Acoustics Company, New York, New York.)

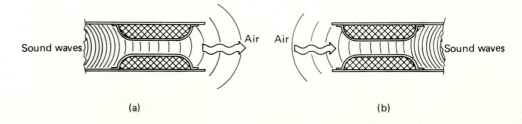

(a)   (b)

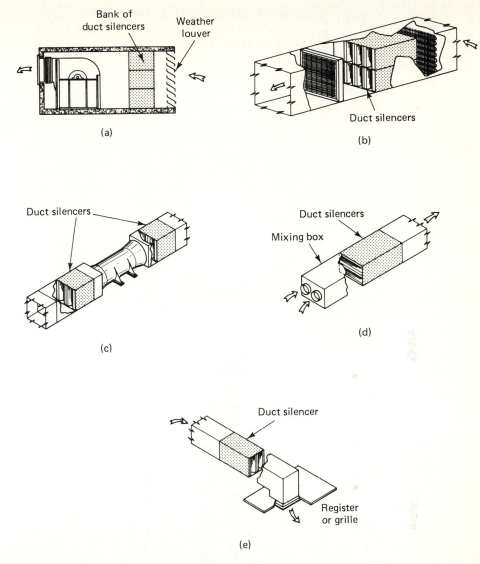

FIGURE 15.47. (a) Equipment room intake silencer behind a weather louver; (b) Silencer bank near coils and filters; (c) intake and discharge silencers for vaneaxial fans; (d) Discharge silencer attached to a mixing box; (e) silencer near an air terminal. (Courtesy of Industrial Acoustics Company, New York, New York.)

15.12 DAMPERS

Volume Dampers

Dampers of various designs and purposes control the flow of air in the distribution ductwork. Intake and exhaust air is regulated by multiblade and opposed-action dampers set in the exterior walls. Those shown in Fig. 15.48 a and b are motor or pneumatically actuated. Within a system *splitter dampers* distribute the supply air proportionally to the branches according to the volume each should receive. Other dampers *shut off* or control the flow in a single duct (Fig. 15.49a to d). Sheet metal firms fabricate splitter and shutoff-type dampers along with the ductwork. Multiblade dampers are usually purchased from manufacturing firms.

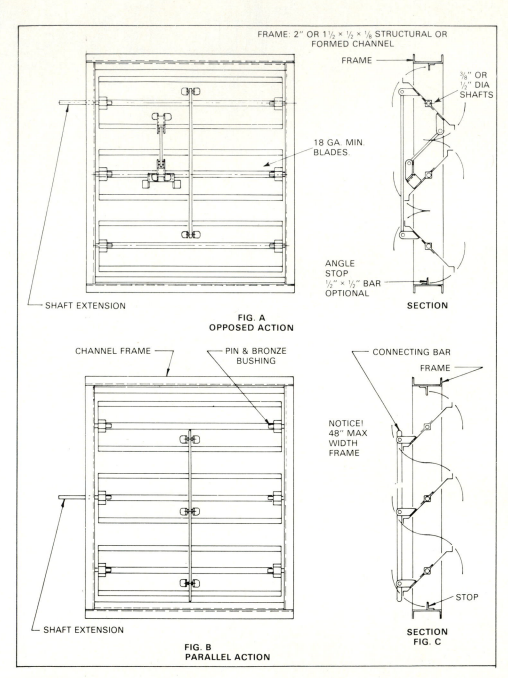

FIGURE 15.48. Multiblade volume dampers. (Courtesy of Sheet Metal & Air Conditioning Contractors' Assoc. Inc.)

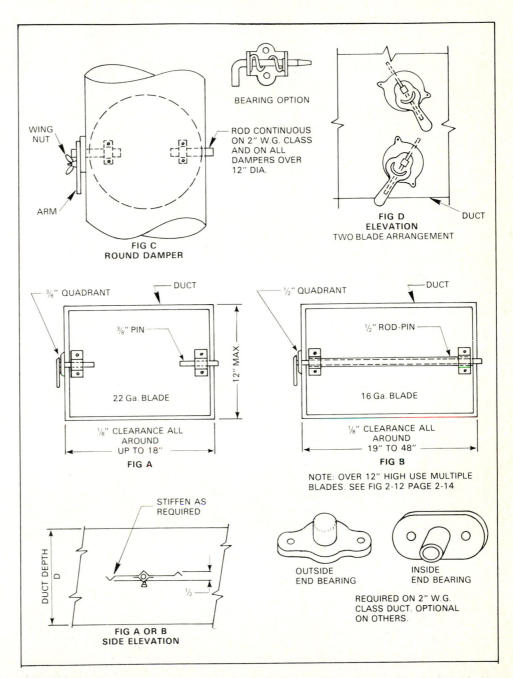

FIGURE 15.49. Volume dampers: single-blade type. (Courtesy of Sheet Metal & Air Conditioning Contractors' Assoc. Inc.)

Fire Dampers

The *single-blade fire damper* for rectangular ductwork shown in Fig. 15.50 successfully passed the Underwriters' Laboratory 1½-hr fire test. The blade is heavily unbalanced about the bearing shaft so that when fire-heat melts the fusible link, the blade rotates to the vertical position and is held closed by the blade catch. The multiblade type operates by gravity in similar fashion. It is used for the larger horizontal ducts. Figure 15.51 shows the round fire damper for ducts 17 in. in diameter and larger. The reliability of the springs is of utmost importance for this type of closure.

FIGURE 15.50. Vertical-mounted single-blade fire damper. (Courtesy of Sheet Metal & Air Conditioning Contractors' Assoc. Inc.)

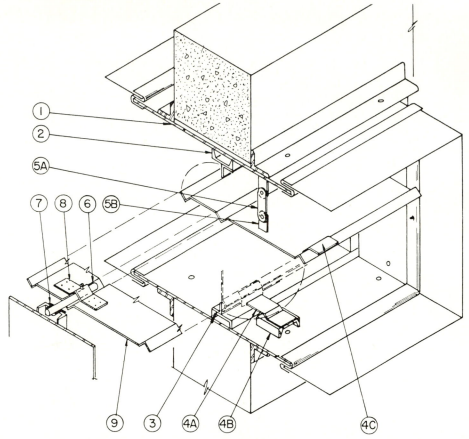

PART NO.	DESCRIPTION	NO. REQUIRED	MATERIAL
1	SLEEVE	1 PER ASSEMBLY	14 GA. HOT ROLLED C-1010 STEEL
2	FRAME	1 PER ASSEMBLY	2"x½"x⅛" CHANNEL H.R. C-1010 STEEL
3	BLADE STOP	2 PER ASSEMBLY	1"x1"x⅛" ANGLE H.R. C-1010 STEEL
4A	BLADE CATCH	1 PER ASSEMBLY	.063 STAINLESS STEEL ½ HARD T-302-SS
4B	BLADE CATCH MOUNT	1 PER ASSEMBLY	2"x½"x⅛" CHANNEL H.R. C-1010 STEEL
4C	STRIKER	1 PER ASSEMBLY	14 GA. HOT ROLLED C-1010 STEEL
5A	FUSIBLE LINK	1 PER ASSEMBLY	VARIOUS U.L. APPROVED
5B	FUSIBLE LINK BRACKET	2 PER ASSEMBLY	⅛"x½" HOT ROLLED C-1010 STEEL
6	BEARING SHAFT	2 PER BLADE	½" DIA. COLD ROLLED C-1010 STEEL
7	BUSHING	2 PER BLADE	SINTERED BRONZE
8	BEARING STRAP W/¼-28x½" SET SCREW	2 PER BLADE	⅛"x1" HOT ROLLED C-1010 STEEL
9	BLADE	1 PER ASSEMBLY	10 GA. HOT ROLLED-SB 10H C-1010 STEEL 14 GA. HOT ROLLED-SB 14H C-1010 STEEL

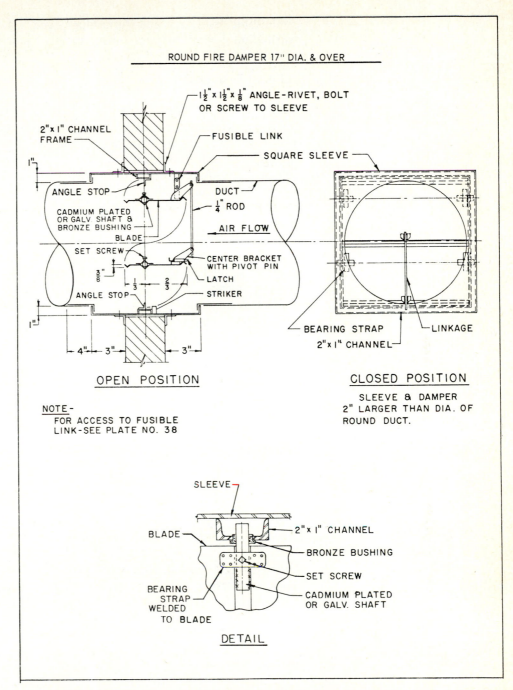

FIGURE 15.51. Fire damper standards. (Courtesy of Sheet Metal & Air Conditioning Contractors' Assoc. Inc.)

15.13 FIRE CONSIDERATIONS Thanks to our building departments, fire departments, insurance underwriters, designers, and engineers, the personal safety of modern building occupants in the United States is well protected by our building codes. Our worst fire losses in terms of loss of life have resulted from violations of our building codes or occupancy laws. The relatively good record has not deluded those concerned with the seriousness of protecting human lives during a fire. Particularly in high-rise buildings, authorities wrestle with some still unsolved problems. These are (1) building evacuation, (2) combustible materials and furnishings, and (3) firefighting. In addition, air distribution systems themselves can present a major problem.

Building Evacuation

Human behavior patterns present a big problem in dealing with building evacuation. Occupants two or more times every day for years on end go to the elevators to exit from their floor. During a fire this may be one of the worst places to go. There are numerous cases of occupants and even firemen entering an elevator which unfortunately stopped at the fire floor. In these cases, once the doors opened the doors stayed open for various fire-related reasons, and the car refused to move, causing the fire to envelope the car interior and the people to perish. A very tragic loss of life related to the habit of always using the elevator occurred on the *second floor* of a Fifth Avenue, New York City, commercial building not many years ago. A flash fire caused by the use of a highly inflammable adhesive enveloped the working space and spread to the hallway. The bodies of three victims were found in the half-closed elevator doorway opening. The fire exit doorway to the fire stairway was only 3 ft away. Firemen came to the conclusion that a "Good Samaritan" was holding the elevator for fellow workers to ride one floor.

The size and height of our inner-city office buildings, which are occupied by thousands of people, make it extremely difficult to rapidly evacuate a high-rise building by the use of fire stairways. The fire floor area creates a substantial hazard for persons from upper floors to overcome. Firefighters use the fire stairways to get at the fire to knock it down. The stairway becomes filled with energetic firemen, their hoses, and equipment. Smoke and heat also pour out of the fire area through the partially open fire doors held open by the hoses, making use of the fire stairway at this level quite difficult. Some authorities recommend upper evacuation to a safe area if one can be found. The rooftop of the building, mechanical equipment rooms, or upper office floor areas might be the safest place, depending on the fire location. Many people have found relative safety on the roofs of high-rise buildings during a fire.

Combustible Materials and Furnishings

It seems ironic to fire chiefs, fire prevention experts, and those who work with building codes that after putting so much effort into building relatively fire-safe buildings, the law permits filling these buildings with combustible drapes, carpets, and furniture. Certainly, drapes and carpeting should be treated with flame retardants. *Highly combustible foamed plastic* overstuffed furniture should be avoided or at least kept to a minimum. Obviously, this is an area where the designer and builder can only make recommendations to the building owners and tenants and hope that their advice will be heeded.

Firefighting

Any fire above the reach of aerial ladders used by firemen to combat fire will be very difficult to fight. Firemen cannot attack a fire at these heights externally. They must go to the fire floor area and combat the fire from the fire stair towers. Initially, they must enter through doorways located at each stair tower. The typical complete curtain wall construction *seals in* smoke, heat, and toxic gases. Firemen like to *vent* a fire, that is, create a flue action to exhaust the smoke and reveal the fire source, which permits them to see how to attack the fire. Modern high-rise construction makes venting difficult since internally it is vital not to let the fire spread upward to other floors, and externally the windows are sealed. Any venting through windows might also allow the fire to reenter the building at the next floor above. In some cases the firefighters have ordered that the return air system be operated to help vent the fire. In these instances apparently the decision was well thought out. A knowledge of the building and the return air system are an important consideration in making this kind of decision.

Distribution Systems and Fires

One of the most disturbing fire safety aspects of the post–World War II high-rise building boom was the unintentional creation of *flues* in the exterior wall construction and a marked increase in the size and number of floor penetrations. *High-velocity perimeter air systems* required numerous holes in the floor for risers at the same time that metal, glass, and precast concrete curtain walls were eliminating the masonry closures at each floor common in the older brick and stone buildings. Curtain walls were erected quickly because a tolerance space was provided between the inside face of the curtain wall and the outside dimension of structural steel or concrete. This created a *flue* from top to bottom of the exterior of the building. In many cases, light-gauge-metal venetian blind pockets provided the only seal from one floor to the next.

Since the high-rise scare of the early 1970s, prudent architects, owners, and contractors have recognized the architectural flue hazard in the exterior wall and the mechanical systems' penetration of floors and walls. The closure at the exterior wall has been recreated to eliminate one source of fire spread from floor to floor. The space between piping and holes through floors and walls has been packed with a compressible fire-resistant substance. Recognizing that fire could penetrate ductwork at one floor and reenter at another, ductwork has been sealed off in fire-resistive materials from floor level to underneath the structural slab.

The hazard of the spread of fire, smoke, intense heat, and toxic gases in supply ductwork has been recognized in the United States for many years. We have seen how dampers shut automatically when low-melting-temperature fusible links melt. Supply system fans can be shut off by smoke detectors and heat-activated devices located in the supply lines near the fan discharge point. It is vital that the fire be denied the supply air, which would only make the fire burn more fiercely.

15.14 CASINGS AND HOUSINGS

The heart of an air conditioning system supplying a particular zone is the unit itself. Where a *package unit* purchased from a manufacturer does not fit the building requirements, which include rigging the unit into the building, the designer selects dampers, filters, coils, and fans which he designs into a casing to house them. Fans, coils, dampers, and filters are usually purchased items, but the casing is shop fabricated and the whole unit is field assembled. The designer arranges and dimensions the unit to connect to the exterior wall and fit the mechanical equipment room provided. Figures 15.52 and 15.53 show a plan view and elevation of a typical job-assembled air conditioning unit casing with five access doors for inspecting and maintaining the filters, coils, and humidifier. The construction of the unit permits replacement of individual sections such as coils (Fig. 15.55). The usual construction practice is to install the coils and fans as a first step, start casing construction and follow with piping to the coils (Fig. 15.54), but there is no set sequence of operations. Many designers place the unit on a 4-in.-high concrete slab called a housekeeping pad. Where space is at a premium, smaller units can be squeezed into upper sections of fan rooms where the entire unit, with all its sections, is hung from the framing above. However, this is usually done only if necessary.

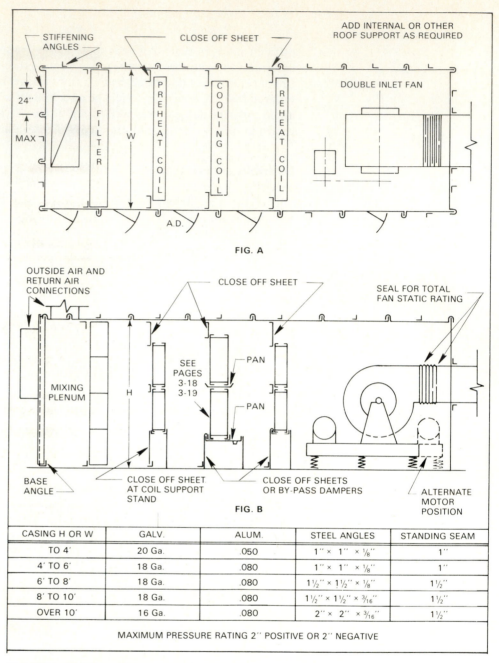

FIG. A

FIG. B

CASING H OR W	GALV.	ALUM.	STEEL ANGLES	STANDING SEAM
TO 4'	20 Ga.	.050	1″ × 1″ × ⅛″	1″
4' TO 6'	18 Ga.	.080	1″ × 1″ × ⅛″	1″
6' TO 8'	18 Ga.	.080	1½″ × 1½″ × ⅛″	1½″
8' TO 10'	18 Ga.	.080	1½″ × 1½″ × ³⁄₁₆″	1½″
OVER 10'	16 Ga.	.080	2″ × 2″ × ³⁄₁₆″	1½″

MAXIMUM PRESSURE RATING 2″ POSITIVE OR 2″ NEGATIVE

FIGURE 15.52. Built-up central station casing. (Courtesy of Sheet Metal & Air Conditioning Contractors' Assoc. Inc.)

FIGURE 15.53. Machine room casing.

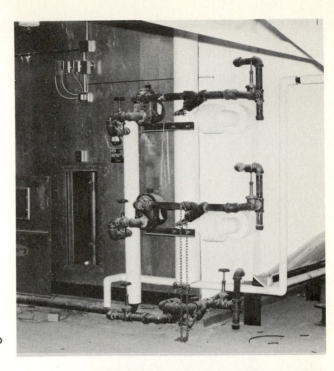

FIGURE 15.54. Piping to coils of casing.

FIGURE 15.55. Typical coils for heating or cooling.

15.15 FILTERS The *filtering section* of an air conditioning system should remove undesirable qualities before the air enters the supply portion of the system. Dust particles, smoke, odors, and fumes present different filtering problems. In most office building installations dust particles are intercepted and tobacco smoke diluted by exhausting some return air and adding fresh air. Dust can be removed by various filtering media, such as fibrous glass, felt, cloth, and cellulose, either as a flat mat or bag (Fig. 15.56). An electrostatic filter attracts the dust from the air to electrically charged plates. One system actually does not filter but depends on neutralizing the electrical charges in airborne pollution. In airport installations the jets produce air ladened with jet-fuel fumes and odor. A charcoal filter is added downstream of the normal filtering units specified to deal with this special problem (Fig. 15.57).

FIGURE 15.56. Metal and plastic filter units inside a casing.

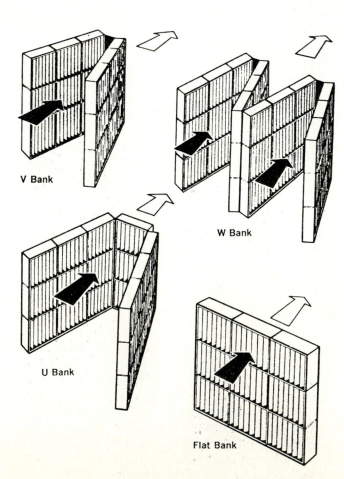

FIGURE 15.57. Four filter configurations. (Courtesy of Barnebey Cheney.)

V Bank

W Bank

U Bank

Flat Bank

15.16 PACKAGE UNITS   The separate design, detailing, purchasing, and construction of a field-assembled air conditioning unit adds to the initial cost of air conditioning. *Package units* offer an economical solution to specific installations such as supermarkets, schools, and industrial buildings. They operate as *single-zone, multizone,* or *dual duct* systems. The package unit comes manufactured complete with coils, fans, filters, and so on (Fig. 15.58). Their size is limited by problems of transportation and placement on the structures. Mammoth cranes and helicopters are used to place them on the structure (Fig. 15.59). They can also be hung from the roof system, particularly in industrial applications. Package units form an integral part of the system's approach to construction of schools. The unit, ductwork, and diffuser are all integrated into a system that maximizes factory manufacture. The school building structure and its HVAC system have been designed and fabricated to fit each other while being manufactured to fixed standards consistent with repetitious mass production.

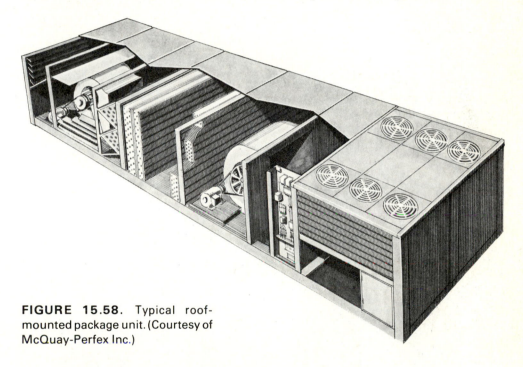

FIGURE 15.58. Typical roof-mounted package unit. (Courtesy of McQuay-Perfex Inc.)

FIGURE 15.59. Rigging a rooftop unit. (Courtesy of McQuay-Perfex Inc.)

16

Solar Energy and Energy Conservation

16.1 INTRODUCTION *Solar energy* is one of those things that is almost too good to be true. The "almost" part—there's the rub. Why is something that we need so badly taking so long to become reality? We understand the scientific principles and a great deal has been accomplished by our research and development organizations. The government provides good financial and moral support for solar research, development programs, and demonstration projects. Many big manufacturers have active solar research and development programs. Over 200 firms in the United States manufacture solar energy hot water equipment. The press constantly publicizes solar energy installations in glowing terms. Why isn't solar energy being eagerly accepted like nylon, cellophane, the zipper, or the pocket calculator? Why has it had so few commercial applications to date?

At the present time our experts have a very good grasp of *solar science*. A large number of government agencies, universities, private companies, and individuals have developed and demonstrated the technical feasibility of numerous ways of utilizing solar energy. Many materials, among them glass, plastic, copper, and aluminum, have been used to manufacture collectors. Rocks, liquid-filled tanks, and ponds have been used as storage media to supply heat during cloudy weather. Systems have been devised using only air or potable water or antifreeze. It seems obvious that we are on sound scientific ground; research and development have made good progress; the technical feasibility of numerous solar applications has been demonstrated; there are manufacturers and contractors involved and very interested in solar progress, and there is a very substantial need that the public, the government, and business wishes to see satisfied.

The "almost" hinges around *cost effectiveness*. Solar hot water heating is one solar application in residential installations that meets the economic test. Depending on:

1. Initial installed cost—the added cost for the solar heating system, including added cost of the structure
2. Interest rates
3. Utility rates or fossil-fuel costs
4. Climate
5. System maintenance costs, if any

savings produced by the sun will pay for the original increase in the price of a home in 7 to 10 years. Unfortunately, solar space heating and solar air conditioning still do not meet this criteria. In commercial buildings there have been very few unsubsidized solar installations. This is due in great part to the necessity to store excess heat in order to carry over building heating during periods of cloudy weather. The nature of this problem can be understood in a recent example of a small all-solar home. A total of 350 tons of rocks at a purchase price of $4 a ton were necessary to create heat storage. To this was added the cost of excavating a space for the rock, enclosing the rock in concrete or masonry block, placing the rock, and related costs. Unfortunately, at this time solar energy is more a good idea than a good solution. Very volatile factors such as economic inflation, interest rates, and energy costs could change this picture and make numerous installations of solar energy a reality. We can also hope for technical breakthroughs to bring costs down, particularly in storage facilities. For the present it is important that we have a good understanding of *residential solar hot water heating*.

16.2 A FUNCTIONAL DESCRIPTION

A solar domestic hot water system functions through the process of the collection and conversion of solar radiation into useful energy. In this process a *collector* absorbs solar energy, *storing* amounts as required, utilizing a *transport* medium as necessary, and *distributing* to the location where it is needed. Automatic or manual devices *control* the performance of each operation. It is deemed prudent to provide an *auxiliary energy system* to supplement the solar output when necessary, and to assume the total energy demand when the solar system cannot function adequately.

With so many competent firms supplying solar system collectors, storage, distribution and transport devices, controls, and backup energy systems, designers have a wide variety of designs, operation, and performance from which to choose. Depending on function, component capability, climatic considerations, desired performance, location, and aesthetics, these components can be arranged in numerous configurations.

Despite the numerous worthwhile ideas that our designers' ingenuity have produced, the relatively simple flat-plate *collector* enjoys the widest application. The flat-plate collector consists of an absorber plate, generally of a metal material, painted black to increase the absorption of the sun's rays. In order to trap the heat properly within the collector and reduce losses through convection, insulation is provided on the underside, while one or more transparent cover plates cover the exposed or sunny side. A *heat transfer medium,* usually air or water, carries the captured heat from the absorber to storage or the point of usage. Liquid is the preferred heat transfer medium for solar hot water systems.

To cover thermal energy demands during the night and other sunless periods, adequate *storage* must be designed and built into the system. A properly designed system stores the energy delivered from the sun and absorbed by the collector which is not being consumed by the users. Transfer of heat from the collector to storage is made (1) by means of a heat exchanger and (2) by direct contact between the heat transfer fluid and the storage medium.

To operate a solar hot water system properly, the *controls* must sense, evaluate, and carry out the required response for the prevailing conditions. When the temperature in the collector rises significantly higher than the storage temperature, the controls cause the stored heat transfer fluid to circulate through the collector and accumulate solar heat.

For dependability, an auxiliary energy system is necessary to overcome periods of severe weather and sunless days. Conventional heating devices and common fuels such as electricity, gas, oil, and wood are used for these auxiliary systems.

Solar systems are generally considered as being active or passive in their operation. In *active solar systems* an energy resource together with the solar energy is used to transfer the thermal energy. Electrically driven pumps responding to the commands of the controls move the heat transfer medium as required. In *passive solar systems* outside energy is not required; heat transfer takes place through convection, radiation, and conduction.

16.3   SOLAR HOT WATER INSTALLATIONS

Either as a separate system or combined with an entire solar heating system, the preheating of domestic water supply has proven to be one of the most effective uses of solar energy. Figures 16.1 and 16.2 are examples of separate systems and Figs. 16.3 and 16.4 show the combined system usage.

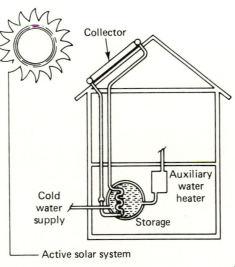

FIGURE 16.1. Separate system: preheating the incoming water supply. [Printed by permission of the National Solar Heating & Cooling Information Center (NSHCIC).]

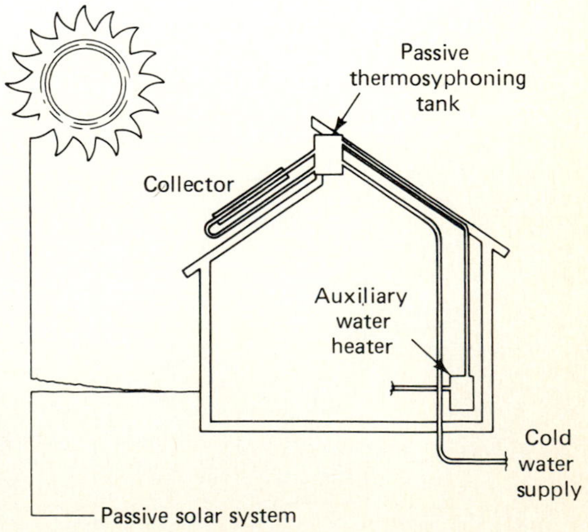

FIGURE 16.2. Separate system: passive preheating of the incoming water supply. [Printed by permission of the National Solar Heating & Cooling Information Center (NSHCIC).]

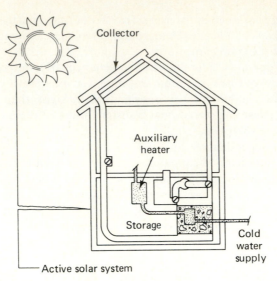

FIGURE 16.3. Combined system: domestic water supply is preheated by the storage area of an active air solar system. [Printed by permission of the National Solar Heating & Cooling Information Center (NSHCIC).]

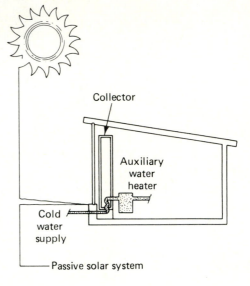

FIGURE 16.4. Combined system: a passive system collector preheats the water supply. [Printed by permission of the National Solar Heating & Cooling Information Center (NSHCIC).]

Solar Collector Orientation and Tilt

For maximum effectiveness the collector should be oriented and tilted to the requirements of the building location. In Fig. 16.5 we can see how tilt relates to the building's latitude. The sun's rays reflecting off an adjacent surface can improve solar performance. In areas where snow accumulates, the collector should be located sufficiently above the reflective surface to ensure proper reflective effect (Fig. 16.6). The collector operates best when facing true south. A deviation of 20° to either side is acceptable (Fig. 16.7).

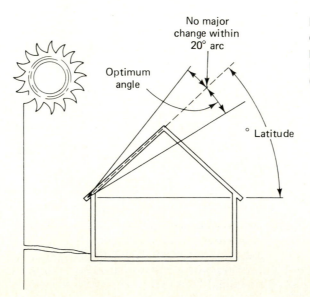

FIGURE 16.5. Collector tilt for domestic hot water. [Printed by permission of the National Solar Heating & Cooling Information Center (NSHCIC).]

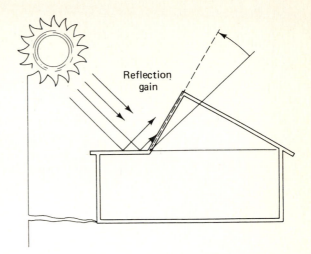

FIGURE 16.6. Solar gain through reflection. [Printed by permission of the National Solar Heating & Cooling Information Center (NSHCIC).]

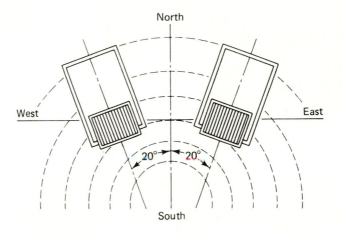

FIGURE 16.7. Orientation recommendation. [Printed by permission of the National Solar Heating & Cooling Information Center (NSHCIC).]

Shading of the Collector*

"Another issue related to both collector orientation and tilt is shading. Solar collectors should be located on the building or site so that unwanted shading of the collectors by adjacent structures, landscaping or building elements does not occur. In addition, considerations for avoiding shading of the collector by other collectors should also be made. Collector shading by elements surrounding the site must also be addressed.

Avoiding all self-shading for a bank of parallel collectors during useful collection hours (9 AM and 3 PM) results in designing for the lowest angle of incidence with large spaces between collectors. It may be desirable, therefore, to allow some self-shading at the end of solar collection hours, in order to increase collector size or to design a closer spacing of collectors, thus increasing solar collection area [Fig. 16.8].

Chimneys, parapets, fire walls, dormers, and other building elements can cast shadows on adjacent roof-mounted solar collectors, as well as on vertical wall collectors. Figure 16.9 shows a house with a 45° south-facing collector at latitude 40° North. By mid-afternoon portions of the collector are shaded by the chimney,

---

*This section is taken from *Solar Hot Water and Your Home,* National Solar Heating & Cooling Information Center, Franklin Research Center, Philadelphia. Reprinted by permission of the National Solar Heating & Cooling Information Center (NSHCIC).

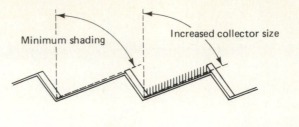

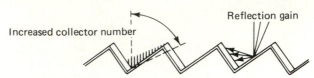

FIGURE 16.8. Self-shading effect. [Printed by permission of the National Solar Heating & Cooling Center (NSHCIC).]

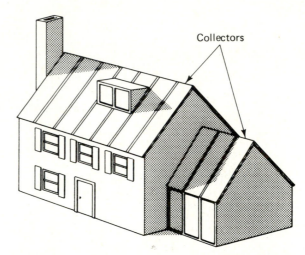

FIGURE 16.9. Shading of a collector by building elements. [Printed by permission of the National Solar Heating & Cooling Information Center (NSHCIC).]

dormer, and the offset between the collector on the garage. Careful attention to the placement of building elements and to floor plan arrangement is required to assure that unwanted collector shading does not occur."

16.4   ACTIVE SYSTEMS*

"Active solar systems are characterized by collectors, thermal storage units, and transfer media, in an assembly which requires additional mechanical energy to convert and transfer the solar energy into thermal energy. The following discussion of active solar systems serves as an introduction to a range of active concepts which have been constructed.

Domestic hot water can be preheated either by circulating the potable water supply itself through the collector, or by passing the supply line through storage enroute to a conventional water heater. Three storage-related preheat systems are shown [in Figs. 16.10 to 16.12]."

---

*Section 16.4 is taken from *Solar Hot Water and Your Home,* National Solar Heating & Cooling Information Center, Franklin Research Center, Philadelphia. Reprinted by permission of the National Solar Heating & Cooling Information Center (NSHCIC).

FIGURE 16.10. Preheat coil in storage. Water is passed through a suitably sized coil placed in storage en route to the conventional water heater. The preheat coil should be of double-wall construction if storage media is toxic. [Printed by permission of the National Solar Heating & Cooling Information Center (NSHCIC).]

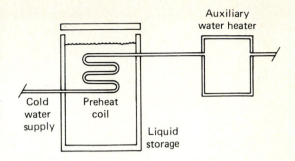

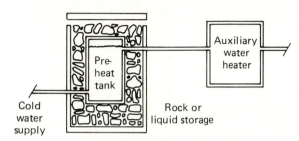

FIGURE 16.11. Preheat tank in storage. Heat from the rock or liquid storage preheats the water supply. Double-wall construction is required if the storage liquid is toxic. [Printed by permission of the National Solar Heating & Cooling Information Center (NSHCIC).]

FIGURE 16.12. Preheat outside of storage. In this preheat method, the heat transfer liquid in storage is pumped through a separate heat exchanger to perform the preheating. Double-wall construction is required if the storage liquid is toxic. [Printed by permission of the National Solar Heating & Cooling Information Center (NSHCIC).]

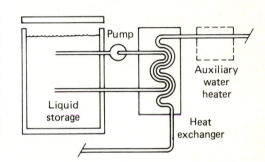

16.5   ENERGY CONSERVATION: THE RUNAROUND SYSTEM

Energy conservation in HVAC systems starts with the recovery of heat from exhaust air. Makeup air (fresh air) must be heated in cold weather while warm spent air is being exhausted. As you can see in Fig. 16.13, the *runaround system* utilizes a *loop pipe* system connected to coils in both the exhaust air ductwork and the incoming air

FIGURE 16.13. Runaround system. (Courtesy of Edison Electric Institute.)

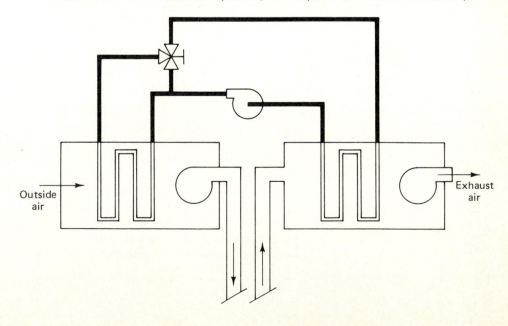

section. The pump circulates an ethylene glycol antifreeze solution through the exhaust section coil, where it recovers heat from the heated exhaust air and gives it up at the other coil to heat the incoming fresh air. Frequently, incoming and exhaust ducts are not located near each other, which is one of the attractive features of the runaround system. The efficiency of this recovery scheme relates to the capacity of the coils to transfer heat. However, fan capacity must overcome the resistance to airflow set up by the coils. If added fan capacity is needed to overcome coil resistance, some of the energy savings will not be realized.

16.6   HEAT WHEELS   Where makeup air and exhaust air ducts pass near each other, the *heat wheel* efficiently recovers exhaust air heat. Air from each duct passes separately through the motor-driven heat wheel, which slowly revolves (Fig. 16.14). The heat wheel is a metal frame cylinder filled with good heat-absorbing fireproof materials such as stainless steel or aluminum mesh or corrugated asbestos. It is designed to avoid significant contamination between the two duct systems.

A heat wheel containing *lithium chloride–impregnated asbestos* will save energy during the summer. Cool dry exhaust air cools and dehumidifies the warm makeup air. The lithium chloride desiccant absorbs moisture as well as heat.

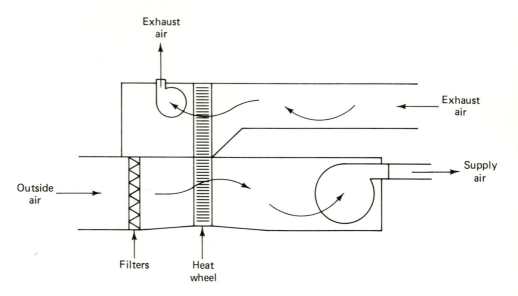

FIGURE 16.14.   Heat wheel. (Courtesy of Edison Electric Institute.)

16.7   RECOVERING LIGHTING HEAT   In the interior of a building *lighting* produces the greatest amount of heat gain. To effect energy savings luminaires are selected containing exhaust slots which pick up a very high percentage of the dissipated heat, as air from the space below passes into the ceiling plenum above. This heated air becomes available for both summer and winter air conditioning operations. In Fig. 16.15 the cold air supply induces the ceiling plenum-heated air to enter the induction box. A thermostat controls the amount of heated plenum air permitted to enter the induction box. Up to 50% of the warm air can be induced in this manner.

One other advantage of this system during cooling periods is the reduction of cooling loads by exhausting this plenum air. Cooler luminaires also produce up to 13% more light output for the same amount of electricity consumed.

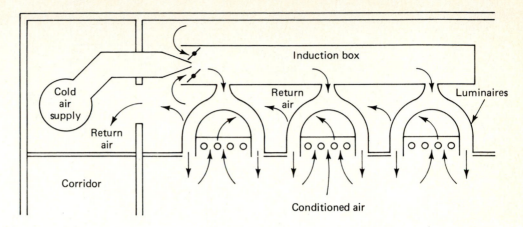

FIGURE 16.15. Recovery of heated air from the space above the ceiling surface. (Courtesy of Edison Electric Institute.)

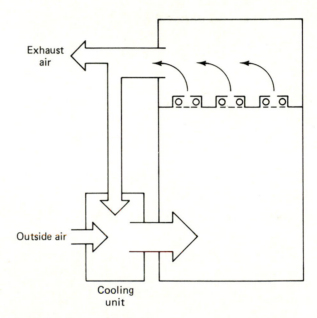

FIGURE 16.16. Total return system. (Courtesy of Edison Electric Institute.)

16.8   TYPES OF RETURN SYSTEMS

Total Return System

As the name implies, in the *total return system* air supplied through ceiling diffusers is all returned through the luminaires (Fig. 16.16). Depending on outdoor temperature and humidity, amounts of air recycled or exhausted vary. In addition to utilizing the return air heat, this system increases the light output from the fluorescent fixtures to their maximum with only a small increase in cooling tonnage.

Bleed-Off System

In the *bleed-off system* (Fig. 16.17), only part of the air to be returned passes through the luminaires; the balance returns through conventional registers and return

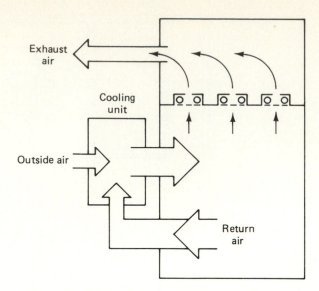

FIGURE 16.17. Bleed-off system. (Courtesy of Edison Electric Institute.)

ductwork. The air passing the luminaires does not reenter the supply system but exhausts to the outside atmosphere. Since it carries off so much luminaire heat, there is perhaps the highest potential for savings in cooling capacity of all air-handling designs. Although not as much cooling of the luminaire occurs, the lighting efficiency again improves with this method.

16.9 HEAT PUMPS   "The hotter it gets, the hotter it gets." This statement is not as inane as it might seem. Every refrigeration device produces heat as it creates cooling. Un-air-conditioned areas such as train platforms become hotter as hot air blasts off air conditioning units cooling the passenger cars. The refrigeration cycle involves heat transfer and makes the *heat pump* an unusual device. Unusual because:

1. It can either heat or cool.
2. It can switch from one to the other automatically.
3. It has the lowest operating cost of any other electric heating/cooling equipment.
4. It can furnish more energy in the form of heat than the electric power put into it because it removes heat from an outside source such as fresh air.

The heat pump is a refrigeration machine and as such has a compressor, condenser, and evaporator. In the reversible cycle type, valves operate to reverse the direction of the refrigeration cycle when changing from one mode to another. As you can see in Figs. 16.18 and 16.19, the condenser becomes the evaporator and the evaporator becomes the condenser. In the summer it takes warm air and produces cool air, and in the winter it takes cold air and produces heat. The heat pump is particularly well suited for residential installations in moderate climates but also has widespread commercial application.

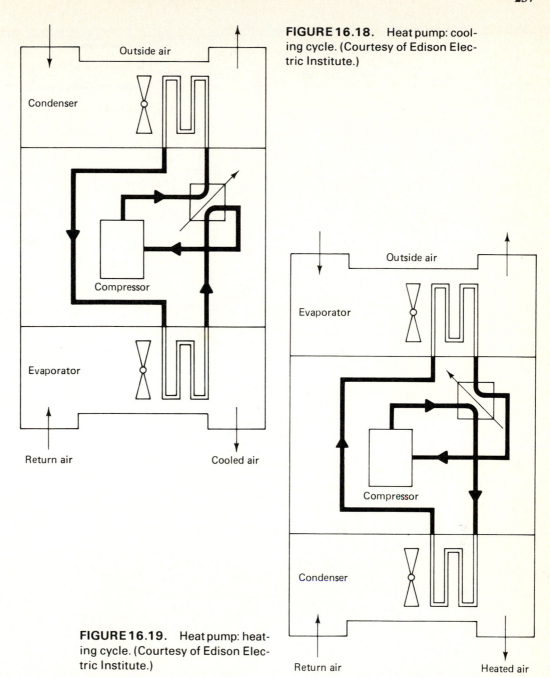

FIGURE 16.18.   Heat pump: cooling cycle. (Courtesy of Edison Electric Institute.)

FIGURE 16.19.   Heat pump: heating cycle. (Courtesy of Edison Electric Institute.)

16.10   THE DOUBLE-BUNDLE CONDENSER

In many buildings at certain times of the year both cooling and heating are required. The chillers provide a good source of heat from the condenser function, which designers utilize rather than exhausting it through the cooling tower. A chiller equipped with two sets of condenser coils, one for the cooling tower system and one for a waste heat utilization system, is referred to as a chiller with a *double-bundle condenser.* Thermostatically regulating the flow of water through each set of coils vitalizes the condenser heat for maximum efficiency. Figure 16.20 shows how the double-bundle system works. The storage tank accumulates the excess heat generated during the day and releases it for heating purposes during the night.

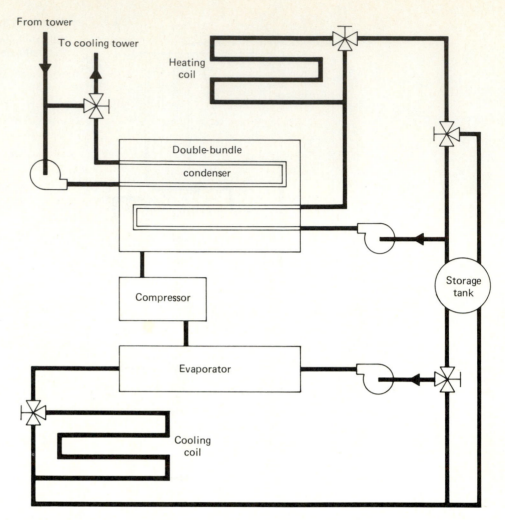

FIGURE 16.20.   Double-bundle condenser. (Courtesy of Edison Electric Institute.)

16.11 CONCLUSION   The examples discussed represent practical applications of energy conservation as they relate to mechanical systems. Many other energy-saving ideas have surfaced in recent years as engineers, owners, architects, contractors, and manufacturers have redirected their thinking along conservation lines. *Energy conservation* has encouraged manufacturers to design chillers and other air conditioning devices on the basis of *consumption* rather than energy-wasteful initial *low cost*.

Energy conservation in buildings is of such magnitude and national importance that the entire design and construction industry has become acutely aware of their heavy energy responsibility. ASHRAE, the American Society of Heating, Refrigeration and Air Conditioning Engineers, quickly faced up to the situation of the seventies and by 1975 developed Standard 90-75, which establishes building component performance standards. This standard sets limits on performance to be met while permitting architects and mechanical and electrical engineers the latitude to select materials, systems, and designs of their choice. "Forty-four states now have energy efficiency codes, most of which are based on the ASHRAE Standard."* The government has established 65° F as the standard for heating and 78° F as the standard for cooling. The Department of Energy has proposed a national *building energy*

* *Engineering News-Record,* June 12, 1980.

performance standard (BEPS), which still has not been passed into law by the Congress, but may be adopted before long. Persons involved with mechanical and electrical installations should keep abreast of developments with the ASHRAE standard and BEPS. The current situation is much too volatile to be stated in a textbook as currently factual.

Our principal interest in this text deals with *constructing* mechanical and electrical systems, but awareness of *operational savings* of HVAC, lighting, and plumbing systems should be one of our continuing concerns. Although much has or can be accomplished in retrofitting existing systems to save energy, we have a tremendous opportunity to save energy by altering the operational pattern of the overwhelming number of buildings already in existence. Many governmental and private organizations now conduct energy audits in their buildings and from these data, which pinpoint where the energy goes, adjust the operation of plumbing, HVAC, electrical, and elevatoring systems. We can only look forward to a greater intensification of our interest in energy consumption since we have a common interest in all aspects of this serious situation.

16.12   GLOSSARY*

"Active solar system.   An assembly of collectors, thermal storage device(s), and transfer fluid which converts solar energy into thermal energy, and in which energy in addition to solar is used to accomplish the transfer of thermal energy.

Auxiliary energy subsystem.   Equipment utilizing energy other than solar both to supplement the output provided by the solar energy system, and to provide full energy backup during periods when the solar domestic hot water systems are not operating.

Collector, flat-plate.   The type of solar collector generally used in solar hot water systems [Fig. 16.21].

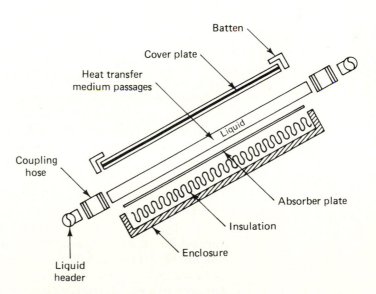

FIGURE 16.21.   Parts of a flat-plate collector. [Printed by permission of the National Solar Heating & Cooling Information Center (NSHCIC).]

*This section is taken from *Solar Hot Water and Your Home,* National Solar Heating & Cooling Information Center, Franklin Research Center, Philadelphia. Reprinted by permission of the National Solar Heating & Cooling Information Center (NSHCIC).

Collector subsystem.   The assembly used for absorbing solar radiation, converting it into useful thermal energy, and transferring the thermal energy to a heat transfer fluid.

Component.   An individually distinguishable product that forms part of a more complex product (such as, subsystem or system).

Contaminants (hazardous).   Materials (solids, liquids or gases) which when added unintentionally (or intentionally) to the potable water supply cause it to be unfit for human or animal consumption.

Control subsystem.   The assembly of devices used to regulate the processes of collecting, transporting, storing and utilizing solar energy.

Design life.   The period of time during which a domestic hot water system is expected to perform its intended function without requiring major maintenance or replacement.

Heat exchanger.   A device for transferring thermal energy from one fluid to another.

Heat transfer fluid.   A liquid which is used to transport thermal energy.

Outgassing.   The emission of gases by materials and components usually during exposure to elevated temperatures or reduced pressure.

Passive solar system.   An assembly of natural and architectural components including collectors, thermal storage device(s) and transfer fluid which converts solar energy into thermal energy in a controlled manner and in which no pumps are used to accomplish the transfer of thermal energy.

Potable water.   Water free from impurities present in amounts sufficient to cause disease or harmful effects to health.

Solar building.   A building which utilizes solar energy by means of an active or passive solar system.

Solar degradation.   The process by which exposure to sunlight deteriorates the properties of materials and components.

Solar domestic hot water system.   The complete assembly of subsystems and components necessary to convert solar energy into thermal energy for domestic hot water in combination with auxiliary energy when required.

Storage subsystem.   The assembly used for storing thermal energy so that it can be used when required.

Subsystem.   A major, separable, functional assembly of a system such as a complete collector or storage assembly.

Thermal energy.   Heat possessed by a material resulting from the motion of molecules which can do work.

Toxic fluids.   Gases or liquids which are poisonous, irritating and/or suffocating."

17

The Electrical Industry

17.1 INTRODUCTION   *Electrical building construction* is an important part of a very large complex industry that has an enormous effect on commerce, transportation, manufacturing, public safety, and on our quality of life. We are greatly affected by the electricity available to us, how it is generated, how much of it we demand, and how much it costs. Low-cost hydroelectric power has a dramatic impact on how we heat our buildings, what products we manufacture, and where we locate our industries. The survival of agricultural projects involving pumped irrigation water, on the other hand, can pivot on the cost of electricity.

The significance of electricity in our lives has twice been emphatically demonstrated to those of us who experienced the two blackouts that hit the New York City and East Coast areas. When all the electric trains and subways stop, when all traffic lights fail to operate, when all ventilation and air conditioning ceases, when all elevators stop running, and almost all lighting goes out, a very serious, vulnerable condition prevails. Even other systems, such as heating and plumbing, fail without electricity, because we do not have electricity for control or the power to pump needed water.

Along with private business there has been for many years a large federal government involvement in electrical generation and transmission: for example, the Tennessee Valley Authority and the Grand Coulee Dam. Big utilities grant huge contracts for long-term fuel supplies and for the construction of mammoth coal, oil, and nuclear generating plants. Electrification of railroads and new mass transit

241

systems improve our transportation systems while increasing our demand for electric power. Large manufacturers dominate the production and sale of electrical items. It is of interest to note, however, that despite its tremendous size, the electrical industry is intimate enough to affect amost every aspect of our lives.

The *electrical construction firms* that serve the building construction industry are by and large the same firms that contract for the electrical work in generating plants, that install *mass transit electrification,* and that construct and maintain *municipal lighting* and *traffic signal control systems.* Although they may have preferences, the nature of electrical contracting is such that the contractors diversify when possible by doing commercial, industrial, or institutional building construction contracts, transmission lines and power plants, or mass transit work.

In the United States one contracting firm occupies a dominant position, doing in one year twice as much work as its nearest competitor. Seven other firms had revenues of over $100 million for 1980. Thirty-seven firms did more than $20,000,000 of construction in a year. Fifty-five firms had completed work in the $10,000,000 to $20,000,000 range.*

To support their common interests the electrical contractors formed and support the *National Electrical Contractors Association* (NECA). In some localities the industry and the electrical unions work together for their common good, setting standards, encouraging education of union members, and promoting the electrical industry. The Joint Industry Board of the Electric Industry in New York City is a good example of this type of labor–management cooperation.

17.2   THE NATURE OF THE WORK

In building construction projects the electrical work generally includes lighting, power, transforming, metering, switching, overload control, grounding, motor control, signal and alarm systems, and perhaps electrical heating. The differences in office building, industrial, and hospital projects would center principally around sizes of power circuits, types of lighting, and particularly with hospitals, degree of sophistication. An office building can contain a complex computer area, an industrial building an electric furnace—both quite different in character from the rest of the project.

Electrical work comprises a surprisingly consistent percentage of the total cost of certain types of buildings; for example:

	Median %		*Median %*
Banks	10.3	Offices—midrise	9.0
College classrooms	10.0	Elementary schools	10.2
Factories	10.8	Junior high schools	9.7
Housing for the elderly	9.5	Senior high schools	10.0
Motels	9.5	Theaters	10.7

Source: This information is copyrighted by Robert Snow Means Co., Inc., and is reproduced from the 1982 edition of *Building Construction Cost Data,* pp. 359–367, with permission.

On the low side are apartment buildings (one to seven stories), 6.8%; and warehouses, 6.7%. On the high side are department stores, 11.0%; hospitals, 14.0%; and supermarkets, 11.8%.

* *Engineering News-Record,* August 27, 1981.

17.3   LICENSING*

"The *National Board of Fire Underwriters* has considerable interest in the licensing of electrical contractors and of journeymen electricians. The *National Electric Code* is the standard set of rules on which most state and municipal electrical codes are based. It is under these rules that electrical licenses are granted. A state or local board of examiners usually conducts examinations for an electrician's license. This board may consist of an architect or engineer, a master or employing electrician, a journeyman electrician, an employee of the public authority having jurisdiction, an employee of the electrical underwriters, and a real estate man representing the public interest. An applicant for a license must have a number of years of experience as a journeyman electrician working with the tools of his trade. There are rules for college or trade school equivalents for part of the practical experience. The applicant must usually pass a written examination and have his background and integrity investigated. An *employer* of electricians must have a *master electrician's license* and must give proof of experience and financial background. The rules are strictly enforced and rightly so. Poorly installed electrical work can cause fires and severe injury or death. The persons who install electrical work must know their business."

17.4   ELECTRICAL CONTRACTING OPERATIONS

Subletting. The industry typifies most specialty contracting. The contractor's personnel do practically all of the electrical work on a particular project. Generally speaking, electrical contractors do not sublet portions of their work.

Purchasing. The industry benefits from good competitive bidding on equipment, materials, and fixtures. The delivery time for certain items often dictates very early ordering of critical items. Special attention should be paid to the market situation for switchgear and certain cable sizes.

Electrical contractors use an unusual number of different small items, such as clamps, fasteners, screws, and so on. A complete and well-stocked inventory makes the work progress efficiently.

Shop Drawings. Electrical shop drawings give many dimensions not shown on design drawings. Much of electrical design, particularly conduit runs, is shown schematically. In schools, offices, homes, and so on, electrical work is buried in walls and floors or hidden in hung-ceiling areas. Since fixture locations are fixed by reflected ceiling design drawings, the electrical and HVAC engineers must detail and *coordinate* their shop drawings to avoid conflicts or undesirable problem solutions. Heavy concentrations of large conduits in hung ceilings can be particularly troublesome. Both ductwork and electric conduit are often made to take circuitous routes simply because they do not have to pitch as do plumbing and heating lines.

Cuts. The industry makes widespread use of cuts as an efficient way of designating items such as disconnect switches, fixtures, and motors (Fig. 17.1).

Sequencing and Scheduling. In scheduling his work the electrical contractor considers his contract in two ways: (1) the work he does in teamwork with others, and (2) his own work, such as installing switchgear, where he is less dependent on others, and vice versa. In multistory structural concrete work the electrician must conform to the general contractor's slab forming and pouring schedule. The electrician is given a definite number of hours in each forming and pouring segment in which to install conduit. He must have the required number of workers when needed. In multistory structural steel work for office buildings, he again must conform to the concrete

---

*Section 17.3 is taken from Laurence E. Reiner, *Handbook of Construction Management,* © 1972, p. 183. Reprinted by permission of Prentice-Hall, Inc., Englewood Cliffs, N.J.

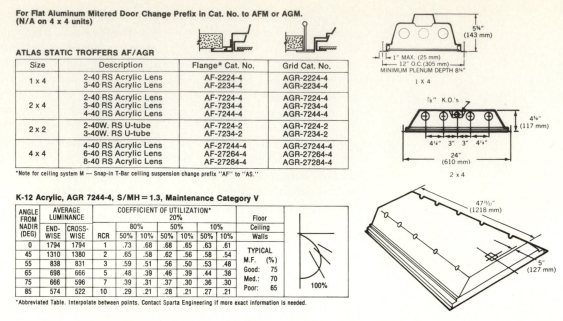

For Flat Aluminum Mitered Door Change Prefix in Cat. No. to AFM or AGM. (N/A on 4 x 4 units)

ATLAS STATIC TROFFERS AF/AGR

Size	Description	Flange* Cat. No.	Grid Cat. No.
1 x 4	2-40 RS Acrylic Lens	AF-2224-4	AGR-2234-4
	3-40 RS Acrylic Lens	AF-2234-4	AGR-2234-4
2 x 4	2-40 RS Acrylic Lens	AF-7224-4	AGR-7224-4
	3-40 RS Acrylic Lens	AF-7234-4	AGR-7234-4
	4-40 RS Acrylic Lens	AF-7244-4	AGR-7244-4
2 x 2	2-40W. RS U-tube	AF-7224-2	AGR-7224-2
	3-40W. RS U-tube	AF-7234-2	AGR-7234-2
4 x 4	4-40 RS Acrylic Lens	AF-27244-4	AGR-27244-4
	6-40 RS Acrylic Lens	AF-27264-4	AGR-27264-4
	8-40 RS Acrylic Lens	AF-27284-4	AGR-27284-4

*Note for ceiling system M — Snap-in T-Bar ceiling suspension change prefix "AF" to "AS."

K-12 Acrylic, AGR 7244-4, S/MH = 1.3, Maintenance Category V

ANGLE FROM NADIR (DEG)	AVERAGE LUMINANCE END-WISE	CROSS-WISE	RCR	COEFFICIENT OF UTILIZATION* 20% 80% 50%	10%	50% 50%	10%	10% 50%	10%
0	1794	1794	1	.73	.68	.68	.65	.63	.61
45	1310	1380	2	.65	.58	.62	.56	.58	.54
55	838	831	3	.59	.51	.56	.50	.53	.48
65	698	666	5	.48	.39	.46	.39	.44	.38
75	666	596	7	.39	.31	.37	.30	.36	.30
85	574	522	10	.29	.21	.28	.21	.27	.21

Floor: Ceiling / Walls

TYPICAL M.F. (%)
Good: 75
Med.: 70
Poor: 65

100%

*Abbreviated Table. Interpolate between points. Contact Sparta Engineering if more exact information is needed.

FIGURE 17.1. Electric fixture "cut." (Courtesy of Thomas Industries Inc.)

schedule, installing floor distribution boxes and header ducts behind the metal deck subcontractor and ahead of concreting operations.

In masonry and drywall work he again works right along with the bricklayers or carpenters, running conduit as the work of others proceeds. In office ceiling work he becomes part of a team of sheet metal duct men and acoustical tile installers, each doing their part of the work in turn.

The electrician should schedule his independent work to blend with the general contractor's structural and architectural progress. He should concentrate effort on the switchgear room and horizontal and vertical runs of large conduits. By following closely behind the structural work, he can usually work with little interference and greater efficiency.

Monitoring Progress. A construction manager or general contractor should monitor electrical work carefully. Because the electrician must work with other trades, he often maintains architectural and structural progress at the expense of other aspects of his work. He can satisfy the other trades by installing conduit and panel box enclosures without pulling wire or installing panel box "guts" and connections. He can fall far behind in his overall contract before others complain of his performance. On one major sports arena project, the electrician had 50 men work five months *after the arena was opened to the public.*

A key date in a major project is the day permanent power is made available. The *owner* works with the *utility company* to bring high-voltage current to the site. That power becomes available to the project when the electrician completes the switchgear room and major distribution circuits. The construction manager or general contractor should schedule this date and monitor the progress of the utility and electrical contractor to see that it is met.

17.5 TEMPORARY POWER AND LIGHT

Construction projects operate on electric power. Even a home builder with one residence to build has electric power brought to the site. The construction manager on the large projects schedules the installation of the temporary light and power system for use before excavation and pumping have been completed (Figs. 17.2 and 17.3). The

FIGURE 17.2. Incoming temporary power source. The utility company and general and electrical contractors plan this installation.

FIGURE 17.3. Temporary power transformer for a hospital project.

electric load he establishes is based on the use of saws, drills, bending machines, pipe threaders, and other small tools. In addition, he provides for lighting all areas of the project plus nighttime safety lighting at stairways and exits. The big loads come from temporary elevators, construction hoists, electric stud welding, regular welding machines, and terrazzo floor grinders. The service should be sized big enough to ensure against dimming lights and underpowered electric motors which curtail work operations.

Temporary light and power should be specified by the architects and engineers or the construction manager/general contractor so that it can be bid *competitively* by the electricians and also so that all the other contractors will understand what temporary electrical services will be available for their use. For lighting and small tools, 120-volt/208 three-phase or 120-volt/240 single-phase should be provided. For the larger loads mentioned, 480-volt service is required (Figs. 17.4 to 17.6).

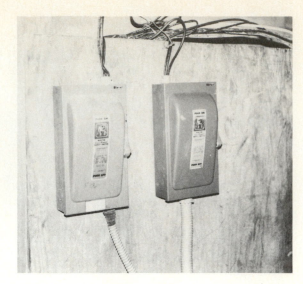

FIGURE 17.4. Disconnect switches are strategically located throughout the project.

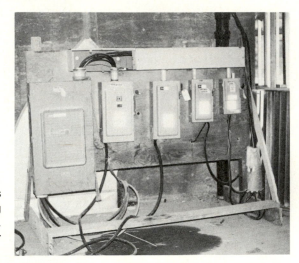

FIGURE 17.5. Power from this installation goes to the climbing crane, hoist, temporary elevators, and small tools. (Courtesy of Turner Construction Company.)

FIGURE 17.6. Temporary power specifications should describe requirements for the personnel–material hoist.

18

Electrical Distribution

18.1 INTRODUCTION The electrical engineer, working with the architect, the utility company, and the owner, designs an electrical distribution system to satisfy the client's requirements. Since client requirements vary greatly, the same type of building, with similar external and internal appearance, may be radically different electrically. One building can depend on one incoming electrical service, whereas a similar building whose owner requires greater electrical dependability will have two incoming service lines, either one capable of satisfying the building's needs should the other fail. Most office building owners insist on an underfloor electric and telephone duct system which provides the capability of bringing power and telephone connections to any desirable desk location with a minimum of time and inconvenience. A few building owners reduce the *initial cost* of a building by eliminating an underfloor distribution system and the versatility and convenience that go with it. They install a system for the original user only. Any change of use or change of tenant requires a completely new or drastically revised installation (Fig. 18.1). This may involve removing the ceiling *below* the floors undergoing change. In all cases the owner and his or her designer decide the degree of dependability, convenience, and provision for future development they wish included in their project within the requirements of the electric code.

The *incoming electric service,* the *distribution systems,* and the *materials, equipment,* and *methods* of distribution are keys to electric construction progress. The construction supervisor or manager will benefit from a knowledge of *electrical distribution* by improving his ability to judge electrical progress and monitor the electricians' performance.

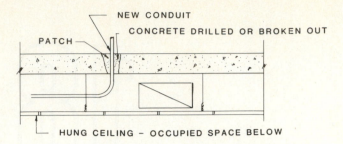

FIGURE 18.1. Tenant changes.

18.2 INCOMING SERVICE

The *utility company* has the responsibility to evaluate the current and future requirements of all its present and potential customers and be prepared to generate and distribute electrical power to satisfy their needs. If an area enjoys a construction boom, the utility must intensify its effort to meet the demand being created. For individual residences and small commercial establishments, connecting into the existing system presents no real problem, but bigger undertakings require evaluation, planning, scheduling, and must involve the owner, designers, and the utility. In bringing service to a customer the utility company usually (1) drops the voltage with a transformer, (2) installs switching which they can control, and (3) installs overhead or underground cables into the building to a service switch which the customer can control.

The building owner pays the cost incurred by the utility in making the service installation. The simplest and cheapest utility system consists of poles and cables (Fig. 18.2), but owners and architects prefer underground installations for aesthetic reasons. Where space permits, the utility's transformers are placed outside the building aboveground in an inconspicuous location. The cables are installed in underground conduit from the utility source to the building (Fig. 18.3).

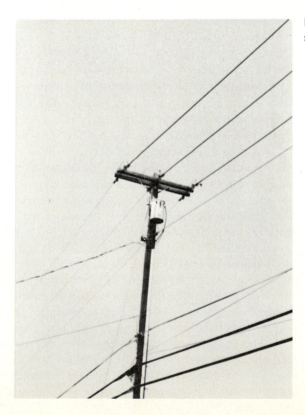

FIGURE 18.2. Overhead utility service.

FIGURE 18.3. Residential service.

In any construction project of appreciable size, whether industrial, commercial, or institutional, the *construction manager* plans and schedules for the availability of permanent power. This construction milestone rates in importance with completing the structural frame or enclosing the building. On open sites, meeting the required date rests largely in the hands of the utility people and the electrical contractor. The transformers are usually remote from the building on a concrete pad with underground conduit run from the utility source into the building. There is virtually no interference with other construction.

In congested city areas concrete vaults are constructed, usually between the curb and property line. Since construction space around an inner-city building is so vital to construction progress and since the vault is often built by the utility, this construction program should be *expedited* and *monitored* by the *building construction manager*. A concerted effort will minimize interference with building construction operations and ensure meeting the projected scheduled date for permanent service. Close liaison with the owner and the utility company will help ensure success.

18.3 THE SWITCHGEAR ROOM

The *switchgear room* performs many important electrical functions, including:

1. Service disconnection
2. Transforming
3. Metering
4. System protection—circuit breaking and fusing
5. Distribution of power to meet requirements of power, lighting, machinery, elevatoring, and so on
6. Emergency usage

Because of its importance to the construction program, the construction manager should build the *switchgear room* early in the project, making sure that it is dry, properly ventilated, warm, and equipped with the necessary hardware for security and emergency exiting. To prevent employees from being trapped by an electrical fire or mishap, these rooms should have two exits and the switchgear should be mounted on 4-in.-high concrete pads to keep them above water level should the adjacent floor area flood. Figure 18.4 shows a typical switchgear installation. From left to right are the

FIGURE 18.4. Switchgear.

main disconnect switch (Fig. 18.5), the well-ventilated transformer section (Fig. 18.6), the meter section over the main circuit breaker, and on the far right eight switches for various services, such as lighting and general power (Fig. 18.7). Depending on the size of the building and the degree of dependability required, the switchgear room could contain one or more of these installations served by at least two separate utility sources. A prudent owner will also insist on an *emergency* power source, switchgear, and distribution system.

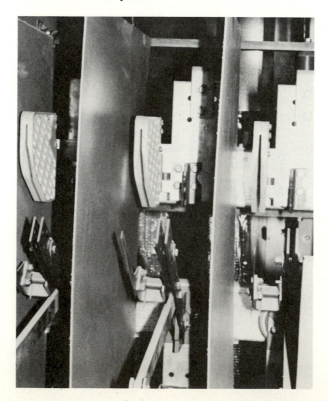

FIGURE 18.5. Main contacts of a medium voltage (5000 to 35,000 volts) air break switch.

FIGURE 18.6.   Dry-type transformer of a switchgear unit.

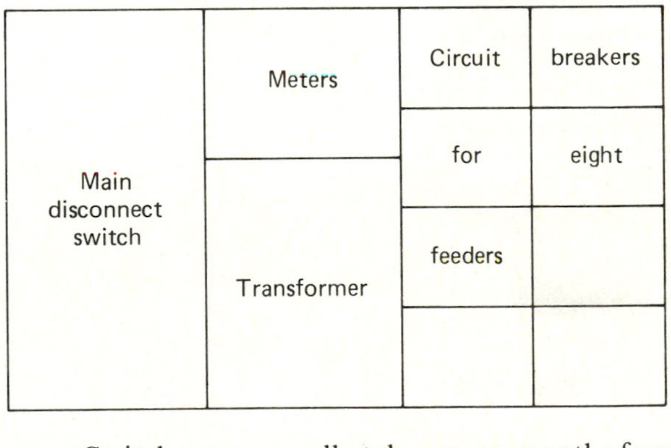

Main disconnect switch	Meters	Circuit	breakers
		for	eight
	Transformer	feeders	

FIGURE 18.7.   Switchgear diagram.

Switchgear generally takes many months for shop drawings, manufacture, and delivery. The construction manager must be alert to see that this vital equipment is ordered in time to meet project schedules. The electrical engineer may have to size and specify the switchgear long before specific design and electrical loads can be pinpointed. Careful analysis and good judgment go into sizing the switchgear in a fast-track project where the design drawings and specifications may be incomplete when purchase of switchgear must be made. When the switchgear arrives on site, it should be protected from water, dust, excessive humidity, and physical damage. Burning a pair of 100-watt bulbs inside the cabinets will greatly aid in keeping the equipment dry.

When the utility company completes its service installation and with the switchgear operational, electrical loads from the temporary service can be hooked into the permanent service, providing greater capacity (Fig. 18.8). The removal of temporary transformers, switchboards, and wiring usually expedites other construction activities.

FIGURE 18.8.   Main lighting and HVAC distribution center in use during construction.

18.4 DISTRIBUTION SYSTEMS: CONSIDERATIONS

The *distribution system* for a building dictates the design of the switchgear equipment. Industrial and office buildings will have versatile distribution systems which provide for relocating machinery or desks. Other buildings, such as schools and apartment houses, have set designs for an unchanging usage. A building is supplied from the switchgear room by *feeders*—the *feeder* can be wire or cable in a conduit, cable laying in tray (Fig. 18.9), or a busway (Fig. 18.10).

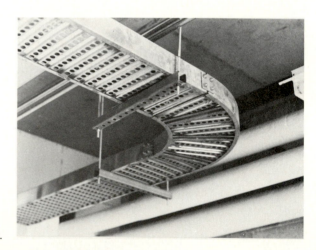

FIGURE 18.9.   Cable rack.

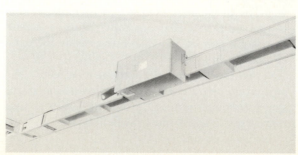

FIGURE 18.10.   Bus duct. (Courtesy of Gould Inc. Distribution & Controls Div.)

Figure 18.11 shows a four-story office building with lighting, power, motor starter, elevator, and air conditioning equipment. Each feeder has its own circuit breaker (Fig. 18.12) to protect itself and the building, plus a manual switch. The phase and voltage carried varies according to the feeder's function. A feeder to a small elevator, for example, might be three-phase 208 volts, which would be supplied by three No. 4 size wires in a 1½-in. conduit.

Hospitals have many special requirements of both a mechanical and an electrical nature. In the electrical distribution system, the sources for incoming service are duplicated to provide continual electrical service should one of the sources fail in some way. This standard of design starting with the incoming service extends to the operating rooms. Each operating room has its own set of isolation transformers (Fig. 18.13), which keep it electrically from affecting other operating rooms and in turn protects it from interference from external mishaps.

Where electrical distribution feeders will be enclosed, the electrician should install the conduit and/or busways right behind the structural work. This is particularly important for vertical risers in high-rise construction, where enclosing architectural work cannot proceed floor by floor until all electrical, plumbing, and HVAC risers are installed. This should present no problem to the electrician provided that he has the proper manpower and coordinates his operations with the other trades.

FIGURE 18.11.   Electrical distribution system.

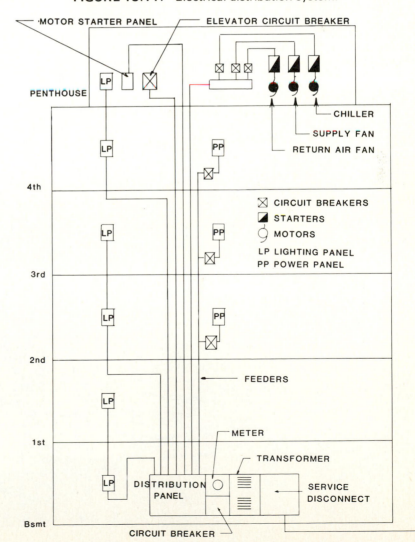

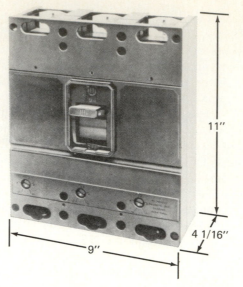

11″

4 1/16″

9″

FIGURE 18.12. Circuit breaker. (Courtesy of Gould I-T-E.)

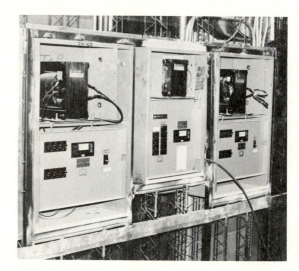

FIGURE 18.13. Isolation transformers for a hospital operating room.

An important practical consideration in distribution systems are the provisions made for pulling wire and cable through the conduit lines. In exterior installations *concrete manholes* and *vaults* simplify the pulling operation. In interior work extensive use is made of *junction boxes* (Fig. 18.14) and *panel boxes* (Figs. 18.15 and 18.16). Switchgear *compartments* allow room for this operation in the switchgear installation.

FIGURE 18.14. Very large junction box.

FIGURE 18.15. Panel box ready for wall construction.

FIGURE 18.16. Panel box ready for connectings.

18.5 UNDERFLOOR DISTRIBUTION SYSTEMS

A medium- to high-quality office building generally utilizes an *underfloor duct system* for low-voltage power and telephone distribution. Since the rentable area of a building increases with a structural steel frame, most high-rise buildings are designed in steel. The floor systems combine steel beams, cellular metal deck, shear studs, and concrete into a structural system that behaves as a *composite unit* (Figs. 18.17 to 18.19). The *composite design* reduces the amount of structural steel required by as much as 20%

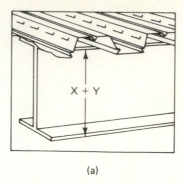

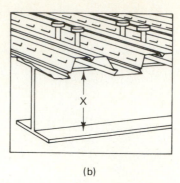

(a)   (b)

FIGURE 18.17.   (a) Noncomposite beam design. (b) Composite beam design. Through cost analysis and field experience composite beam design has demonstrated reduction in steel costs of 15 to 20%. This design economy is achieved by making the floor slab work in conjunction with the steel beam. This permits shallower beams, lighter beam sections, or both. Total building height can be reduced. (Courtesy of H.H. Robertson Co.)

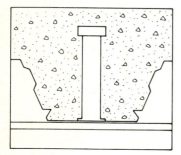

FIGURE 18.18.   Section through the shear stud and beam. The composite floor slab acts as a cover plate for the steel beam. The key element in composite design is the shear stud, which assures the proper structural bond between the steel beam and the floor slab. (Courtesy of H.H. Robertson Co.)

FIGURE 18.19.   Shear stud connectors field welded to beams below.

and makes the metal deck available for underfloor distribution. The *metal deck* blends active cells with pure structural decking (Fig. 18.20). The system is fed from *electrical and telephone risers* located in closets through a *trench header duct* divided into electric and telephone sections and run perpendicular to the metal deck cells (Figs. 18.21 to 18.23). The metal deck is prepunched for easy installation of the outlet units,

FIGURE 18.20.   Cross section of a metal deck. (Courtesy of H.H. Robertson Co.)

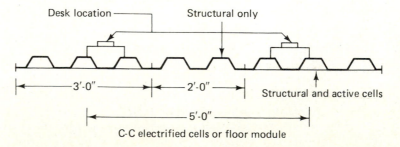

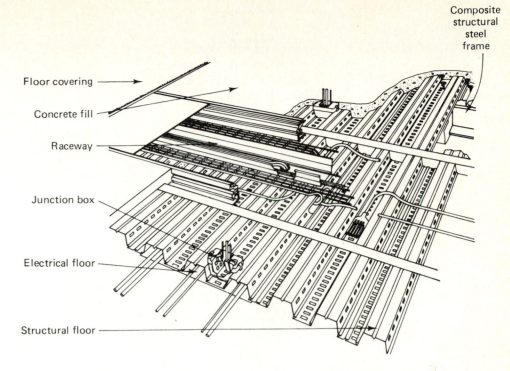

Composite structural steel frame

Floor covering

Concrete fill

Raceway

Junction box

Electrical floor

Structural floor

FIGURE 18.21. Diagram of the intersection of a header duct with cellular ducts. (Courtesy of H.H. Robertson Co.)

FIGURE 18.22. Installing a header duct for electrical and telephone service.

FIGURE 18.23. Plan view showing how the system is fed from closet to trench header to cells. (Courtesy of H.H. Robertson Co.)

Trench header

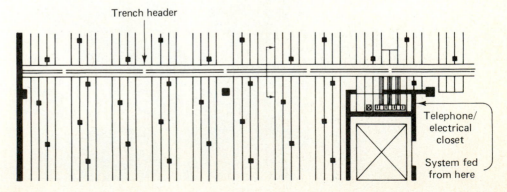

Telephone/electrical closet

System fed from here

which are cast into the concrete floor 5 ft in each direction (Fig. 18.24). Any outlet unit can be activated by pulling the necessary wire through the trench header and through the cell to the unit. At the unit a small amount of concrete is broken out (Fig. 18.25) after the carpeting has been slit (Fig. 18.26). The breakout pan is then removed and the electrical and telephone connections are made (Fig. 18.27). The telephone and electric wires penetrate the carpet without leaving unsightly holes.

FIGURE 18.24.   Steps in the installation of a junction box in a metal deck.

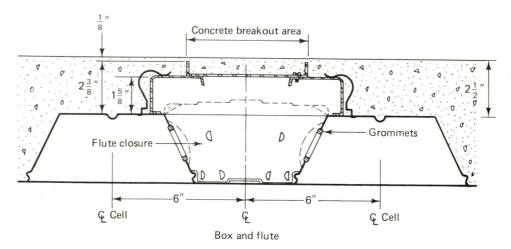

FIGURE 18.25.   Section through cells showing the area of concrete broken out for access to cells. (Courtesy of H.H. Robertson Co.)

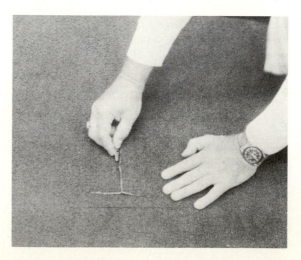

FIGURE 18.26.   "H"-shaped cut leaves carpet unharmed. (Courtesy of H.H. Robertson Co.)

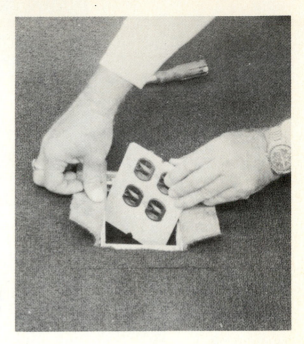

FIGURE 18.27. Installing two duplex receptacles. (Courtesy of H.H. Robertson Co.)

The construction supervisor will schedule concrete pouring immediately behind the placing and securing of metal deck. The electrical contractor must keep pace with this operation by installing the trench header duct and securing the outlet units to the decking, a significant amount of work.

The system just described is one of many metal deck distribution systems. Others vary in metal gauge, capacity, outlet spacing, prepunching, cost, and so on.

The general, metal deck, and electrical contractors share the responsibility of creating a structurally sound, electrically safe installation. Metal deck is often bent and deformed prior to pouring of concrete by careless placing of excessive construction loads on the structurally vulnerable deck. Any deformation or cutting of the metal to fit the decking must be corrected by welding and grinding to make sure that the duct through which the wire will be pulled is smooth and free from any sharp surfaces that might injure the wire during the pulling operation.

Underfloor distribution can also be provided utilizing separate units of rectangular ducts. Although cellular deck has largely supplanted this method, it has excellent specific uses:

1. For office space on slabs on ground (Figs. 18.28 and 18.29)
2. In use with simple corrugated metal forming
3. When fill and finish cover a structural slab (Figs. 18.30 and 18.31)

This method adapts readily to the recycling of old office buildings.

FIGURE 18.28.   Slab-on-ground underfloor distribution system.

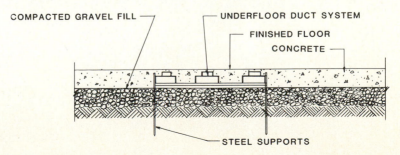

FIGURE 18.29. Cellular duct slab on the ground. (Courtesy of Walker Parkersburg Textron.)

FIGURE 18.30. Header duct and cellular units ready for concrete fill.

FIGURE 18.31. New underfloor system for recycling old office buildings.

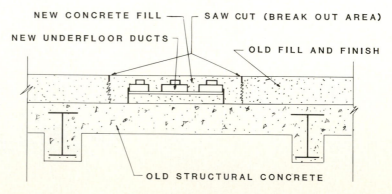

NEW CONCRETE FILL — SAW CUT (BREAK OUT AREA)

NEW UNDERFLOOR DUCTS — OLD FILL AND FINISH

OLD STRUCTURAL CONCRETE

During the first half of this century high-rise buildings were constructed with concrete structural slabs over which was poured 4 in. or more of fill and finish. By cutting *channels* in the old concrete fill the rectangular ducts fit neatly into the old floor system and provide excellent underfloor duct capabilities (Fig. 18.31). This type of underfloor system has two disadvantages that the cellular metal deck system overcame:

1.   To tap the duct, a small core of concrete must be drilled out (Fig. 18.33).
2.   A small section of floor covering is cut out.

However, it provides an excellent solution where cellular deck cannot be used.

Figure 18.34 shows an underfloor distribution system utilizing standard conduit

FIGURE 18.32. Junction boxes facilitate wire pulling. (Courtesy of Walker Parkersburg Textron.)

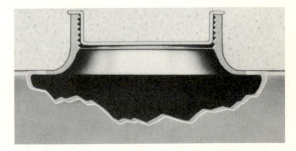

FIGURE 18.33. Break-out area cap protects threads. (Courtesy of Walker Parkersburg Textron.)

FIGURE 18.34. Floor boxes and conduit. (Courtesy of Midland Ross.)

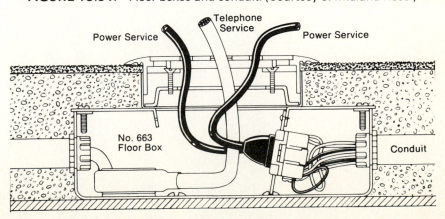

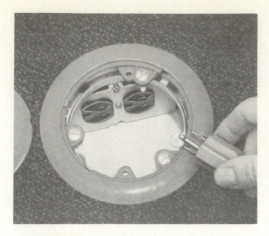

FIGURE 18.35. Receptacle below floor level. (Courtesy of Midland Ross.)

FIGURE 18.36. Nearly flush floor condition. (Courtesy of Midland Ross.)

and special electrical boxes designed for underfloor installation (Fig. 18.35). Although lacking the versatility for tenant changes of other systems, this method has these advantages:

1. It is adaptable to a wide range of structural concrete installations.
2. It is very economical.
3. It provides an attractive method of bringing the wiring out of the floor (Fig. 18.36).

18.6   CONDUITS   In nonresidential building construction work, *conduits* make up the largest portion of the electrical distribution system. They carry the service into the building and in most cases the feeders to panel boxes, elevators, motor control centers, and so on (Fig. 18.37). Used underground, they are made from *polyvinyl chloride, galvanized steel,* or *asphaltic products* encased in concrete for protection (Figs. 18.38 to 18.41). In concrete and masonry slabs and walls, galvanized steel gives the best results since it can be readily bent and resists the corrosive action of portland cement and moisture.

FIGURE 18.37.   Conduits spaced and supported ready for forming and concrete.

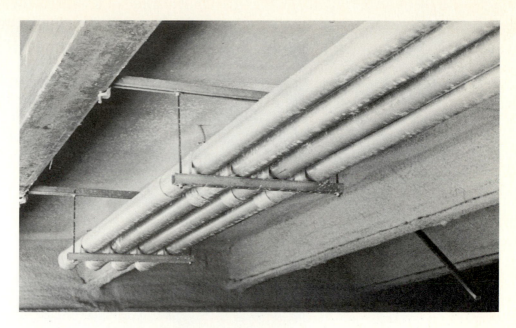

FIGURE 18.38. Conduit carried on a trapeze hanger.

FIGURE 18.39. Conduit connected into a horizontal wire trough and panel boxes.

FIGURE 18.40. The electrician works along with the bricklayer installing conduit, boxes, and panel boards.

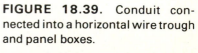

FIGURE 18.41.   Forms, reinforcing steel, and conduit for a concrete slab.

Aluminum and galvanized steel are both used in exposed areas such as riser shafts or in dry, covered areas such as ceilings and drywall partitions (Figs. 18.42 and 18.43).

Metal conduit is joined together and into electrical boxes and compartments with *threaded* or *friction clamping* connections. Both standard and thin-wall weights of conduit are used, depending on code allowances. On the project site machines cut, bend, and thread metal conduit, but much of the bending of ½- to 1-in.-size conduit is done by hand using a *hickey* (Fig. 18.44). Although most sizes of aluminum conduit cost more to purchase than rigid galvanized steel, the installed price is lower due to lightweight and easier cutting and bending. The 3- to 6-in.-sizes produce the biggest savings.

FIGURE 18.42.   Maze of small conduits in a lobby ceiling.

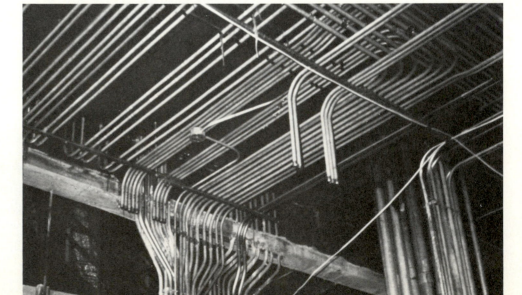

FIGURE 18.43. Conduit and switch box for drywall construction.

FIGURE 18.44. Bending conduit with a hickey.

Polyvinyl chloride (PVC) conduit installed in concrete encasement is approximately 80% cheaper than rigid galvanized steel (Fig. 18.45). In certain industrial applications (Fig. 18.46) PVC conduit performs better than metallic products. The National Electric Code requires that PVC conduit be encased in 2 in. of concrete when carrying voltages above 600 volts. Some local codes still do not permit its use in general building construction work. PVC conduit costs less to purchase and install. It can be readily cut, joined, and bent, as shown in Figs. 18.47 and 18.48.

FIGURE 18.45. Bank of PVC conduits. (Courtesy of Florida Power & Light Co.)

FIGURE 18.46. PVC conduit: external application. (Courtesy of Florida Power & Light Co.)

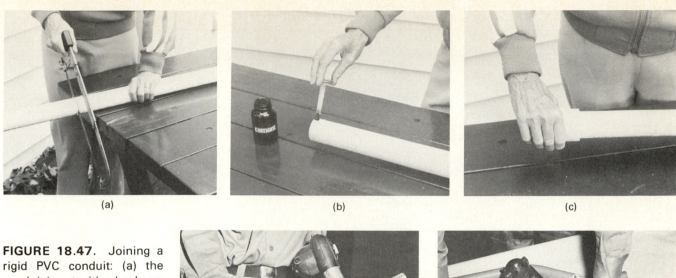

(a)   (b)   (c)

FIGURE 18.47. Joining a rigid PVC conduit: (a) the conduit is cut with a hacksaw or power saw; (b) Solvent is applied to the clean surface; (c) the conduit is joined and twisted for best adhesion; (d) heating the PVC conduit in the bend area; (e) bending the PVC conduit. (Courtesy of B.F. Goodrich Chemical Co.)

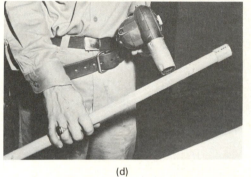

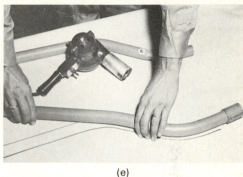

(d)   (e)

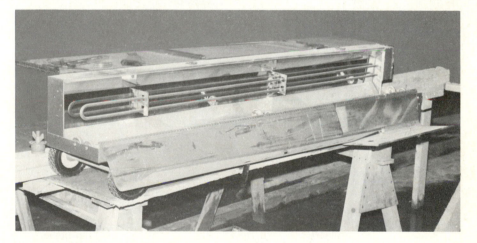

FIGURE 18.48. Electric oven for bending a plastic conduit.

18.7 WIRE AND CABLE   Underfloor distribution systems, junction boxes, conduit, and wiring troughs provide a protected route for carrying the conductors. The electrical contractor has completed a substantial part of his contract when this type of roughing has been completed. *Preparing* the system for wire and cable pulling is a joint effort with the architectural and structural trades. *Pulling* cable and wire is an independent electrical operation.

The term *wire* applies to a single conductor of any size not larger than No. 8 AWG (American Wire Gauge). *Cable* refers to a single No. 6 AWG and larger, or two or more conductors regardless of size when combined. Sizes larger than No. 4 in diameter are expressed in MCM, thousand circular mils. Copper is superior to aluminum for use as conductors. To resist heat, moisture, and oil various insulation materials cover the wire or cable. Among the most popular are rubber, thermoplastic, and cross-linked synthetic polymer. The eight examples listed in Table 18.1 are among 30 designated for various uses by the National Electrical Code®.

TABLE 18.1 CONDUCTOR APPLICATION AND INSULATIONS[a]

Trade name	Type letter	Max. operating temp.	Application provisions	Insulation	Thickness of insulation AWG or MCM	Mils	Outer covering
Heat-resistant rubber	RH	75°C 167°F	Dry locations	Heat-resistant rubber	14-12[b]	30	Moisture-resistant, flame-retardant, nonmetallic covering[c]
	RHH	90°C 194°F	Dry locations		10	45	
					8-2	60	
					1-4/0	80	
					213-500	95	
					501-1000	110	
					1001-2000	125	
Moisture and heat-resistant rubber	RHW	75°C 167°F	Dry and wet locations; for over 2000 volts, insulation shall be ozone-resistant	Moisture and heat-resistant rubber	14-10	45	Moisture-resistant, flame-retardant, nonmetallic covering[c]
					8-2	60	
					1-4/0	80	
					213-500	95	
					501-1000	110	
					1001-2000	125	
Heat-resistant latex rubber	RUH	75°C 167°F	Dry locations	90% unmilled, grainless rubber	14-10	18	Moisture-resistant, flame-retardant, nonmetallic covering
					8-2	25	
Moisture- and heat-resistant thermoplastic	THW	75°C 167°F	Dry and wet locations	Flame-retardant, moisture- and heat-resistant thermo-plastic	14-10	45	None
		90°C 194°F	Special applications *within* electric discharge lighting equipment; limited to 1000 open-circuit volts or less (size 14-8 only as permitted in Section 410-26)		8-2	60	
					1-4/0	80	
					213-500	95	
					501-1000	110	
					1001-2000	125	
	THWN	75°C 167°F	Dry and wet locations	Flame-retardant, moisture- and heat-resistant thermo-plastic	14-12	15	Nylon jacket
					10	20	
					8-6	30	
					4-2	40	
					1-4/0	50	
					250-500	60	
					501-1000	70	
Moisture- and heat-resistant cross-linked synthetic polymer	XHHW	90°C 194°F	Dry locations	Flame-retardant cross-linked synthetic polymer	14-10	30	None
					8-2	45	
					1-4/0	55	
		75°C 167°F	Wet locations		213-500	65	
					501-1000	80	
					1001-2000	95	

Trade name	Type letter	Max. operating temp.	Application provisions	Insulation	AWG or MCM	(A) Mils	(B) Mils	Outer covering
Moisture-, heat-, and oil-resistant thermoplastic	MTW	60°C 140°F	Machine tool wiring in wet locations as permitted in NFPA Standard No. 79 (see Article 670)	Flame-retardant, moisture-, heat-, and oil-resistant thermo-plastic	22-12	30	15	(A) None
					10	30	20	(B) Nylon jacket
					8	45	30	
					6	60	30	
		90°C 194°F	Machine tool wiring in dry locations as permitted in NFPA Standard No. 79 (see Article 670)		4-2	60	40	
					1-4/0	80	50	
					213-500	95	60	
					501-1000	110	70	

[a] For insulated aluminum and copper-clad aluminum conductors, the minimum size shall be No. 12.
[b] For 14-12 sizes RHH shall be 45 mils thickness insulation.
For insulated aluminum and copper-clad aluminum conductors, the minimum size shall be No. 12.
[c] Outer covering shall not be required over rubber insulations which have been specifically approved for the purpose.

Source: Extracted by permission from NFPA 70-1978, National Electrical Code®, copyright © 1977, National Fire Protection Association, Boston.

Construction supervisors should observe and report on the start of cable and wire pulling. With proper manpower and a well-coordinated electrical effort, wire and cable should be sticking out of panel and junction boxes a few weeks after their installation (Figs. 18.49 to 18.51). The electrical contractor should not be permitted to let this operation slide with the hope of making up lost progress later in the project even though he appears on schedule and does not hold up the work of other trades. Many projects have run into an electrical progress crisis near their completion with all kinds of work that only an electrician can do.

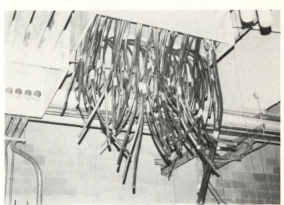

FIGURE 18.49.   Ceiling pull box.

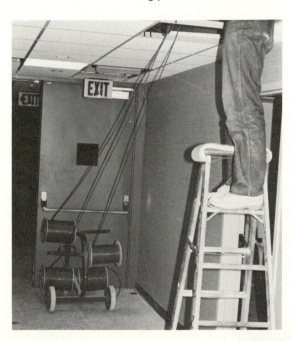

FIGURE 18.50.   Wire pulling.

FIGURE 18.51.   Wire pulling cart.

18.8 BUSWAYS   In place of conduit and cable, *busways* are employed in industrial plants (Fig. 18.52) and high-rise buildings to carry large amounts of power from the main switchgear to locations of heavy usage. The design of busways makes them electrically more efficient and easier and simpler to plug in or otherwise tap the energy source (Figs. 18.53 and 18.54). Heavy bars of copper or aluminum (busbars) enclosed in a metal enclosure carry the 600-volt (or less) current (Fig. 18.55). The electricians bolt the 10-ft sections together and firmly fasten the busway to the structure. All manner of fittings, such as tees, elbows, trolleys, and plugs, are available to make busway design fit the structure.

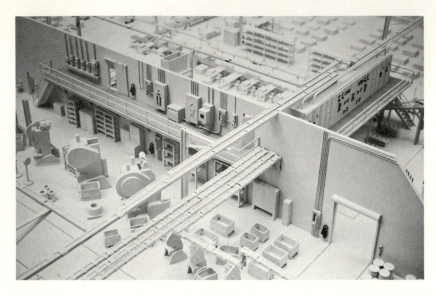

FIGURE 18.52. Model of an industrial building utilizing a bus duct. The main switchgear is located on the mezzanine to the right. (Courtesy of Gould Inc. Distribution & Controls Div.)

FIGURE 18.53. Typical riser layout for a nine-floor building. (Courtesy of Gould Inc. Distribution & Controls Div.)

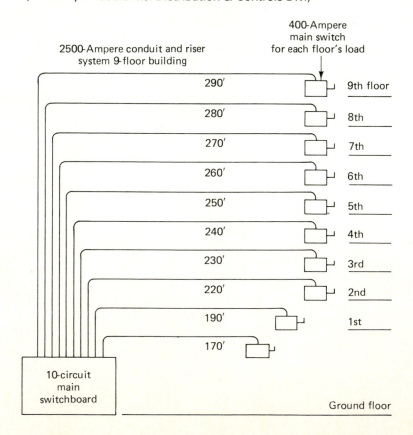

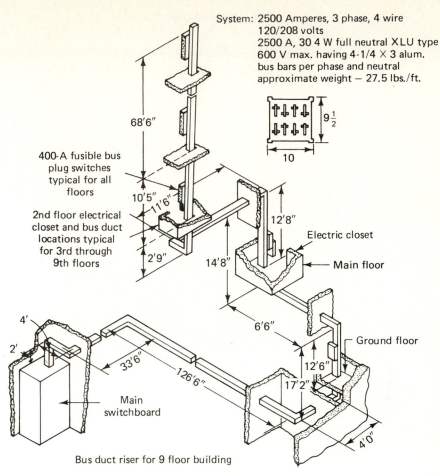

System: 2500 Amperes, 3 phase, 4 wire
120/208 volts
2500 A, 30 4 W full neutral XLU type
600 V max. having 4-1/4 × 3 alum.
bus bars per phase and neutral
approximate weight — 27.5 lbs./ft.

400-A fusible bus plug switches typical for all floors

2nd floor electrical closet and bus duct locations typical for 3rd through 9th floors

68'6"

10'5"

11'6"

2'9"

12'8"

14'8"

6'6"

Electric closet

Main floor

Ground floor

12'6"

17'2"

4'0"

4'

2'

33'6"

126'6"

Main switchboard

9½

10

Bus duct riser for 9 floor building

FIGURE 18.54. A more economical bus duct riser for the same nine-floor building. (Courtesy of Gould Inc. Distribution & Controls Div.)

FIGURE 18.55. Cutaway view of bus duct construction. (Courtesy of Gould Inc. Distribution & Controls Div.)

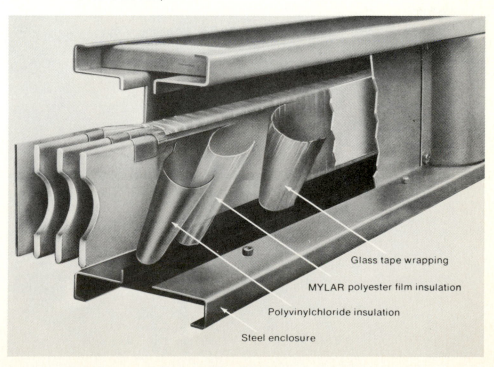

Glass tape wrapping

MYLAR polyester film insulation

Polyvinylchloride insulation

Steel enclosure

The *feeder type* is used in both industrial and high-rise applications where low-voltage drop between two principal points is essential (Figs. 18.56 and 18.57). The *plug-in type* has the advantage of versatility and is particularly adapted to industrial applications where production requirements dictate machinery relocations (Figs. 18.58 and 18.59). The plug-in type, as its name implies, has openings spaced along the length of the busway which readily permit *power taps*. The plug-in type fills the need in high-rise buildings for power taps at various floor levels (Figs. 18.60 and 18.61).

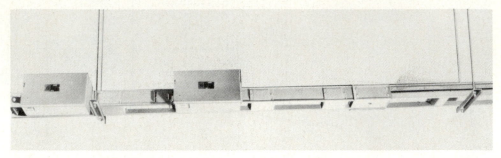

FIGURE 18.56   A bus duct riser is supported on hangers ten feet apart. (Courtesy of Gould Inc. Distribution & Controls Div.)

FIGURE 18.57   Installing a bus duct on trapeze hangers. (Courtesy of Gould Inc. Distribution & Controls Div.)

FIGURE 18.58   A bus duct riser is supported at each floor by angle supports. (Courtesy of Gould Inc. Distribution & Controls Div.)

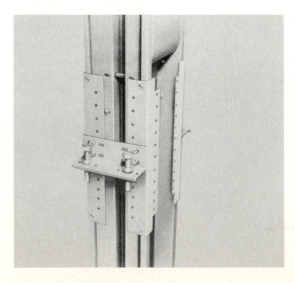

FIGURE 18.59 Installing a bus duct riser. (Courtesy of Gould Inc. Distribution & Controls Div.)

FIGURE 18.60. Installing a plug-in device. (Courtesy of Gould Inc. Distribution & Controls Div.)

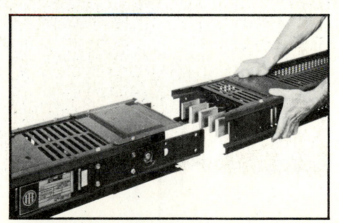

FIGURE 18.61. Joining a feeder-type bus duct. (Courtesy of Gould Inc. Distribution & Controls Div.)

During construction periods the dry and dust-free conditions needed for safe electrical installations are difficult to attain and maintain. Considerable care should be exercised in utilizing busways during construction for permanent machinery power or temporary applications before the building has been completed. A very serious electrical fire occurred some years ago in a New York City high-rise building under construction. The fire started in the busway and traveled up and down the busway, spreading the conflagration to many floors, ultimately destroying the busway and heavily damaging construction materials and parts of the building.

18.9   FUSES   Electrical devices and systems protect themselves and the building they serve with *fuses* and *circuit breakers.* Fuses are rated according to voltage, amperes, and interrupting and overload ability.

Ratings

Voltage Rating. The *voltage rating* of a fuse must be equal to or greater than the voltage of the circuit in which the fuse is applied. The voltage rating is the ability of the fuse to extinguish the arc quickly after the fuse element has melted and to prevent the system open-circuit voltage from restriking across the opened fuse element. The fuse voltage rating is not a measure of its ability to withstand a specified voltage while carrying current. A fuse should never be applied where the system voltage exceeds the fuse voltage rating.

Ampere Rating. Every fuse is designated by an *ampere rating* which has been determined under specified test circuit conditions. In selecting a fuse ampere rating, consideration must be given to the type of load and code requirements for the specific application. To provide reliable overload and short-circuit protection, the ampere rating of the fuse should normally not exceed the current-carrying capacity of the circuit. However, there are some specific circumstances where the fuse ampere rating is permitted to be greater than the current-carrying capacity of the circuit. A typical example is a motor circuit, where dual-element fuses are generally permitted to be sized at 175% and non-time-delay fuses sized at 300% of the motor full-load amperes.

Interrupting Rating. The *interrupting rating* is the maximum available short-circuit current that an overcurrent protective device can safely interrupt without damage to itself. This is a very important rating when applying overcurrent protective devices. Sections 110-9 and 230-98 of the National Electrical Code® require sufficient interrupting ratings wherever overcurrent protective devices are applied. Modern current-limiting fuses have an interrupting rating of 200,000 amperes, which is sufficient for virtually all applications.

Overloads

An *overload* is a current, larger than normal, but flowing within the normal current path. Overloads are usually in the range of one to six times normal. Rapid removal of overloads is not generally necessary except for certain delicate components such as semiconductors. The chief damage expected under overload conditions is thermal and most often appears as insulation deterioration and failure.

Time Current Characteristics

A 100-ampere fuse does not instantly open at 101 amperes, or even at 200 amperes. When an overcurrent exceeds a fuse's ampere rating, the opening time of the fuse is dependent on the type of fuse and the magnitude of overcurrent. In most circuits, *temporary overloads* can be tolerated by circuit components without damage. When the fuse is properly selected, the delay in opening under overload conditions is desirable. An overload current condition may be temporary in nature and the current may subside to normal current conditions before opening the fuse. Yet fuses will open to protect components if an overload persists for too long. A typical harmless overload current surge is encountered whenever most motors are started. Motor-starting currents of four to six times full-load current occur for a period of up to several seconds. Built-in fuse time delay can permit the motor to start without fuses opening unnecessarily.

Two broad fuse characteristic types are (1) *dual-element, time-delay* fuses, and (2) *non-time-delay* fuses. Each characteristic type has certain attributes suitable for specific applications. In general, the dual-element, time-delay fuses are the most widely used for general-purpose applications, motor circuits, transformers, and other circuits. The non-time-delay current-limiting fuses are used where a fast speed of response with little or no intentional overload delay is desired; a typical application is for protection of low-interrupting-rating circuit breakers.

Short Circuits*

"A *short-circuit* current is distinguished from an overload by virtue of the current flowing outside of the normal path. Usually this current is much larger than overload values ranging from tens to hundreds of times normal. Great damage can be expected quickly under short-circuit conditions and takes the form of thermal, magnetic, and arcing damage. Time is crucial in the removal of short-circuits from the system.

Another form of short-circuit can occur, the low level or ground fault. These faults are difficult to detect and correct as their magnitude lies in the same range as normal or overload currents. These faults are best prevented rather than corrected."

18.10   OPERATING PRINCIPLES OF FUSES *

"Modern high interrupting rated, current limiting fuses provide safe and reliable protection for electrical distribution systems. They can be used with complete confidence in their performance. Fuses operate from increased fuse element temperature caused by an overcurrent flowing through them. Since the functioning of fuses does not depend on the operation of intricate moving mechanical parts, fuse performance characteristics are reliable. Fuses remain safe and accurate since age does not increase their current carrying capacity nor lengthen their opening time. After a modern fuse has been called upon to clear a short-circuit or overload, it is replaced by a new fail safe, factory calibrated unit, as accurate and dependable as the original. The two basic types of modern current-limiting fuses are single element fuses and dual-element fuses. Each type has a simple, reliable operating principle.

Single-Element Fuses

A single-element fuse consists of a link or several links contained in a tube filled with an arc-quenching filler [Fig. 18.62a].

Overload Operation. If an overload current of more than rated current is continued for a sufficiently long period of time, the fuse link melts. An arc across the resulting gap burns back the metal, lengthening the gap [Fig. 18.62b]. The arc is quenched and circuit interrupted when the break is of sufficient length. The surrounding filler aids the arc interrupting process [Fig. 18.62c].

*This section is taken from *Electrical Protection Handbook,* Bulletin SPD-78, Bussmann Division, McGraw-Edison Company, St. Louis, Mo. Reprinted with permission by Bussmann Division, McGraw-Edison Company.

*Section 18.10 is taken from *Electrical Protection Handbook,* Bulletin SPD-78, Bussmann Division, McGraw-Edison Company, St. Louis, Mo. Reprinted with permission by Bussmann Division, McGraw-Edison Company.

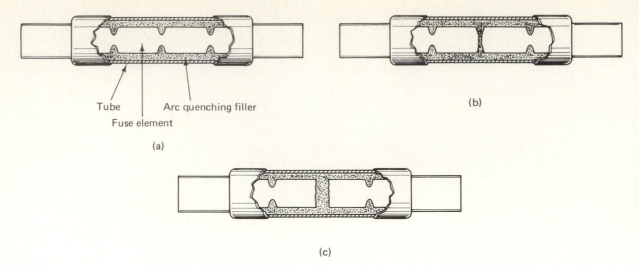

Tube   Arc quenching filler

Fuse element

(a)

(b)

(c)

FIGURE 18.62.   Single-element fuse: (a) cutaway view; (b) melting of the fuse element; (c) circuit interrupted. (Reprinted with permission by Bussmann Division, McGraw-Edison Company.)

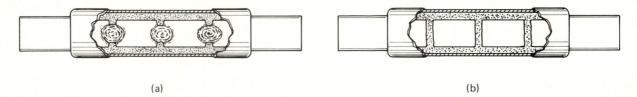

(a)   (b)

FIGURE 18.63.   Single-element fuse: (a) melting of several sections; (b) circuit interrupted. (Reprinted with permission by Bussmann Division, McGraw-Edison Company.)

Short-Circuit Action. On a short-circuit, several sections of the link melt instantly [Fig. 18.63a]. Arcing across the vaporized portions of the link commences, and the arc extinguishing filler quickly aids in quenching these arcs, and thus clearing the circuit [Fig. 18.63b].

Although the fuse operation is caused by an increase in the link temperature, the fusing operation is safely contained in the fuse casing and the circuit is interrupted in a fraction of a second.

Dual-Element Fuses

Dual-element, time-delay fuses [Fig. 18.64a] provide time-delay in the low overload range to eliminate needless opening on harmless overloads and transient conditions; yet these fuses are extremely fast opening and current-limiting on short-circuit currents. This beneficial time-current characteristic is obtained by using two fusible elements connected in series and contained in one tube: (1) the thermal cutout element and (2) the fuse link element surrounded with arc extinguishing filler (thus the name dual-element). The magnitude of the overcurrent determines which element functions. The thermal cutout element is designed to open on overcurrents of up to approximately 500% of the fuse ampere rating. The fuse link or short-circuit element is designed to open on heavier overloads and short-circuit currents.

Overload Operation. On a low overload current, the short-circuit fuse link remains entirely inactive. The overload element consists of a center mass of copper on which is mounted a spring and a short connector. This connector is held in place by a

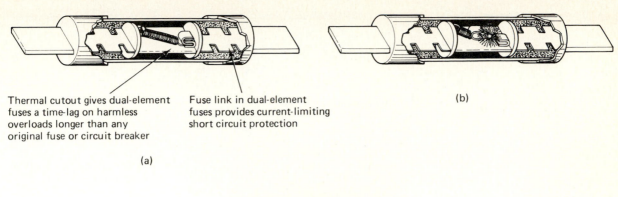

Thermal cutout gives dual-element fuses a time-lag on harmless overloads longer than any original fuse or circuit breaker

Fuse link in dual-element fuses provides current-limiting short circuit protection

(b)

(a)

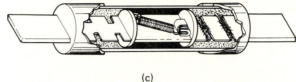

(c)

FIGURE 18.64. Dual-element fuse: (a) cutaway view (thermal cutout gives dual-element fuses a time lag on harmless overloads longer than any ordinary fuse or circuit breaker; the fuse link in dual-element fuses provides current-limiting short-circuit protection); (b) overload operation; (c) short-circuit operation. (Reprinted with permission by Bussmann Division, McGraw-Edison Company.)

low melting point solder and connects the center mass of copper to the fuse link. When an overload current flows long enough to raise the temperature of the center mass to the melting point of the solder, this connector is pulled out of place by the spring, thereby opening the circuit.

[An] *opening overload* [is shown in Fig. 18.64b]. The thermal cutout element has a built in time-delay which permits momentary harmless inrush currents such as motor starting current to flow without opening the fuse. If the overload current persists too long, the thermal cutout element clears the circuit thus protecting the components and equipment. These fuses are ideal for motor, transformer, and other circuits with harmless startup surges.

Short-Circuit Operation [Fig. 18.64c]. On [an opening] short-circuit, the fuse link acts as described for a single element fuse. A portion of the link vaporizes and the surrounding filler aids in extinguishing the arc. This fuse element has current-limiting ability for short-circuit currents thereby reducing the mechanical, thermal, and arcing stress which system components must withstand."

18.11 CIRCUIT BREAKERS

Circuit breakers have the distinct advantage over fuses of providing electrical protection *without the necessity of replacing the protective device.* When the breaker trips, the cause is determined, the condition corrected, and the breaker reset. The electric distribution system presented in Fig. 18.4 shows eight circuit breakers protecting the main switchgear, power panels, the elevator service, and circuits to HVAC equipment. Circuit breakers have replaced fuse boxes in modern residences. They are used extensively in industrial applications in combination with starters and power taps to busways. One of the leading manufacturers offers over 800 breakers from which to choose. In Fig. 18.65 the functioning of a circuit breaker is shown under normal, overload, and short-circuit current conditions. The breaker is designed with the inverse time element concept, in which the time between the occurrences of an overload and the moment of tripping varies inversely with the magnitude of the overload current. This provides optimum overload protection without nuisance tripping from momentary higher overload currents.

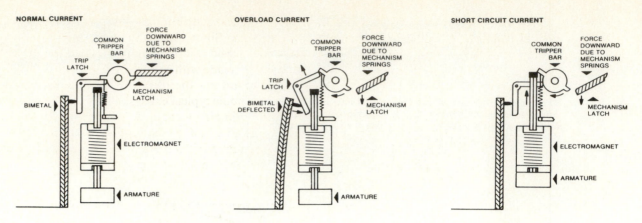

NORMAL CURRENT   OVERLOAD CURRENT   SHORT CIRCUIT CURRENT

FIGURE 18.65. Operation of a thermal magnetic circuit breaker. (Courtesy of Gould I-T-E.)

18.12 MOTOR CONTROL CENTERS

As the name implies, the *motor control center* performs a number of functions (Fig. 18.66). Power from the basic switchgear is brought to a motor control center located in a local mechanical equipment area. The motor control center contains an externally operable circuit disconnect, a fusible switch, or a molded-case circuit breaker and a magnetic starter with an overload relay in the motor lines. The MCC can be fed by either a busway or conduit service.

Motor control centers are contained in neat and attractive cabinets usually located adjacent to the machinery they control. This important piece of equipment is essentially a custom-designed free-standing assembly of standard electrical equipment bolted rigidly together. The MCC should be designed to permit future additions, changes, or regrouping of the units by the purchaser. It should be specified and ordered early in the project. The designers should carefully designate what they require, what contractor will supply the MCC (the HVAC or electrical contractor, for example), and the exact point where the electrician's work *terminates* and the other contractor *takes over*. The construction manager or general contractor should check this item to avoid an estimating or bidding gap.

FIGURE 18.66. Motor control center.

18.13 CONNECTORS AND TERMINALS

A very important link in the electrical distribution system takes place as the various parts of the system join together. Since electricity develops *thrusts, vibrations, expansion,* and *contraction,* the method of joining the separate parts is of considerable importance. Most connections are made with *pressure connectors* of the mechanical screw or compression type using a tool and die. Connectors for every conceivable need are available. Figure 18.67 shows some typical examples. Proper torque is important (Fig. 18.68). Figure 18.69 shows crimping technique. Mechanical connectors known as "pigtails" are shown in Fig. 18.70, and mechanical terminals are shown in Fig. 18.71. Joining copper and aluminum requires more care than if these two metals were not combined. Figures 18.72 and 18.73 show two examples of this type of joint.

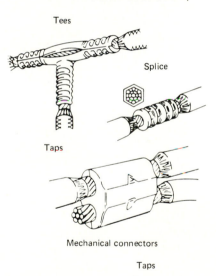

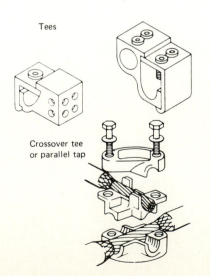

FIGURE 18.67. Compression connectors. (Reprinted from the 1978 edition of the "Aluminum Building Wire Installation Manual and Design Guide" with permission from the Aluminum Association.)

FIGURE 18.68. Proper torque is important. Overtightening may sever the wires or break the fitting; undertightening may lead to overheating and failure. (Reprinted from the 1978 edition of the "Aluminum Building Wire Installation Manual and Design Guide" with permission from the Aluminum Association.)

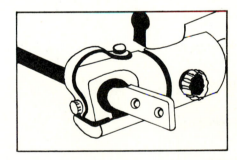

FIGURE 18.69. Crimping tool must be fully closed to avoid unsatisfactory and weak joints. (Reprinted from the 1978 edition of the "Aluminum Building Wire Installation Manual and Design Guide" with permission from the Aluminum Association.)

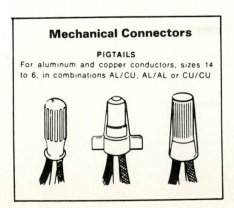

Mechanical Connectors

PIGTAILS
For aluminum and copper conductors, sizes 14 to 6, in combinations AL/CU, AL/AL or CU/CU

FIGURE 18.70. Mechanical connectors: pigtails. (Reprinted from the 1978 edition of the "Aluminum Building Wire Installation Manual and Design Guide" with permission from the Aluminum Association.)

Mechanical Terminals

SCREW-TYPE TERMINAL LUGS

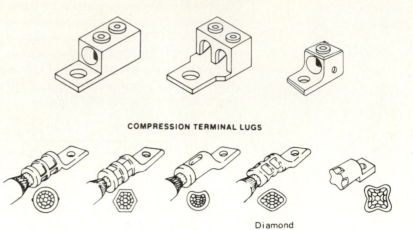

COMPRESSION TERMINAL LUGS

Circumferential   Hexagonal   Indented   Diamond Compression   Versa-Crimp

FIGURE 18.71. Mechanical terminals.

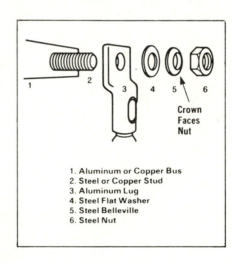

1. Aluminum or Copper Bus
2. Steel or Copper Stud
3. Aluminum Lug
4. Steel Flat Washer
5. Steel Belleville
6. Steel Nut

Crown Faces Nut

FIGURE 18.72. Belleville washer is used to make an aluminum to copper or steel joint. (Reprinted from the 1978 edition of the "Aluminum Building Wire Installation Manual and Design Guide" with permission from the Aluminum Association.)

FIGURE 18.73. Method for connecting large aluminum connectors to equipment studs or terminal pads made of copper. (Reprinted from the 1978 edition of the "Aluminum Building Wire Installation Manual and Design Guide" with permission from the Aluminum Association.)

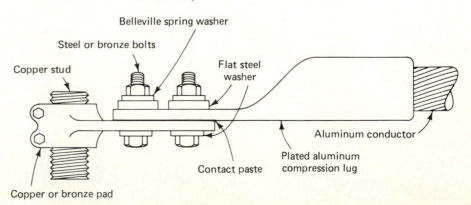

Belleville spring washer

Steel or bronze bolts

Copper stud

Flat steel washer

Aluminum conductor

Copper or bronze pad

Contact paste

Plated aluminum compression lug

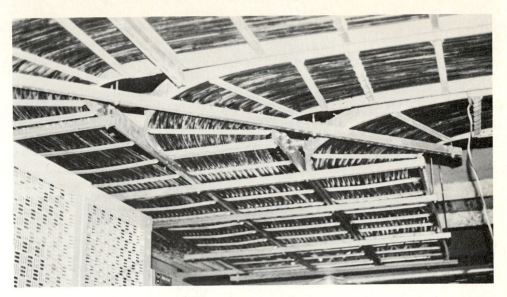

FIGURE 18.74.   Cable trays.

18.14   CABLE TRAYS   *Cable trays* are used in open areas to carry large numbers of protected cables in a safe, efficient, and economical manner (Fig. 18.74). Trapeze hangers support the trays, which have closely spaced struts that keep the cable from drooping. Clips firmly affix the cables to the tray.

18.15   EMERGENCY GENERATOR SYSTEMS   Prudent owners have been installing *emergency generators* since dependency on electrical power began (Fig. 18.75). Airline ticket reservation computer centers and hospital operating rooms are prime examples of this need. During the first widespread blackout along the East Coast, many buildings equipped with emergency generators continued to operate satisfactorily. The best results were experienced by those buildings that periodically conducted emergency drills. One big building operated

FIGURE 18.75.   Temporary electric generator, generally powered by a diesel motor. Gasoline and propane types are also used.

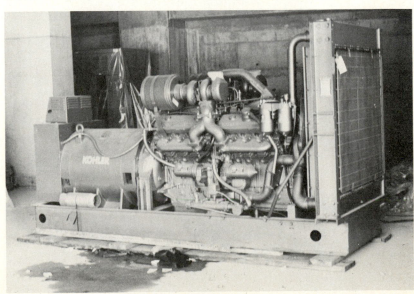

satisfactorily until the house tanks drained down and no water was available for sanitary purposes. Pumping water has since been added to the emergency system. Generating emergency power is readily accomplished with machines operating on diesel, gasoline, or propane gas fuel. The exhaust line sometimes causes a bit of a comparatively expensive problem, but the installation basically goes routinely. The *design* of the emergency system requires real thought and attention in covering requirements for lighting, elevator service, pumping water, ventilation, control, and alarm systems.

19

Lighting

19.1 ENERGY CONSERVATION AND OTHER CONSIDERATIONS

Until the energy crunch of the 1970s, illumination levels in schools, offices, and work places kept rising and rising. The *new trend* in design follows the recommendation of the ASHRAE Standard 90-75, "Energy Conservation in New Building Design," a standard used to modify state and national model codes. The standard does not dictate how to design electrically but utilizes a lighting power budget as part of an overall building energy package. The new proposals tend to limit the recommended design wattage for electric lighting systems to approximately 50 to 65% of that used in typical commercial and institutional buildings designed in the 1960s. General lighting levels proposed drop down to the 20- to 35-footcandle level in office areas. Lighting to meet the new standards will also discriminate between areas requiring high levels of illumination and those that do not. Lighting will be analyzed on the basis of (1) *task lighting,* (2) *general lighting,* and (3) *noncritical lighting.* In addition, emphasis is being placed on greater use of reflective surfaces, lower-wattage luminaires, and more cleaning of fixtures (see also section 19.5).

Despite the fact that the construction manager and electrical contractor can install the lighting fixtures specified without much thought about the illumination objectives, there is merit to their understanding of illumination. Designers use three kinds of lighting: *incandescent, fluorescent,* and *high intensity,* their choice based on economy, efficiency, lamp life, starting characteristics, fixture cost, and other considerations. Features of the three types are compared in Table 19.1.

TABLE 19.1   COMPARISON OF TYPES OF LIGHTING

	Incandescent	*High-intensity*	*Fluorescent*
Requirements	Simple socket, simple wiring	Ballast, simple socket, simple wiring,	Ballast, two sockets
Luminaire size	Small	Medium	Large
Luminaire weight	Light	Heavy	Medium to heavy
Efficiency (lumines/watt)	Least—10–15	60–110	60–80
Lamp life (hr)	750–1000	24,000	10,000–20,000
Starting	Instantaneous	Delayed—2 to 5 min	Almost instantaneous
Dimability	Yes	No for most types	Yes only for the 4-ft rapid-start type
Fixture cost	Low	Varies with location	Very much more than incandescent
Color	Warm	Mostly unattractive	Warm or cool

19.2   INCANDESCENT LIGHTING

In the *incandescent lamp* a coiled tungsten wire glows "white hot" in a vacuum. This "burning" of the metal causes blackening of the bulb from the vaporization of the tungsten element and loss of illumination. The lamps are of comparatively very short life, 750 to 1000 hours, the first one burning out as quickly as 300 hours and 50% burning out between 375 and 500 hours. Lamp life varies considerably with voltage, as shown in the following comparison, where 120 volts is taken as the standard:

Voltage	*Illumination*	*Lamp life*
120	100%	100%
100	70%	400%
132	140%	35%

In hard-to-reach high ceilings where lamp replacement costs are excessive, a *reduced-voltage* system makes good sense. A *high-voltage* system would be selected for lighting a parking lot of short-term usage, where reducing the number of poles would reduce costs. Finally, incandescent light produces a continuous color spectrum, an important consideration to many clients interested in a pleasing lighting effect and better color rendition.

19.3   FLUORESCENT LIGHTING

Fluorescent lighting has certain characteristics which have made it a very popular form of lighting. Designers use fluorescent lighting extensively in schools, supermarkets, office buildings, and industrial plants because it produces five times as much light per watt as an incandescent lamp and has a lamp life 10 to 20 times longer (Figs. 19.1 to 19.8).

The fluorescent lamp produces ultraviolet light by carrying an *arc* through *mercury vapor* in an *inert gas* (argon, krypton, or neon) at a low pressure. The tubes are coated with phosphors which react with the invisible ultraviolet rays to produce visible light. The combination of red and green phosphors produces all colors, which the human eye sees as white.

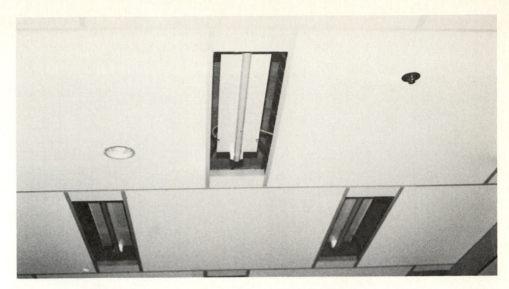

FIGURE 19.1. A 40-watt single-tube fluorescent luminaire.

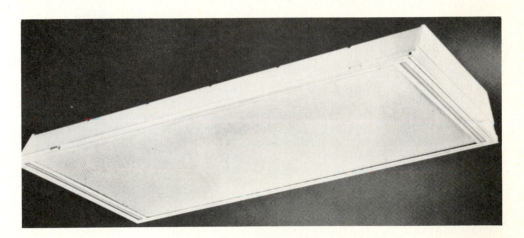

FIGURE 19.2. Heat removal troffor. (Courtesy of Thomas Industries Inc.)

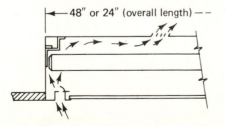

FIGURE 19.3. Return air path. (Courtesy of Westinghouse Electric Corp.)

FIGURE 19.4. End view of heat removal. The air train directs the flow across the lamp filament area for maximum heat pickup. (Courtesy of Thomas Industries Inc.)

285

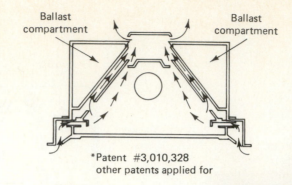

Ballast compartment · Ballast compartment

*Patent #3,010,328
other patents applied for

FIGURE 19.5. Convection flow heat removal design. The heat produced by the lamp develops a natural convection effect, drawing cooler air upward through the air gap provided by the double-wall construction of the reflector. (Courtesy of Thomas Industries Inc.)

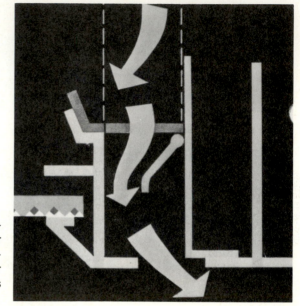

FIGURE 19.6. Side view of supply air. An optional adjustable air pattern vane in housing trim provides for horizontal or vertical air pattern control. (Courtesy of Thomas Industries Inc.)

FIGURE 19.7. Suspension-type fixture. (Courtesy of Thomas Industries Inc.)

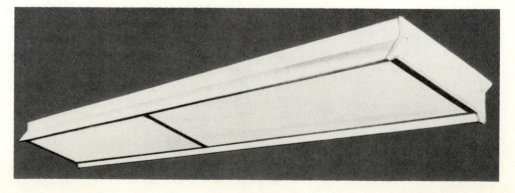

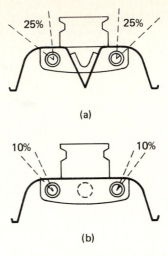

FIGURE 19.8. Cross section of a fluorescent luminaire showing the amount of light designed to reflect off the ceiling: (a) 25% upward component unit with a crosswise shielding of 30°, available only in two-lamp construction; (b) 10% upward component unit with a crosswise shielding of 13°, available in two- and three-lamp construction. (Courtesy of Thomas Industries Inc.)

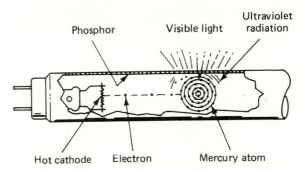

FIGURE 19.9. How light is produced in a typical hot cathode fluorescent lamp. (Courtesy of GTE Lighting Products.)

"The lamp is essentially a coated and evacuated tubular bulb containing a small quantity of mercury and inert gas. A specially treated electrode, called "hot cathode," is sealed in either end. Figure 19.9 shows how visible light is generated in a typical hot cathode fluorescent lamp."* (page 3, 0-341)

Electrodes

"The *electrode* at each end of a fluorescent lamp is generally a coated coiled-coil or triple-coil tungsten wire [Fig. 19.10]. The coating on the tungsten wire is an emissive material (barium, strontium and calcium oxide) which emits electrons when heated to an operating temperature of about 950° C. At this temperature, electrons are given off freely with only a small wattage loss at each cathode. This process is called thermionic emission because the heat is more responsible for the emission of electrons than is the voltage. An electrode of this design is called a "hot cathode." This type of cathode lowers the starting voltage required to strike the arc." (page 4, 0-341)

---

*This quotation and the remainder of Section 19.3 through "Starters" is taken from *Fluorescent Lamps*, GTE Sylvania Engineering Bulletin 0-341, Sylvania Lighting Center, Danvers, Mass., courtesy of GTE Lighting Products.

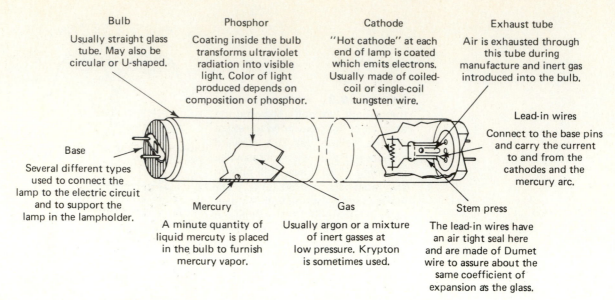

Bulb
Usually straight glass tube. May also be circular or U-shaped.

Phosphor
Coating inside the bulb transforms ultraviolet radiation into visible light. Color of light produced depends on composition of phosphor.

Cathode
"Hot cathode" at each end of lamp is coated which emits electrons. Usually made of coiled-coil or single-coil tungsten wire.

Exhaust tube
Air is exhausted through this tube during manufacture and inert gas introduced into the bulb.

Lead-in wires
Connect to the base pins and carry the current to and from the cathodes and the mercury arc.

Base
Several different types used to connect the lamp to the electric circuit and to support the lamp in the lampholder.

Mercury
A minute quantity of liquid mercuty is placed in the bulb to furnish mercury vapor.

Gas
Usually argon or a mixture of inert gasses at low pressure. Krypton is sometimes used.

Stem press
The lead-in wires have an air tight seal here and are made of Dumet wire to assure about the same coefficient of expansion as the glass.

FIGURE 19.10.   Basic parts of a typical hot cathode fluorescent lamp. (Courtesy of GTE Lighting Products.)

Energy Distribution

"Approximately 60 percent of the input energy in a cool white fluorescent lamp is converted directly into ultraviolet, with 38 percent going into heat and 2 percent into visible light, as shown in Fig. 19.11. The phosphor changes about 21 percent of the ultraviolet into visible light with the remaining 39 percent becoming heat. The 23 percent conversion of energy into light for a 40-watt fluorescent lamp is approximately

FIGURE 19.11.   Energy distribution of a typical 40-watt Cool White fluorescent lamp. (Courtesy of GTE Lighting Products.)

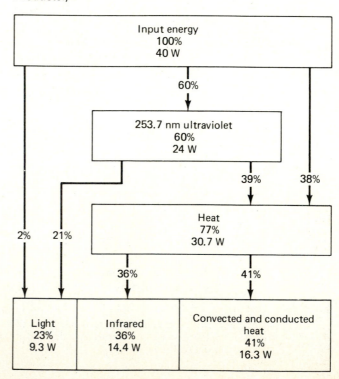

twice the percentage of a 300-watt incandescent lamp, which changes only 11 percent of the input energy into light. The production of 36 percent infrared compares with 69 percent for a 300-watt incandescent." (page 5, 0-341)

Types of Fluorescent Lamps

"The original fluorescent lamps introduced in 1938 were of the preheat type, requiring separate starters. The starter supplies several seconds of current flow through the cathodes to preheat them between the time the lamp is turned on and the time the lamp lights. The cathodes are preheated to emit electrons to aid in the striking of the arc at a lower voltage. The starter is usually of an automatic type which applies current to the cathodes for sufficient time to preheat them and then automatically opens to stop the current flow and cause the full voltage with an inductive voltage spike to be applied across the two cathodes, thus striking the arc. There are some preheat systems, such as fluorescent desk lamps, in which the starting is done by depressing a manual start button for a few seconds and releasing it to start the lamp." (page 7, 0-341)

[Other types of fluorescent lamps are: *instant start, rapid start,* and *rapid start high output.* In addition, there are many different kinds of special usage fluorescent lamps, such as outdoor or plant growth types.]

The Ballast

"Fluorescent lamps, in common with all arc discharge lamps, must be operated with an auxiliary called a *ballast* which limits the current and provides the required starting voltage. The more the current in the arc increases, the more the resistance of the arc decreases. Thus, the arc in a fluorescent lamp would "run away with itself" and draw so much current that it would destroy the lamp if it were not held back. Limiting the current is the most important function of the ballast, whether it be a choke coil, reactor, a capacitor, or a resistance. Each fluorescent lamp requires a ballast that is designed especially for its electrical characteristics, the type of circuit in which it is to be operated and the voltage and frequency of the power supply.

While limiting the current in a fluorescent lamp is the most important function of a ballast, the ballast must also provide adequate voltage for starting the lamp and supply low voltage to heat the cathodes.

Although fluorescent lamps may be ballasted by *inductance, capacitance,* or *resistance,* the most practical and widely used of the three is *inductance.* In most cases, the fluorescent lamp ballast includes an inductive device such as a choke coil or an autotransformer to limit the current. Use is also made of the series combination on inductive coil and capacitor.

All ballasts produce an inherent sound, commonly described as a "hum." This will vary with the type of ballast, from a nearly inaudible sound to a noticeable noise. Most manufacturers give their ballasts a sound rating from A to F, as an aid in the selection of ballasts. An "A" ballast will usually have the least hum and should be used in quiet areas, such as study halls and homes. The most audible hum is produced by an "F" ballast, which should be satisfactory for street lighting and noisy factory areas.

Because of the losses within the ballast, they consume a small amount of wattage which must be added to the lamp wattage to obtain the total wattage of the lighting equipment." (page 10, 0-341)

Starters

"The principal functions of a *starter* are to close the starting circuit of a preheat lamp while the cathodes heat up, and then to open the circuit to start the lamp. If the arc fails to strike, the starter keeps trying until the lamp starts. A further function in protective starters is to disconnect a lamp from the starting circuit when it fails to start after several attempts. Starters may be of the *thermal* or the *glow-switch* type, with the latter being in much more common use." (page 12, 0-341)

Lamp Life *

"*Life* of a fluorescent lamp is dependent to a major degree on the cathode coating being able to supply the electrons necessary to sustain the low pressure mercury arc discharge in the lamp. The rate of erosion of this coating during operation is dependent on the lamp gas fill, gas pressure, and cathode temperature. In lamp design these factors are considered carefully to assure the meeting of rated average life figures.

Another important factor in determining *lamp life* is the manner in which the lamp is started by the ballast. Each time the lamp is started, no matter how perfectly the start is made, a certain quantity of coating is dislodged from the cathode during the first few cycles of operation.

Since lamp rating figures are generally based on a three hour burning cycle, the ratings reflect the effects of both starting and burning. Therefore any changes in the hours burned per cycle will affect life" [Fig. 19.12].

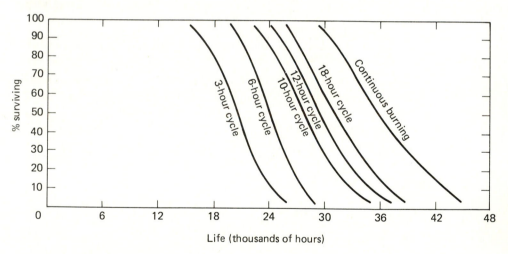

FIGURE 19.12. Life in thousands of hours. Typical mortality curves as a function of burning cycles for 40-watt rapid start lamps with a rated life of 20,000 hours. (Courtesy of GTE Lighting Products.)

19.4 HIGH-INTENSITY LIGHTING

This form of lighting, called *H.I.D.* (for *high-intensity discharge*) provides us with the most efficient of light sources (Fig. 19.13). Coupled with the best performance for lamp life, H.I.D. lighting has many excellent applications (Figs. 19.14 to 19.16). Depending on the type of lamp, there are varying degrees of color spectrum problems. The biggest disadvantage, however, to the H.I.D. lighting system is the very slow starting time of 2 to 5 minutes, compared with almost instantaneous incandescent and fluorescent lighting. Three forms of H.I.D. lighting commonly in use are (1) *mercury,* (2) *metal-halide,* and (3) *sodium.*

---

* This section is taken from *Effect of Burning Period on Fluorescent Lamp Life,* GTE Sylvania Engineering Bulletin 0-335, Sylvania Lighting Center, Danvers, Mass., courtesy of GTE Lighting Products.

FIGURE 19.13. High-pressure sodium luminaire. (Courtesy of Holophane Division of Manville Products Corporation.)

FIGURE 19.14. Maximum vertical footcandles, required for reading labels, tags, and so on, are achieved by a special prismatic glass reflector–refractor combination which bends and redirects the light at precise angles onto the stacks. (Courtesy of Manville Products Corporation.)

FIGURE 19.15. Comparison of HID versus fluorescent. (Courtesy of Manville Products Corporation.)

FIGURE 19.16. Comparison of HID versus fluorescent. (Courtesy of Manville Products Corporation.)

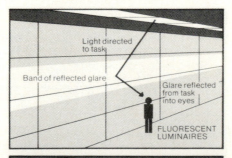

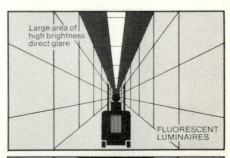

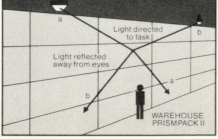

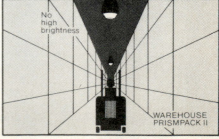

The Mercury Lamp*

"The *mercury* lamp belongs to the classification known as high intensity discharge (H.I.D.) lamps. In lamps of this type, light is produced by the passage of an electric current through a gas or vapor under pressure instead of through a tungsten wire as in the incandescent lamp.

The electrical circuit of a typical mercury lamp is shown schematically in Fig. 19.17. Ballasts of the correct size and type are required to operate mercury lamps on any standard electrical circuit to convert the distribution voltage of the lighting circuit to the required starting voltage for the lamp and to control the current during operation of the lamp. This current control is necessary because the mercury lamp, like all discharge light sources, has a "negative resistance" characteristic. Once started, the arc will "run away" with itself and draw an excessive current that will destroy the lamp if not controlled by a ballasting device.

When the line switch is turned on, the starting voltage of the ballast is impressed across the gap between the operating electrodes at opposite ends of the arc tube and also across the small gap between the operating electrode and the starting electrode. This ionizes the argon gas in the starting gap, but the current is limited to a small value by the starting resistor. When there is sufficient ionized argon and mercury vapor distributed throughout the arc tube, an arc strikes between the operating electrodes. This vaporizes more mercury, and the lamp quickly warms up to a stable condition. After the main arc strikes, the starting resistor causes the potential across the starting gap to be too low to maintain that discharge, and the lamp current flows between operating electrodes.

The ions and electrons which comprise the current flow, or "arc discharge," are set in motion at tremendous speeds on the path between the two operating electrodes at opposite ends of the arc tube. The impact of the speeding electrons and ions on the surrounding gas or vapor briefly changes their atomic structure. Light is produced from the energy given off by the affected atoms as they change back to their normal structure."

Efficacy. "*High light output* is one of the important advantages of mercury lamps [Fig. 19.18]. The initial efficacy (at 100 hours operation) ranges from 30 to 63 lumens per watt, depending on the wattage and color of the lamp. This does not include the ballast losses which should be added to the lamp watts when comparisons are made with other light sources."

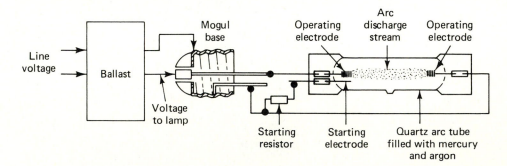

FIGURE 19.17.   Electrical circuit of a typical mercury lamp. (Courtesy of GTE Lighting Products.)

---

*This section is taken from *Metalarc Lamps,* GTE Sylvania Engineering Bulletin 0-344, Sylvania Lighting Center, Danvers, Mass., courtesy of GTE Lighting Products.

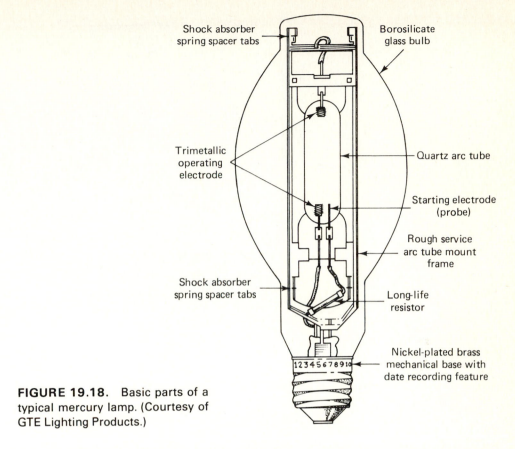

Shock absorber
spring spacer tabs

Borosilicate
glass bulb

Trimetallic
operating
electrode

Quartz arc tube

Starting electrode
(probe)

Rough service
arc tube mount
frame

Shock absorber
spring spacer tabs

Long-life
resistor

Nickel-plated brass
mechanical base with
date recording feature

FIGURE 19.18. Basic parts of a
typical mercury lamp. (Courtesy of
GTE Lighting Products.)

Spectral Energy Distribution. "The spectrum of a mercury lamp contains strong
lines in the ultraviolet and visible regions. The pressure in the quartz arc tube is
responsible to a large degree for the mercury lamp's characteristic spectral energy
distribution. The exact spectral distribution varies greatly with the pressure at which
the arc tube operates" [Figs. 19.19 and 19.20].

FIGURE 19.19. Mercury vapor luminaire. (Courtesy of Thomas Industries Inc.)

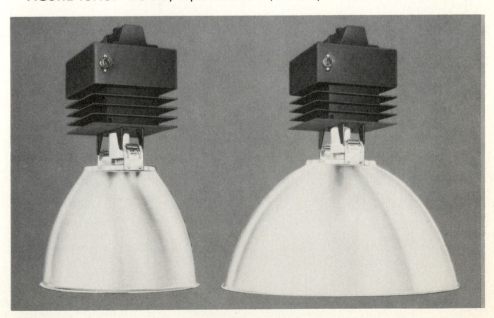

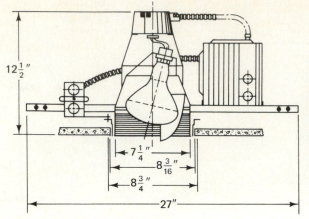

FIGURE 19.20. Wall washer. In these units a mercury vapor lamp is combined with a half-ellipsoidal type reflector, a black baffle, and a scoop trim. The wall is flooded with an even sheet of light while the aperture remains dark. (Courtesy of Lighting Products Division, McGraw-Edison Company.)

The Metal-Halide Lamp*

"The *metal-halide* lamp has a quartz arc tube which is slightly smaller than that of the same wattage mercury lamp [Fig. 19.21]. The arc tube contains argon gas and mercury plus thorium iodide, sodium iodide and scandium iodide. These three materials are responsible for the outstanding performance of this remarkable light source. The ends of the arc tube have a heat retention coating that raises the temperature of the ends during operation, thus assuring adequate evaporation of the metal iodides. The arc tube harness includes spring supports at the neck and dome, which make the mount structure very durable and resistant to rough service and vibration. The bimetal shorting switch in the metal-halide lamp closes during lamp operation, providing a short circuit between the starting electrode and the adjacent main electrode. This prevents the flow of electric current through arc tube materials that may collect

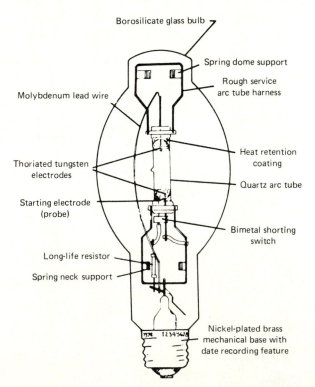

FIGURE 19.21. Construction of a 400-watt, base-up-burning metal-halide lamp. (Courtesy of GTE Lighting Products.)

---

*This section is taken from *Metalarc Lamps,* GTE Sylvania Engineering Bulletin 0-344, Sylvania Lighting Center, Danvers, Mass., courtesy of GTE Lighting Products.

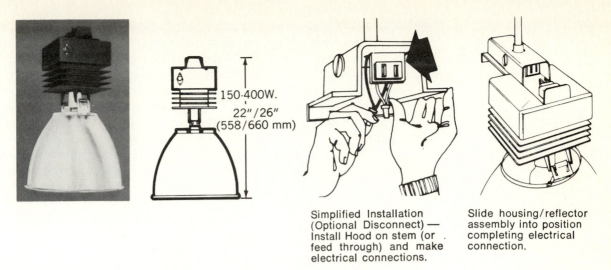

150-400W.
22"/26"
(558/660 mm)

Simplified Installation
(Optional Disconnect) —
Install Hood on stem (or .
feed through) and make
electrical connections.

Slide housing/reflector
assembly into position
completing electrical
connection.

FIGURE 19.22.   Metal halide 250 watt/1000 watt luminaire. (Courtesy of Thomas Industries Inc.)

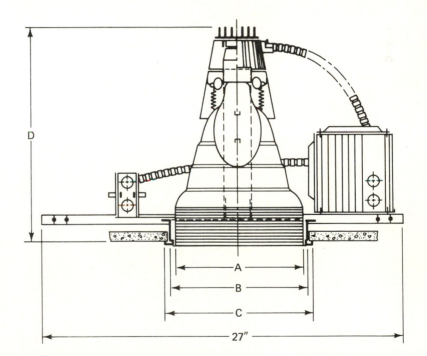

FIGURE 19.23.   Ceiling-mounted HID fixture. (Courtesy of Lighting Products Division, McGraw-Edison Company.)

between the starting and main electrode and eliminates electrolytic failure of the molybdenum foil within the seal" [Figs. 19.22 and 19.23].

The Sodium Vapor Lamp*

"The *sodium vapor* lamp is one of the high intensity discharge (H.I.D.) lamp family. Its physical, electrical and photometric characteristics are quite different from those of

*This section is taken from GTE Sylvania Engineering Bulletin 0-348, Sylvania Lighting Center, Danvers, Mass., courtesy of GTE Lighting Products.

other H.I.D. lamps. However, light is produced in a similar manner by the passage of an electric current through a vapor under pressure in an arc tube [Fig. 19.24].

The electrical circuit of a sodium vapor lamp is shown schematically in Fig. 19.25. The sketch indicates that a starting electrode and starting resistor are not required as in mercury and metal-arc lamps. A ballast of the correct size and type is required to operate the lamp. The function of the ballast as with other H.I.D. lamps is to provide the necessary starting voltage and to control the current during lamp operation.

The principal radiating element in the arc of the lamp is sodium. Mercury is added for color and voltage control. Xenon gas is also introduced to facilitate starting. Since the arc tube does not conveniently allow the use of a starting electrode (probe), a special starting circuit is used to supply a short high voltage pulse on each cycle or half cycle. This pulse is of sufficient amplitude and duration to ionize the xenon gas and to initiate the starting sequence of the lamp.

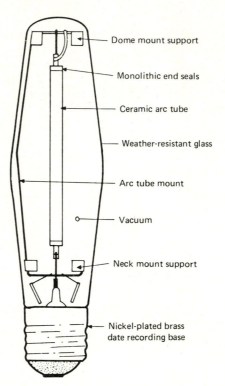

Dome mount support

Monolithic end seals

Ceramic arc tube

Weather-resistant glass

Arc tube mount

Vacuum

Neck mount support

Nickel-plated brass date recording base

FIGURE 19.24. Basic parts of a sodium vapor lamp. (Courtesy of GTE Lighting Products.)

FIGURE 19.25. Electrical circuit of a sodium vapor lamp. (Courtesy of GTE Lighting Products.)

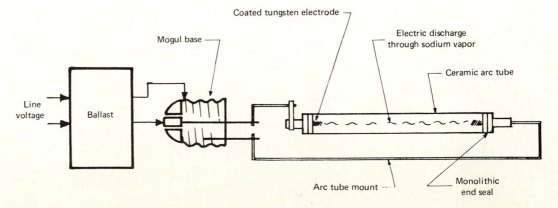

Coated tungsten electrode

Electric discharge through sodium vapor

Ceramic arc tube

Mogul base

Line voltage

Ballast

Arc tube mount

Monolithic end seal

FIGURE 19.26. Sodium vapor lighting. (Courtesy of Holophane Division of Manville Products Corporation.)

FIGURE 19.27. Gasketed types of outdoor HID fixtures. (Courtesy of Thomas Industries Inc.)

1¾" PENDANT
2" STANCHION
1¾" WALL
1½" CEILING

14½"

16"

1¾" PENDANT
2" STANCHION
1¾" WALL
1½" CEILING

13¾"

12"

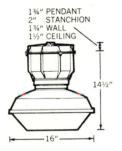

The warm-up period for the lamp to reach full brightness is about 3 to 4 minutes—somewhat less than that of a mercury or metal-arc lamp. During warm-up, there are several changes in the color of the light. Initially, there is a very dim, bluish-white glow produced by ionized xenon which is quickly replaced by a typical blue, brighter mercury light. With an increase in brightness, there is a change to monochromatic yellow which is characteristic of sodium at low pressure and temperature. Then as the pressure in the arc tube increases, the lamp comes to full brightness with a golden white light. Should there be a momentary interruption of power, the restrike time is approximately one minute" [Fig. 19.26].

H.I.D. Lighting Applications

Mercury and *sodium vapor* lamps are generally found in outdoor uses such as street lighting and industrial storage areas (Figs. 19.27 and 19.28). The *metal-halide* lamps are very well suited for industrial uses that require relatively high mounting of the

FIGURE 19.28. Outdoor HID luminaire. (Courtesy of Thomas Industries Inc.)

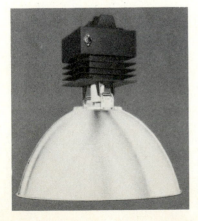

luminaires (Figs. 19.29 and 19.30). One example is the use of 900 of the 400-watt metal-halide lamps mounted at 20 ft that provide 80 footcandles for the dirty atmosphere of a metal stamping and forming plant. At Cape Kennedy's Jet Propulsion Laboratories Spacecraft Assembly area, *500* footcandles of white light give faithful color rendition using 1000-watt metal-halide lamps. In chemical and petroleum plants where color identification of pipes, circuits, and safety devices is extremely important, metal-halide lamps are being specified by increasing numbers of designers.

FIGURE 19.29. Metal halide lighting. (Courtesy of Holophane Division of Manville Products Corporation.)

FIGURE 19.30. High-intensity lighting. (Courtesy of Holophane Division of Manville Products Corporation.)

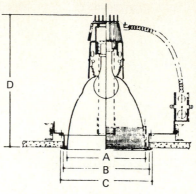

FIGURE 19.31. Incandescent downlight. These incandescent units are designed as auxiliary units for HID installations. These units can be either wired into an emergency system or installed to operate as night lights or when the HID lamps are starting up. (Courtesy of Lighting Products Division, McGraw-Edison Company.)

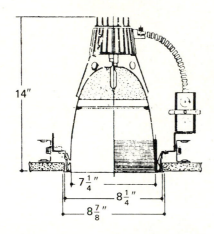

FIGURE 19.32. Quartz downlight, for use as emergency or standby lighting units. The tungsten-halogen lamps used in these fixtures permit use of a higher wattage and greater efficiencies than the A lamp incandescent units. (Courtesy of Lighting Products Division, McGraw-Edison Company.)

Because they offer high efficacy and good color rendition, metal-halide lamps are well suited for sports arenas and stadiums. The special high-intensity-light demands of color TV make this an excellent choice in these locations. Particularly in industrial plants, gymnasiums, and sports arenas, the 2 to 5 minutes of starting time or restarting time is unacceptable. In most cases an incandescent or fluorescent system is used as a backup, providing minimum lighting (Figs. 19.31 and 19.32).

19.5 AUTOMATIC ENERGY CONTROL*

"One of the most significant developments of the last decade in H.I.D. lighting is the *Automatic Energy Control* (AEC) system. Extending some of the concepts and equipment first developed for H.I.D. dimming, this new option to H.I.D. lighting systems can save dollars and energy for the system owner [Fig. 19.33].

It uses a photocell sensor to read the illumination level in a given area. When that level is too high, the AEC system automatically reduces power to lamps in the system. If the level is low, power is increased automatically.

It maintains exactly the illumination level desired. The owner doesn't use or pay for energy not actually needed. And since total system power relates directly to lamp power, the savings are indeed significant.

During system installation, the AEC monitor is set for the desired illumination level. When the system is first turned on, the AEC equipment immediately turns down lamp power and thereby reduces the energy being consumed. As lamp lumen falls off with use and as dirt effects reduce the total illumination level, the photocell sensor "calls for" more light and the AEC system ups lamp power.

---

*Section 19.5 is taken from Bulletin 0212-0180, *Wide-Lite's Modular Dimming Systems*, Wide-Lite Corporation, San Marcos, Tex., courtesy of Wide-Lite Corporation.

TYPICAL SAVINGS WITH AUTOMATIC ENERGY CONTROL
kWh Saved Per Luminaire Over Conventional Operation

Lamp Type	Annual Savings[1] AEC Only		Annual Savings[2] AEC/Dimming Cycle		Annual Savings[3] AEC/Dimming Cycle/Daylight	
	kWh	Cost	kWh	Cost	kWh	Cost
1000w DX mercury vapor	1946	$ 97.00	2636	$132.00	3128	$156.00
1000w metal halide	620	$ 31.00	1291	$ 65.00	1771	$ 89.00
1000w HPS	2031	$102.00	2644	$132.00	3081	$154.00

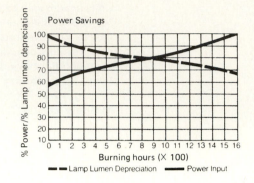

Power Savings

[1] Based on continuous operation in dirty environment at $0.05/kWh. Savings result from constant level illumination compensation for LLD and LDD.

[2] As above, plus savings from 4 hr/day operation dimmed to 30% light output.

[3] As above, plus savings from 25% daylight contribution to illumination level 8 hr/day.

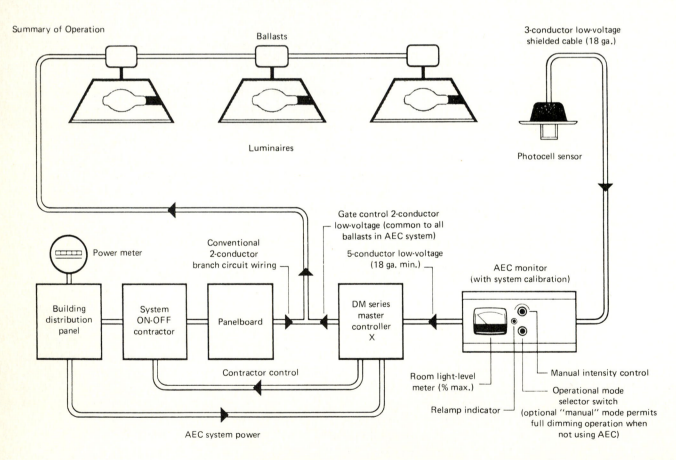

FIGURE 19.33. Automatic energy control. (Courtesy of Widelite Corporation.)

When the system has increased lamp power to maximum value, an indicator lamp comes on to warn that it is time to relamp.

If the area receives supplemental light—such as from daylight—the AEC simply reduces lamp power accordingly to yield further energy and cost savings!

Major Factors Affecting Savings with AEC

Long Burning Hours. An AEC system operating 18 hrs/day saves about twice as much energy as a 9 hrs/day operation. AEC is ideal for typical industrial schedules.

High Power Rates. Considers not only electricity cost but also the demand charge and projected rate of increase.

Daylighting. Takes advantage of daylight hours from windows, skylights.

Multi-level Operation. Makes use of manual or computer-programmed dimming to custom tailor lighting to the task. For shift changes, clean-up, restocking, or maintenance, use lower light levels for significant savings.

High Lamp Wattage. Takes advantage of the greater efficacy (lumens per watt); set and maintain the luman output as required for the task for maximum economy."

19.6 LUMINOUS CEILINGS

An outgrowth of indirect lighting, *luminous ceilings* have great aesthetic appeal. Architects and illumination engineers have many different artistic ideas to select from, but the basic scheme produces a ceiling of diffused or broken up light using attractive metal and glass designs. Generally, the light source reflects downward from white painted surfaces or strip lighting with reflectors. Sometimes a separate reflecting surface of drywall materials is constructed under the structural floor slab. The illumination level usually runs fairly high since luminous ceilings are used primarily in lobbies and other areas of high-occupancy concentration. Since luminous ceilings are generally custom designed and made, early detailing, sample submission, and approval will help ensure on-time construction. Fluorescent lighting produces special effects where it provides the illumination for interior stained glass windows and imitation sunlight for basement areas.

19.7 FIXTURES: ORDERING, DELIVERY, AND STORAGE

The electrical portion of a project requires considerable care in providing all the materials and fixtures at the right time and place. Whether the electricians order conduit materials, wires and cables, or the fixtures, there are many different items that make up the order list. The designers usually provide a "Lighting Fixture Schedule" which lists the types of fixtures, a description, the number and type of lamps, and manufacturer information. Figure 19.34 is an example of a schedule for a medium-size four-story office building. Thirty-five types of lamps are shown.

The 1 × 4 ft, 2 × 4 ft, and 4 × 4 ft fluorescent fixtures take up a great deal of space in the manufacturing plant and on the project. In any sizable quantity they are palletized for easier shipment, unloading, hoisting, and project storage. Manufacturers have given up putting each fixture in a separate corrugated paper box, which was time consuming, costly, and buried the bigger projects in volumes of highly inflammable waste material. By use of compressible padding and polyethylene covering the fixtures are banded to wood pallets for easy mechanical and semimechanical handling. Care should be exercised in selecting on-site storage areas to protect against physical damage from the weather, scaffolding, debris, and general construction activity.

| TYPE | DESCRIPTION | LAMPS | | MANUFACTURER | |
		NO.	TYPE	NAME	MODEL NO.
AH2	1' × 4' fluorescent-recessed	2	40 W fl. C.W.	Day-Brite	4FC21-4
AH3	2' × 4' fluorescent-recessed	3	40 W fl. C.W.	Day-Brite	4FC32-4
AH4	2' × 4' fluorescent-recessed	4	40 W fl. C.W.	Day-Brite	4FC42-4
AH44	4' × 4' fluorescent-recessed	4	40 W fl. C.W.	Day-Brite	4FC44-4
AH46	4' × 4' fluorescent-recessed	6	40 W fl. C.W.	Day-Brite	4FC64-4
AH6	1' × 4" fluorescent-recessed	3	40 W fl. C.W.	Day-Brite	4FC31-4
B2	1' × 4' fluorescent sur. mtd.	2	40 W fl. C.W.	Curtis Electro.	DX-5212M
B44	4' × 4' fluorescent sur. mtd.	4	40 W fl. C.W.	Curtis Electro.	DX-5444
B4W	2' × 4' fluorescent sur. mtd.	4	40 W fl. C.W.	Benjamin	CO-7244-4
B44W	4' × 4' fluorescent sur. mtd.	4	40 W fl. C.W.	Benjamin	CO-27244-4
A2	1' × 4' fluorescent-recessed	2	40 W fl. C.W.	Day-Brite	4-T21WF-818
D	Incandescent downlight	1	75 W par 38	Marco	D4-T411J
D1	Incandescent downlight	1	150 W par 38	Marco	D4-T411J
DC	Incandescent downlight	1	150 W par 38	Devine Lighting	HCD-2074
DW1	Incandescent downlight	1	150 W R40	Marco	A7-T173
F	Incandescent square	1	100 W	Marco	S1-T152J
F1	Incandescent square	1	150 W	Marco	S1-T152J
G	Incandescent square	1	150 W	Marco	D415-A54/P15
L22	Wall fluorescent	2	20 W fl. C.W.	Tech. Lite Co.	Profile C-227
L42	Wall fluorescent	2	40 W fl. C.W.	Tech Lite Co.	Profile C-227
VT	Vapor tight	1	100 W	Russell · Stoll	4940A
M22	Mirror light	2	20 W fl. C.W.	Curtis-electro-"M Series"	M-220-T3H
M42	Mirror light	2	40 W fl. C.W.	Curtis-electro-"M Series"	M-240
OC	Special	2	Up 150 W/dn. 150 W	McPhilben	Special
VE21	Dual pole light	2	Each 150 W par	Stonco	40 L with 207 guard
PL	Special	1	150 W	Sterner	B4080
S21	Strip light	1	20 W fl. C.W.	Curtis-electro	B120TSL
S41	Strip light	1	40 W fl. C.W.	Curtis-electro	B140RS8
S32	Strip light	2	30 W fl. C.W.	Curtis-electro	B230RSH
UC41	Undercounter strip	1	40 W fl.	Lightolier	10247
UC31	Undercounter strip	1	30 W fl.	Lightolier	10245
3R42	Strip light w/reflector	2	40 W fl. C.W.	Curtis-Electro	B240SE W/848255 REF
VW	Wall light	1	150 W	Marco	F615-V13
Exit signs		2	25 W	McPhilben "50 Line"	50W-6MG(wall), 50C-6MG(Cig.)
OF	Outside flood	1	Q300T3	Stonco	8501

FIGURE 19.34. Lighting fixture schedule.

19.8 CEILING SEQUENCING

In an area ready for tenant work consisting of dividing partitions, ceilings, and finishes, it is usually best to start with the sheet metal ductwork. Ductwork is bulky, requires a lot of area for storage, and starting at the building core winds to the building perimeter. There are no hard and fast rules as to who goes first except the obvious need of one trade to provide for the next and a striving by the construction manager to have all trades work at their overall greatest efficiency. Sprinkler work on main and branch lines can proceed with the lighting conduit network, boxes, flexible cable, and wire pulling. When the initial roughing has progressed sufficiently, the metal-carrying system for the acoustical ceiling goes in. The roughing also might include the plumber, HVAC control lines, and the insulator. The electrician follows the ceiling support system progress, placing the fixtures into the grid system (Figs. 19.35 and 19.36). If the supply air enters the room through the fluorescent light fixture, the sheet metal workers make the connection between the lighting fixture and the rectangular ductwork with flexible ductwork. The electrician wires the fixture into the electrical system and the area is ready for installation of the acoustic material.

FIGURE 19.35.   Fixture installed in a ceiling support system.

FIGURE 19.36.   Flexible duct air conditioning connection: metal duct to lighting fixture. (Courtesy of Wiremold Company.)

19.9   PREFABRICATION   Some projects offer opportunities for *prefabrication* by the electrician. Conduit work for repetitive structural slabs in high-rise buildings and lighting roughing for apartment and office buildings are two examples. Figures 19.37 to 19.40 depict a series of simple tasks which greatly cut costs when performed as a repetitive production-oriented operation. In this case, the lighting roughing assemblies are prefabricated, ready to be hung from the structural slab.

FIGURE 19.37.   Roughing prefabrication: preassembly of junction boxes. (Courtesy of Truland Systems Corporation.)

FIGURE 19.38.   Roughing prefabrication: the precut wire (left center) has been pulled through prepared cable (left) ready for final assembly. (Courtesy of Truland Systems Corporation.)

FIGURE 19.39.   Roughing prefabrication: joining boxes, conduit, and cable. (Courtesy of Truland Systems Corporation.)

FIGURE 19.40.   Roughing prefabrication: assembled units taped together for shipping and job handling, ready for delivery to the project. (Courtesy of Truland Systems Corporation.)

20

Elevators and Escalators

20.1 THE ELEVATOR INDUSTRY

Two very large corporations and four other big companies tend to dominate the elevator industry. These companies manufacture, construct, and maintain the elevators they sell. Sixty or more companies do not manufacture or only make small portions of the elevators they install. The industry employs personnel trained through an education program conducted by the *National Elevator Industry Inc.* These workmen belong to the *International Union of Elevator Constructors,* but unlike many of the other AFL-CIO building trades members who change employers frequently, elevator construction personnel usually work many years for one firm. With some local exceptions they also install, adjust, and maintain both the mechanical and electrical parts of an elevator as the regular work of their craft. As a result, elevator constructors are considerably more versatile than most other construction workers. Unlike other specialty contractors, who avoid maintenance work like a plague, elevator companies depend more on maintenance than on new construction for their profit. Although the importation of elevators and escalators is on the increase, the foreign companies have not made a significant impact on the industry at this time.

20.2 ELEVATORS AND THE PROJECT

An elevator construction program for a major high-rise building has many facets which interest the construction manager, the owner, and the tenants. On these large projects the construction manager usually needs all the vertical transportation he can get. It frequently takes a half-hour morning and afternoon to carry every worker to his assigned floor, a frightful loss in production and money. The construction manager looks to the elevator program for

1. *Temporary elevators* with *temporary motors* and *cables*
2. *Permanent platforms* and *motors* brought to an acceptable degree of readiness and with *temporary cabs* used as temporary elevators and/or construction hoists
3. *Completed elevators* protected and used as *temporary elevators* and *hoists*

Elevators for this use will expedite the removal of platform hoists, climbing, and ground-operated cranes, which in turn expedites completion of the building exterior. As the project nears completion the owner wants to control as many elevators as possible. Owners do not relish sharing elevator usage with construction workers, nor the dirt they track into the elevator and the lobby. Owners want elevators for use by building maintenance personnel and for moving tenants into the building.

Tenants also do not want to share elevators with the construction people. They look to the owners to live up to the lease agreement and provide the elevator service agreed upon.

Most owners consider high-rise buildings too expensive to lie idle until the whole building has been completed. As a result, sections of the building are occupied while the balance of the project goes through the completion stages. This produces an extended period where the three parties mentioned above all clamor for elevators. To have a successful high-rise project, it is essential to have a successful elevator program.

20.3 PASSENGER ELEVATOR CONSIDERATIONS

Our big cities are, to a great extent, the product of our building codes and the laws of economics. The size of a city's blocks help dictate the height of high-rises built there. The combination of block size, code, and most efficient use of a piece of property has resulted in the building of hundreds of inner-city buildings 30 to 40 stories high. The potential gross earning power of a building can be measured in net rentable area, the theoretic area a tenant could use. A very important decision an owner has to make in high-rise buildings is the quality of elevator service he or she plans to provide. The more elevators the happier the building occupants, but the more elevators the less the rentable area. In addition, the more elevators the higher the building cost. There are many factors that tenants consider in reaching a decision on renting a property, particularly location and rental per square foot, but elevator service is one of the most important.

A major building usually has passenger and freight elevators of both the traction and hydraulic types, one or more escalators, and perhaps a dumbwaiter. The elevators are usually grouped into banks of four to eight elevators, each bank serving approximately 12 floors. The banks are in the core of the building and the core usually, but not always, is in the center of the building. The floor space above the low-level bank elevator machine room can be developed as office space. Elevators serving the upper areas—for example, above the 14th floor—travel in enclosed shafts which occupy valuable space. The areas *between* these shafts can be utilized for toilet rooms or service areas.

20.4 PASSENGER ELEVATOR DESIGN

A passenger elevator must have big door openings for rapid loading and unloading, must accelerate and decelerate quickly without uncomfortable human sensation, travel at high speed (1600 ft per minute in the tallest buildings), be completely safe, and have an intricate control system to handle the greatest number of passengers in the shortest possible time. Figure 20.1 shows a typical installation for an apartment

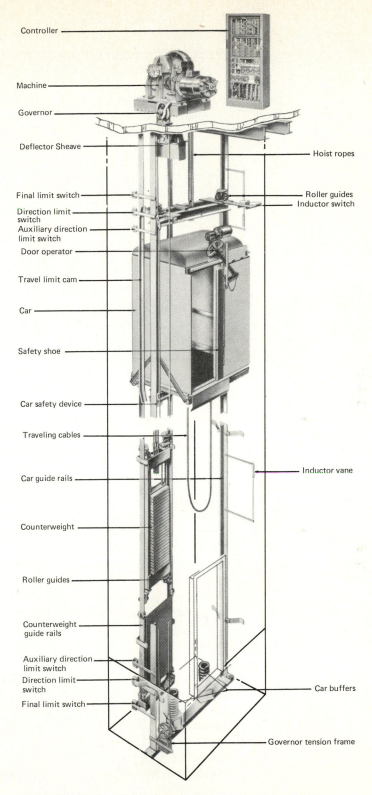

FIGURE 20.1. Typical installation: geared machine, ac resistance control. (Courtesy of Otis Elevator Co.)

building, a slightly simpler arrangement than in a big office building. The essential parts of an elevator system are depicted. Note that the cab mounts on a platform supported on a steel frame or sling which hangs from the hoist ropes. The elevator machine in this example is a gearless ac (alternating current) machine for slower and lighter duty and therefore does not require a motor generator set to provide dc (direct current) to the elevator machine as in the bigger office building installations. The single door would also not be adequate for a large office building population. An elevator is a balanced machine—the cab, platform, and sling balanced against counterweights traveling on a set of counterweight guide rails (Fig. 20.2). Both car and counterweights travel smoothly on roller guides (Fig. 20.3). The rails are carefully machined to fit together with tongue-and-groove joints and are firmly held in place after careful alignment (Figs. 20.4 to 20.6).

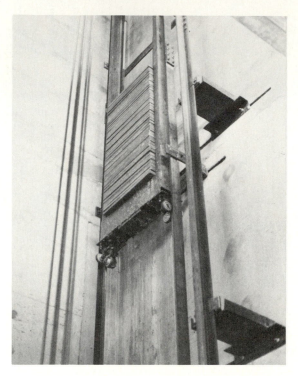

FIGURE 20.2. Counterweight, frame, and counterweight guide rails.

FIGURE 20.3.   Roller guides. (Courtesy of Otis Elevator Co.)

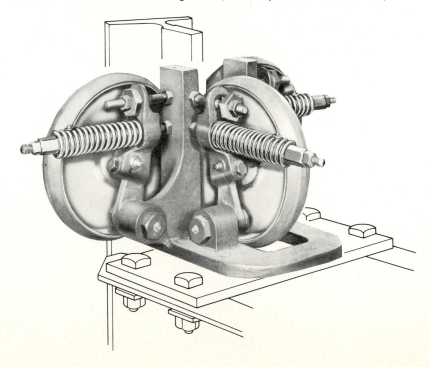

FIGURE 20.4. Setting rail brackets, one of the first steps in elevator construction.

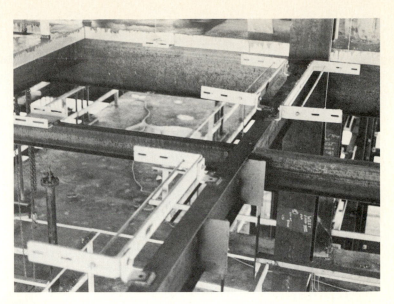

FIGURE 20.5. Rails stored on the site.

FIGURE 20.6. Connecting rails.

20.5 TRACTION ELEVATOR MACHINES

Both *ac* and *dc motors* are used in elevatoring and both *geared* and *gearless* machines. For high-speed, large-capacity cars for tall buildings a *gearless traction machine* (Fig. 20.7) operating on *direct current* from a *motor generator set* (Fig. 20.8) gives the highest performance. The dc motor has excellent acceleration characteristics as well as smooth motor-controlled slowing. However, in addition to the cost of the motor generator, this installation sometimes requires a two-level elevator machine room, which adds to the initial cost. Geared traction machines with motor generator sets are used on both freight and passenger elevators (Figs. 20.9 and 20.10). The mechanical advantage of gearing produces a more powerful but somewhat slower elevator. Figure 20.11 shows a *geared traction machine* with an ac motor. This less expensive machine drives low-speed freight and passenger cars. To improve the drive action between the elevator motor drive sheave and the wire ropes, nonmetallic inserts are fastened to the sheaves (Fig. 20.12). Manufacturers claim that they also double the life of the hoist ropes.

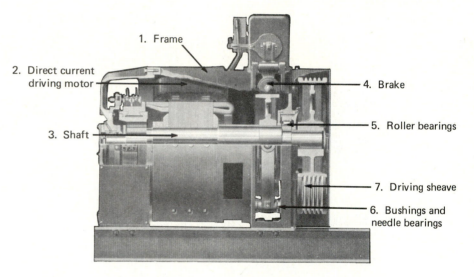

1. Frame
2. Direct current driving motor
3. Shaft
4. Brake
5. Roller bearings
7. Driving sheave
6. Bushings and needle bearings

FIGURE 20.7.   Gearless elevator machine (items 1 to 7). (Courtesy of Otis Elevator Co.)

FIGURE 20.8.   Motor generator set. (Courtesy of Otis Elevator Co.)

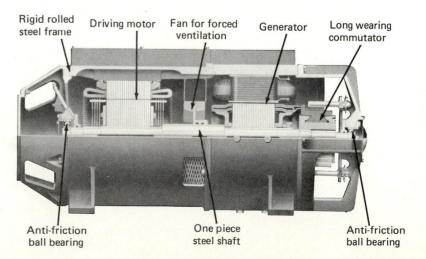

Rigid rolled steel frame
Driving motor
Fan for forced ventilation
Generator
Long wearing commutator
Anti-friction ball bearing
One piece steel shaft
Anti-friction ball bearing

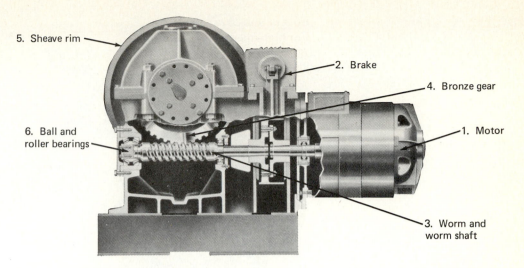

5. Sheave rim

2. Brake

4. Bronze gear

6. Ball and roller bearings

1. Motor

3. Worm and worm shaft

FIGURE 20.9. Geared traction machine with an ac motor. (Courtesy of Otis Elevator Co.)

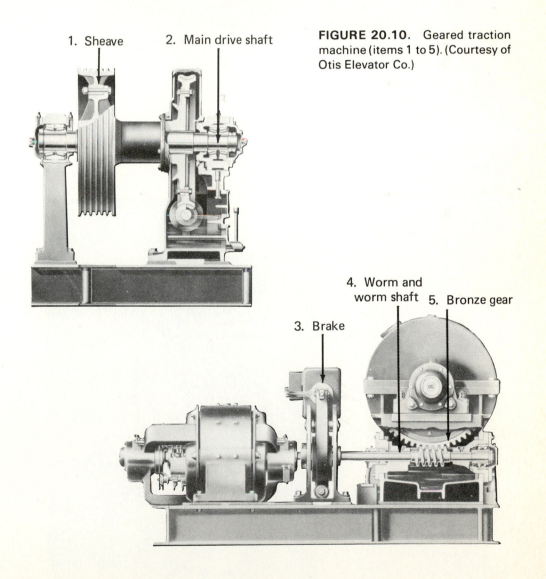

1. Sheave

2. Main drive shaft

FIGURE 20.10. Geared traction machine (items 1 to 5). (Courtesy of Otis Elevator Co.)

4. Worm and worm shaft

5. Bronze gear

3. Brake

FIGURE 20.11. Geared traction machine with unit multivoltage control. Geared traction machines are used on both freight and passenger elevators. Machines of this type transmit the power developed by the driving motor to the drive sheave through a worm and gear, and transmit motion to the hoist ropes by the traction between the ropes and the grooved driving sheave. (Courtesy of Otis Elevator Co.)

FIGURE 20.12. Nonmetallic inserts on the sheaves prolong cable life. (Courtesy of Otis Elevator Co.)

20.6   THE SAFETY SYSTEM

"*The safety system* protects the elevator car from over-speeding downward [Fig. 20.13]. Completely independent of the elevator machine and brake, the system consists of a speed governor, car safeties, safety switch, governor rope and tension sheave. The steel governor rope, attached at both ends to the lift rods of the car safeties, passes over the drive sheave of the governor at the top of the hoistway and under a tension sheave at the bottom of the hoistway.

The safety system has no influence on the operation of the elevator, unless the elevator speed reaches or exceeds the pre-set tripping speed of the governor. Should an overspeed occur, the governor "trips" and clutches the governor rope to "set" the safeties by bringing the rollers into contact with the rails. Simultaneously, the safety switch cuts off power to the elevator machine, applying the service brake.

After the car has stopped, the safeties continue to grip the rails and hold the car stationary. When the cause of the overspeed has been corrected, the safety may be reset by running the car a few inches in the UP direction, reversing the wedging action of the clamps and returning the rollers to the "Ready" position, thus eliminating manual re-setting. Roll type safeties cannot be released from within the car unless the elevator machinery is in condition to lift the car. No wrenches, levers or special tools are required to release these safeties and car floors are free of wrench holes or cover plates.

Roll type safeties [Fig. 20.14] are designed to stop the car quickly and independently of speed action by the governor in the event of sudden, rapid acceleration in the down direction. Should such acceleration occur, the inertia of the governor rope system resists this motion with sufficient force to raise the lift rods and set the safety. This action is known as *Inertia Tripping* of the safety system."*

A similar type of safety device is the *flexible guide clamp* (Fig. 20.15), which functions in the same manner as the roll type. Each clamp has two steel jaws to grip the

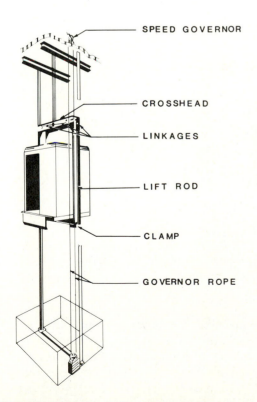

SPEED GOVERNOR

CROSSHEAD

LINKAGES

LIFT ROD

CLAMP

GOVERNOR ROPE

FIGURE 20.13. Safety system. (Courtesy of Otis Elevator Co.)

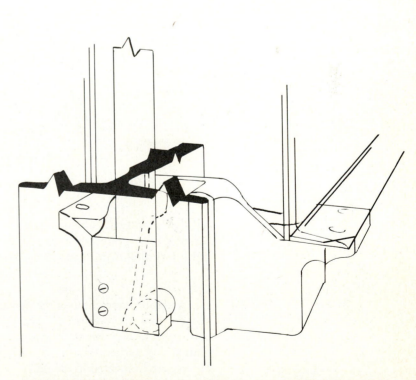

FIGURE 20.14. Roll-type safeties. (Courtesy of Otis Elevator Co.)

*Otis Elevator Company, Bulletins B-600 [0780] and B-406 [8630].

FIGURE 20.15.   Flexible guide clamp safety. (Courtesy of Otis Elevator Co.)

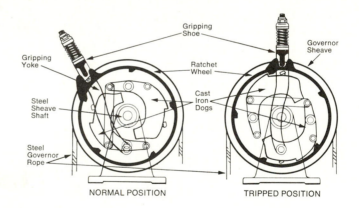

FIGURE 20.16.   Safety governor. (Courtesy of Otis Elevator Co.)

rail and a heavy flexible spring to regulate the pressure exerted by the jaws. When actuated by overspeed, both clamps operate simultaneously to bring the car to a smooth sliding stop. Figure 20.16 shows the tripping action of the safety governor.

20.7   THE SHAFT   The machine room houses the elevator machine, the motor generator set if one is needed, and the controller. The shaft, including the pit, contains the sling and platform; the cab, which may or may not be furnished by the elevator company; the shaft doors and entrances; the rails and counterweights; the safety system; and the buffer system located in the pit (Figs. 20.17 and 20.18). The structural elements that create the shaft, together with the enclosing architectural cement masonry units, drywall, or brick, should be built to a tolerance of ±½ in. There is a definite relationship between the structural parts of the shaft, the rails, and the hallway entrances. Figure 20.19 illustrates the hallway door construction.

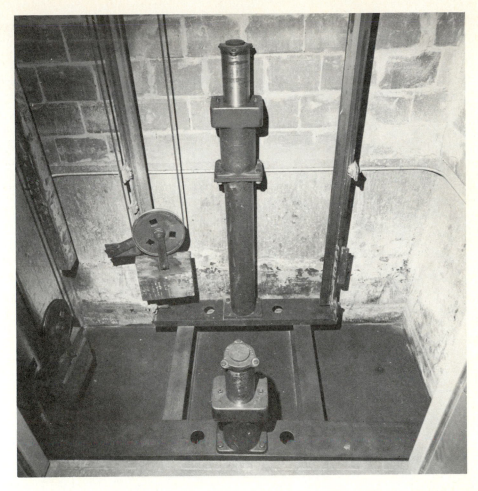

FIGURE 20.17.   Elevator pit, showing the location of buffers. (Courtesy of National Elevator Industry Education Program.)

FIGURE 20.18.   Buffers for pits stored on the job site.

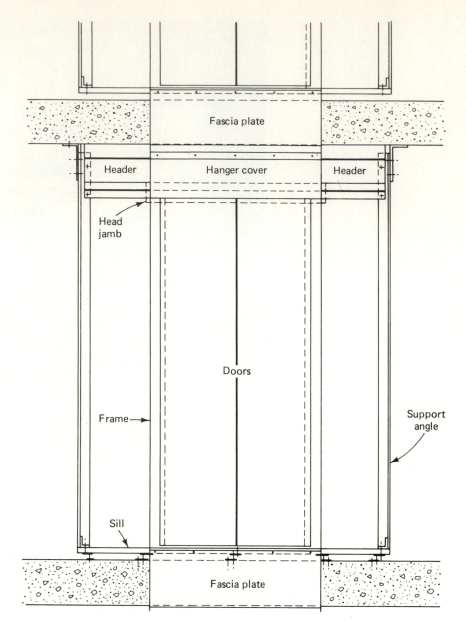

FIGURE 20.19.   Hoistway view. (Courtesy of Otis Elevator Co.)

20.8 TYPES OF ELEVATORS

The Freight Elevators

The *freight elevators* in a high-rise building are very important to the functioning of the completed building. They operate at a moderately high speed and can carry rather large and heavy loads. They are particularly useful since they have a good-sized platform, stop at every floor, and are accessible from the off-street trucking area. This type of elevator is depicted in Fig. 20.20. Note that a *geared traction machine* driven by a *dc motor* and a *motor generator set* provides excellent lifting capacity with moderate speed. Note also that the car is roped 2:1, with larger sheaves used at the top of the car and at the counterweight. This cuts the travel speed in half and practically doubles its

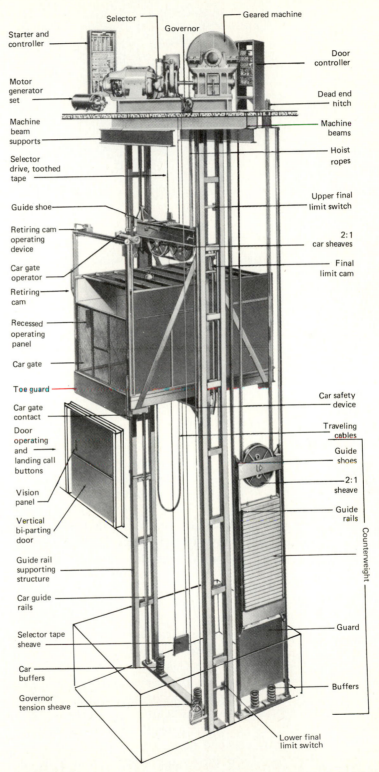

Selector

Starter and controller

Governor

Geared machine

Door controller

Motor generator set

Machine beam supports

Dead end hitch

Machine beams

Selector drive, toothed tape

Hoist ropes

Guide shoe

Upper final limit switch

Retiring cam operating device

2:1 car sheaves

Car gate operator

Final limit cam

Retiring cam

Recessed operating panel

Car gate

Toe guard

Car safety device

Car gate contact

Traveling cables

Door operating and landing call buttons

Guide shoes

2:1 sheave

Vision panel

Guide rails

Vertical bi-parting door

Counterweight

Guide rail supporting structure

Car guide rails

Selector tape sheave

Guard

Car buffers

Governor tension sheave

Buffers

Lower final limit switch

FIGURE 20.20. Heavy-duty freight elevator. (Courtesy of Otis Elevator Co.)

FIGURE 20.21. Freight elevator car. (Courtesy of Otis Elevator Co.)

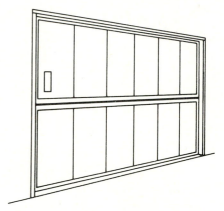

FIGURE 20.22. Freight elevator hoistway door. (Courtesy of Otis Elevator Co.)

lifting capacity. Although the cab (Fig. 20.21), car gate, and hall doors (Fig. 20.22) are simple and utilitarian, the freight elevator very nearly resembles the passenger type. Perhaps the greatest difference comes from simplifying the elevator programming.

The Double-Deck Elevator

Some of the newer and tallest buildings in North America, such as the Time-Life and Standard Oil of Indiana buildings in Chicago, the John Hancock building in Boston, and the Commerce Court in Toronto, have utilized the *double-deck* elevatoring method. In this method the elevator has two cabs, one a story above the other, riding on the same sling and operated by the same motor and controls. At the lobby level passengers board the car at the level serving odd-numbered floors or the level serving even-numbered floors (Fig. 20.23). The elevator traveling upward stops only at every second floor. Traveling in the down direction the elevator circuitry switches to a different mode and stops at both odd- and even-numbered floors to allow the passenger to travel from an odd to an even floor, or vice versa.

It is estimated that double decking increases passenger handling by 25 to 50%, permitting reducing the number of elevators and hoistways by one-third. Figure 20.24 shows the potential savings that double decking can produce.

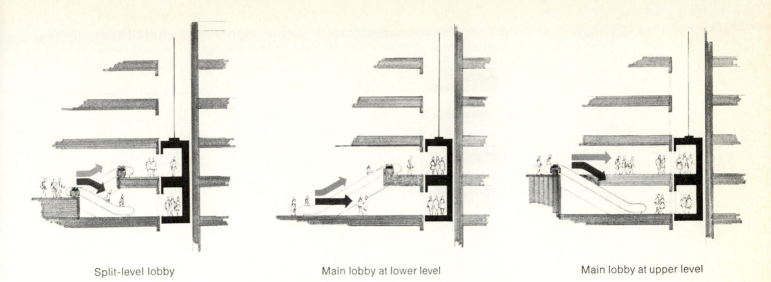

Split-level lobby   Main lobby at lower level   Main lobby at upper level

FIGURE 20.23. Split-level lobby: the main lobby at lower and upper levels. (Courtesy of Otis Elevator Co.)

FIGURE 20.24. Space saving using double-deck elevatoring. (Courtesy of Otis Elevator Co.)

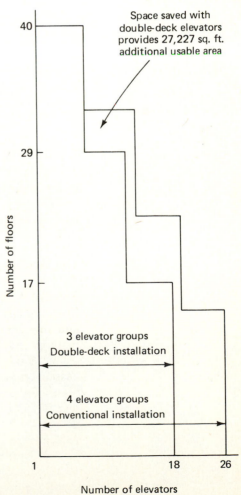

"Potential savings with double-deck elevators, as demonstrated by the following, results from a recent survey of a 40 story office building, with both systems meeting the same service criteria.

	Double-deck	Conventional
Elevator groups	3	4
Elevator units	18	26
Hoistway entrances	264	301
Total elevator core area required (including walls, lobbies and machine rooms)	56,192 sq. ft.	83,419 sq. ft.
Total area saved	27,227 sq. ft.	

"For this building, the conventional installation requires 48% more elevator core area than the double-deck installation. Even greater savings are possible, but each building must be considered individually."*

The Hydraulic Elevator

Hydraulic elevators are used extensively for freight and passengers in low-rise buildings of all kinds and to serve various basement levels of tall buildings. They come in all capacities since the hydraulic principle seems to have no physical limitations. Some buildings have 30-ton hydraulic truck elevators to lower and raise tractor trailer rigs to subbasement loading docks. Low cost, minimum building space, and very little maintenance make them popular choices. Unlike the usual traction installation, they do not need a machine room rising above the top floor of a building. This appeals economically and aesthetically to architects and owners. Their short traveling distance (± 42 ft) and slow speed (125 fpm compared to 500 fpm or greater for traction) are their principal disadvantages.

Drilling the hole for an hydraulic elevator should occur during the excavation and foundation stage of construction whenever possible. However, it can be done during the structural phase (Fig. 20.25), and there is even a drilling method in which the drill mechanism operates from the installed car rails. For exceptional locations a holeless hydraulic elevator can be specified. The generally simpler installation makes hydraulic elevators easier to build from the construction manager's viewpoint.

FIGURE 20.25. Drilling hole for hydraulic elevator casing. (Courtesy of Otis Elevator Co.)

*Otis Elevator Co.

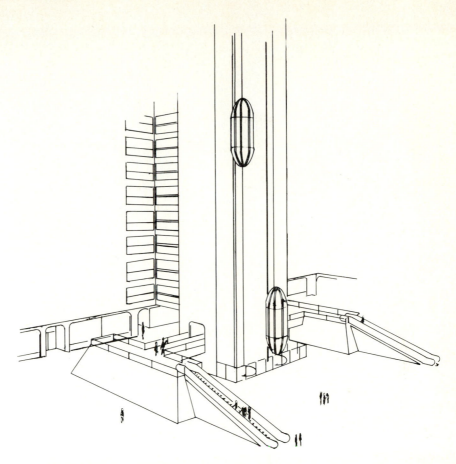

FIGURE 20.26. Observation elevators. (Courtesy of Otis Elevator Co.)

The Observation Elevator

Exposed elevators add new and exciting dimensions to elevator construction (Fig. 20.26). These installations require special care and attention to ensure against the intrusion of rain, wind, and snow. The visual and emotional impact of these elevators on passengers and the public, the participants and the spectators, has been quite dramatic. Technically just another elevator, one wonders to what extent glass-walled observation elevators will be selected for buildings in the future.

20.9 ESCALATORS

An *escalator* is quite a bit like a sand and gravel bucket conveyor in which a motor drives a chain to which are attached many separate work-producing units. The step treads in an escalator take the place of the buckets, each one capable of conveying one or more human beings between two levels. Escalators (Figs. 20.27 and 20.28) are ideally suited to handle large volumes of people in subway and railroad stations, sports arenas, airport terminals, department stores, and office buildings. Their location and importance to the functioning of the area they serve make them a vital part of a construction program. Since they are installed in areas with fancy marble or terrazzo floors, beautiful walls and ceilings, and handsome lighting, they require careful attention during construction, particularly in the sequencing of job operations and protection from plaster, terrazzo grindings, or other debris. The principal parts of an escalator are:

FIGURE 20.27. Twin design escalators. (Courtesy of Otis Elevator Co.)

FIGURE 20.28. Sectional view of an escalator. (Courtesy of Otis Elevator Co.)

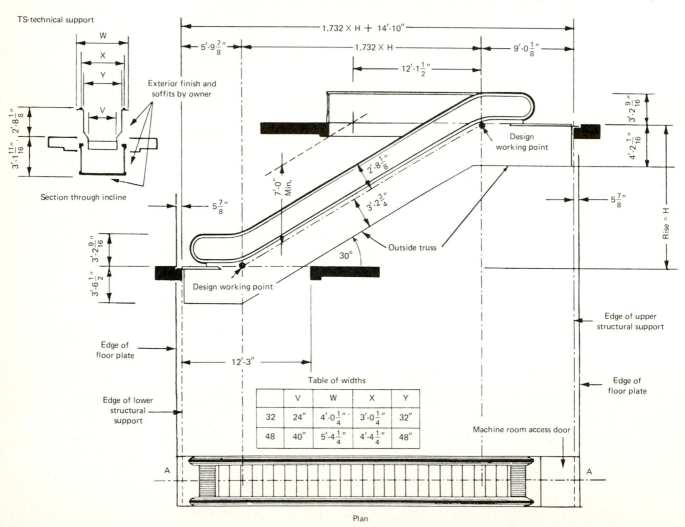

TS-technical support

Exterior finish and soffits by owner

Section through incline

$1.732 \times H + 14'\text{-}10''$

$1.732 \times H$

$5'\text{-}9\frac{7}{8}''$

$9'\text{-}0\frac{1}{8}''$

$12'\text{-}1\frac{1}{2}''$

$3'\text{-}2\frac{9}{16}''$

$4'\text{-}2\frac{1}{16}''$

Design working point

$2'\text{-}8\frac{1}{8}''$

$3'\text{-}2\frac{3}{4}''$

$5\frac{7}{8}''$

$7'\text{-}0''$ Min.

$3'\text{-}1\frac{11}{16}''$, $2'\text{-}8\frac{1}{8}''$

$5\frac{7}{8}''$

$3'\text{-}6\frac{1}{2}''$ $3'\text{-}2\frac{9}{16}''$

Rise = H

Outside truss

30°

Design working point

Edge of floor plate

Edge of lower structural support

Edge of upper structural support

Edge of floor plate

$12'\text{-}3''$

Machine room access door

Table of widths

	V	W	X	Y
32	24"	$4'\text{-}0\frac{1}{4}''$	$3'\text{-}0\frac{1}{4}''$	32"
48	40"	$5'\text{-}4\frac{1}{4}''$	$4'\text{-}4\frac{1}{4}''$	48"

A   A

Plan

1. The steel U-shaped truss (Fig. 20.29)
2. The step chain
3. Steps and risers
4. The worm gear machine
5. The handrail
6. The emergency stop button
7. The main drive sprocket and brake

Some of our major sports arenas have as many as 50 escalators, which make the elevator–escalator contract a very important part of the project. Both elevators and escalators after assembly are turned over to adjusters, specialists in tuning the machinery until it operates smoothly and accurately. During busy periods it is advisable for the construction manager to secure commitments from the elevator–escalator contractor as to the availability of adjusters.

FIGURE 20.29.   Preassembled escalator truss frame. (Courtesy of Otis Elevator Co.)

20.10 OTHER CONVEYORS   Moving Walks

Occasionally pedestrians traveling through long corridors in airports and other public places are transported by *moving walks*. Their use is usually optional, since elderly and handicapped people sometimes prefer to decline their use. They are very similar to an escalator and a bit easier to construct. Figure 20.30 shows some sectional views.

FIGURE 20.30.   Moving sidewalk: sectional view. (Courtesy of Otis Elevator Co.)

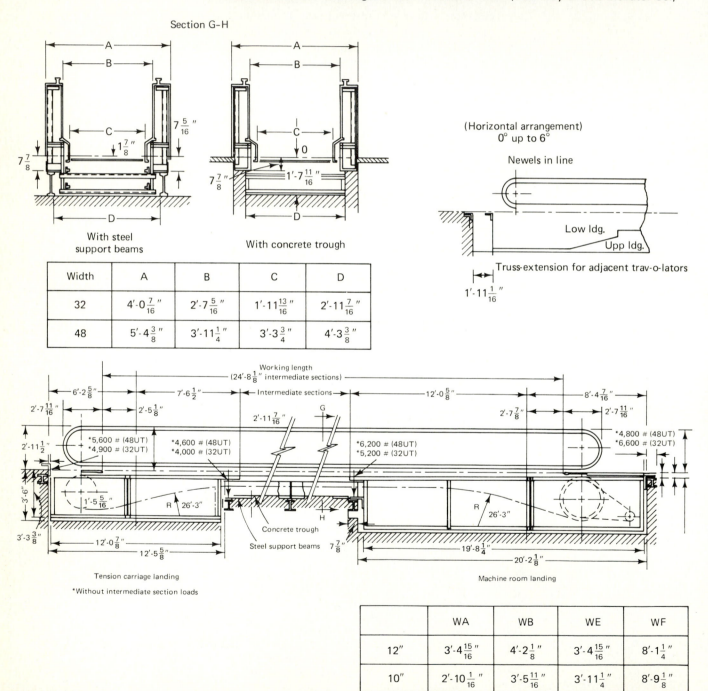

Width	A	B	C	D
32	$4'\text{-}0\frac{7}{16}''$	$2'\text{-}7\frac{5}{16}''$	$1'\text{-}11\frac{13}{16}''$	$2'\text{-}11\frac{7}{16}''$
48	$5'\text{-}4\frac{3}{8}''$	$3'\text{-}11\frac{1}{4}''$	$3'\text{-}3\frac{3}{4}''$	$4'\text{-}3\frac{3}{8}''$

	WA	WB	WE	WF
12"	$3'\text{-}4\frac{15}{16}''$	$4'\text{-}2\frac{1}{8}''$	$3'\text{-}4\frac{15}{16}''$	$8'\text{-}1\frac{1}{4}''$
10"	$2'\text{-}10\frac{1}{16}''$	$3'\text{-}5\frac{11}{16}''$	$3'\text{-}11\frac{1}{4}''$	$8'\text{-}9\frac{1}{8}''$

Dumbwaiters

Dumbwaiters (Figs. 20.31 and 20.32) are simple lifting devices for restaurants, hospitals, motels, apartments, and so on. They lift 75 to 500 pounds 50 to 150 fpm serving two to six landings. Standard operation is automatic call–send.

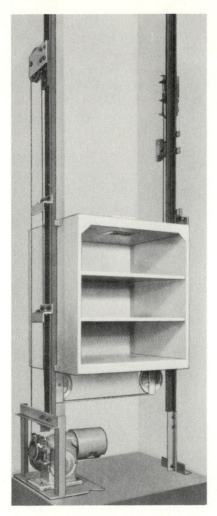

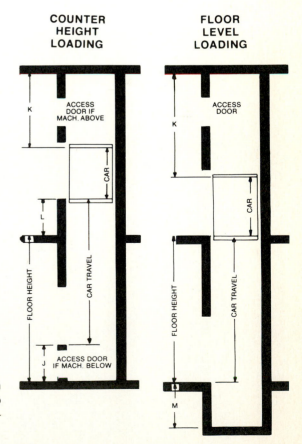

FIGURE 20.31. Dumbwaiter. (Courtesy of Montgomery Elevator Co.)

FIGURE 20.32. Section through a dumbwaiter shaft, showing two loading modes. (Courtesy of Montgomery Elevator Co.)

20.11 ELEVATOR OPERATION

Elevator companies have developed a widely ranging system of elevator operating mechanisms, from very simple to highly complex. The more sophisticated ones are based on the principle of handling the greatest number of people in the shortest possible time without "forgetting" anyone for an undue time. This solid-state equipment has been invented and perfected by the elevator companies. The construction manager purchases the expertise of elevator control when he contracts for the entire elevator installation. Below are two car operation examples.

Single-Car Operation. The first call for service establishes the elevator's direction of travel. The elevator stops at any car call or hall call registered, in the sequence that floors are passed. At the highest or lowest car or hall call, the car reverses and proceeds to answer calls in the opposite direction (Fig. 20.33).

When the car stops at a floor, the automatic door system opens and closes the doors. If no additional calls are registered, the car remains parked at that floor until further calls for service.

Multiple-Car Operation. The multiple zoning system divides the building into traffic zones. Here, three zones are illustrated in a two-car system. One car is assigned to the "top" and one to the "lobby." Both cars respond to all car and hall calls within their assigned zone. When a car leaves its zone to complete a call, all calls registered in that zone are answered by the second car. Similarly, hall calls registered behind a traveling car are answered by the second car (Fig. 20.34).

When either car travels through the "middle" zone, it responds to calls in that zone. After servicing all calls, each car parks in its assigned zone until needed. Hall calls registered in the "middle" zone while both cars are parked are answered by the "lobby" car.

FIGURE 20.33.   Single-car operation. (Courtesy of Otis Elevator Co.)

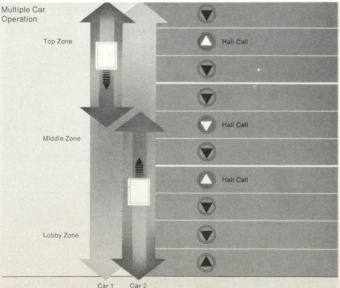

FIGURE 20.34.   Multiple-car operation. (Courtesy of Otis Elevator Co.)

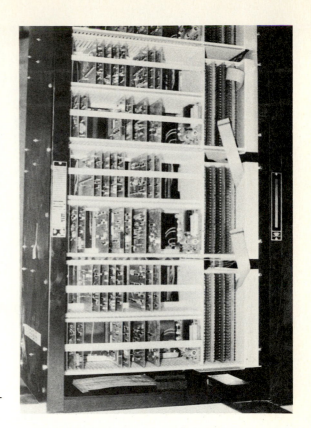

FIGURE 20.35.   Controllers: construction phase.

Machine Controllers

Economical, simple, and slow machines can be controlled by rheostat, a variable-resistance device operating either an ac or dc machine. For installations in the moderate- to high-speed range unit multivoltage systems are used. A unit multivoltage controller (Fig. 20.35) contains all the equipment required to start and stop the elevator machine, determine the direction of travel, regulate the rate of acceleration and deceleration, and control running speed. It also controls the opening and closing of car and hoistway doors and the starting and stopping of the motor generator set.

20.12 ELEVATOR CONSTRUCTION

Elevator construction for a large high-rise project with 20 to 40 elevators requires careful planning, coordination, and integration into the overall construction program. As soon as the schedule for the occupancy of the building becomes known, objectives can be identified and target dates established for particular elevators. General objectives would be:

1. One car in each bank as soon as practical for temporary use.
2. One freight car as soon as practical.
3. Partial concentration on the bank of elevators in the initial occupancy area.
4. One car in each bank for permanent use.
5. A house car (a house car is one of the cars in the highest bank. It stops at every floor, including basement and subbasement levels).

These general objectives would be part of an overall program in which construction on all elevators would proceed to a degree. For example, rails, sills, and hallway entrance work would be initial operations that would continue uninterrupted until all shafts were complete. In similar fashion, car slings and platforms, at the ground floor; roping;

FIGURE 20.36. Installation of a secondary sheave. (Courtesy of National Elevator Industry Education Program.)

buffer work in the pits; and motor generator and elevator machine setting in the machine rooms would be separate repetitive programs (Fig. 20.36). The cooperation of the construction manager and the other trades is an important consideration. The top two or three floors are usually a bit of a headache for many trades. Pouring the machine room slabs and building the enclosing walls when best for the elevator constructor may require real cooperation from other contractors. Fortunately for the elevator contractor, *everyone* wants elevators, both temporary and permanent, and cooperation is not too difficult to obtain. By decking over the shafts with 3-in. timbers at the third-floor level, work can proceed safely at the machine room level, in the shaft, and on the platform and cabs at the same time.

Safety

Injuries and deaths occur in all phases of mechanical and electrical construction. However, although elevator work has many of the hazards of other types of construction, it has some peculiar to elevator work. Shafts are a particular concern. For a lengthy period while rails and sills are being installed, the shaft has protective railings (Fig. 20.37). Later, the door frames are in place but not the doors. Up to the

FIGURE 20.37. Removable elevator shaft guard rails.

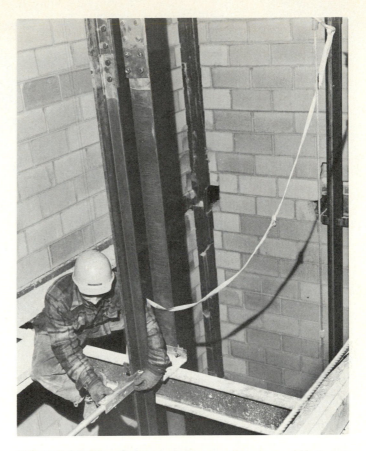

FIGURE 20.38. Safety belt in use. (Courtesy of National Elevator Industry Education Program.)

time the hallway doors are in place, the guard rails frequently have been removed by the elevator constructors and not replaced. The elevator constructor does much of his work in the shaft (Fig. 20.38) and runs the risk of falls or being struck by lumber, bolts, or bricks that inadvertently fall into the shaft.

Later, long before completion, cars move in the shaft. No one should work in or next to a shaft with a moving elevator, but unfortunately this is not an unusual elevator accident. It is advisable for the construction manager to pay particular heed to the elevator part of his safety program.

20.13 ESCALATOR CONSTRUCTION

Escalators are generally located in an area of finely finished materials. It is therefore necessary to sequence the various operations—structural, mechanical, electrical, and architectural—to constitute an orderly program. After the structural opening has been prepared, the assembled truss (Fig. 20.29) can be rigged into place together with the motor sprockets and chain drive (Figs. 20.39 and 20.40). The work may then have to be interrupted while architectural, mechanical, and electrical trades work on ceilings, adjacent walls and floors, and the escalator well plastering, setting marble, grinding terrazzo, installing architectural metal, roughing for lighting, and so on. The escalator should be covered over with polyethylene or other covering during this stage to keep plaster, mortar, and construction debris out of the chain links, motors, bearings, and sprockets. When work resumes, practically all the work involves finished materials: step treads, traveling hand-rail, comb plates, and interior and exterior skirting material (Fig. 20.41). The general area will be in a very advanced stage of architectural completion while this operation proceeds. The escalator people at this stage must be careful to protect adjacent floors from oil and grease stains.

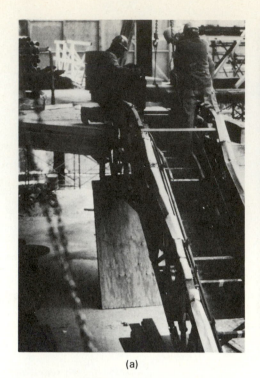 

(a)   (b)

FIGURE 20.39. Rigging the main drive sprocket into A and/or B position. (Courtesy of Otis Elevator Co.)

FIGURE 20.40. Part of a lower-chain-drive assembly. (Courtesy of Otis Elevator Co.)

FIGURE 20.41.   Setting a glass skirt. (Courtesy of Otis Elevator Co.)

20.14 TEMPORARY ELEVATORS

In high-rise construction, structural steel work progresses at the rate of two floors every six working days. Metal decking and concrete slabs are expected to match this rate of speed. The work of many other trades installing mechanical and electrical systems and exterior cladding follows closely behind this structural work. All this means hundreds of workers to be carried to all floors of the structure. During the structural steel phase temporary elevators are rented and erected on the outside of the building. These elevators consist of a self-contained structural frame which attaches to the building and on which the car rides, the car motor, and safety devices. They are expensive to rent and often have a low capacity and slow rate of travel. They also tend to slow the completion of the exterior of the building.

As soon as practical, the construction manager schedules temporary elevator service with the elevator contractor as a superior method of personnel transport. One solution is to rent a *complete portable unit* from the elevator companies, which operate from the *permanent rails* and utilize the car frame and platform (Fig. 20.42). This unit can be installed in a shaft running to the top of the building. Initially, the unit would operate from say the 12th floor, then jump to the 24th floor, then to the 36th, and finally to the 48th. The car is operated by an attendant using a simple forward and reverse type of control.

The *portable temporary elevator* is designed to provide temporary elevator service in buildings under construction, utilizing permanent elevator hoistways, and requires no major revisions to final building plans. Better elevator service is afforded construction workers, with resultant savings in time, since the "portable unit" can be easily relocated to a higher location, keeping pace with steel or concrete construction on the site.

Designed to serve up to 100 floors, portable units are available in three machine sizes, depending on the hoistway dimensions and elevator duty required. Elevator guide rails must be installed to a level above the portable unit for both the initial installation and subsequent moves. The system utilizes the permanent car frame, platform, and rails, which facilitates installation and operation of the permanent machine room equipment.

Each relocation of the portable unit may be scheduled over a weekend, with the elevator operating from its new location on the next working day, ensuring no interruption in elevator service for the work force.

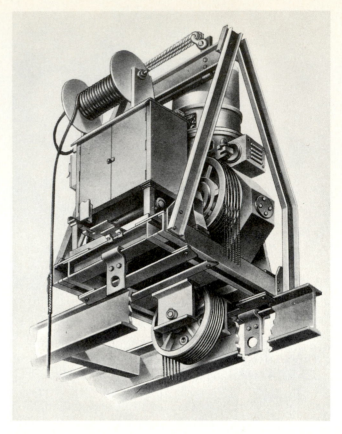

FIGURE 20.42. Relocatable temporary elevator machine. (Courtesy of Otis Elevator Co.)

Contractors' requirements in utilizing this equipment are usually limited to:

1. Furnishing hoist facilities for all temporary equipment
2. Furnishing power to each portable unit location
3. Furnishing machine beam supports at each location
4. Furnishing a protected hoistway (permanent or temporary)
5. Furnishing a fire- and water-resistant enclosure of substantial construction

Responsibility includes:

1. Installation of the portable unit equipment
2. Relocating the portable unit equipment
3. Servicing the equipment
4. Removal of equipment

More often the construction manager arranges for a shaft to be prepared, by the elevator contractor, with rails, machine, motor generator set, buffers, frame, and platform. The construction manager builds a 2 × 4 and plywood car with an old-fashioned used accordian gate. The shaft is enclosed in wire fencing with plywood door and hook-and-eye hardware (Fig. 20.43). The car is controlled by an elevator constructor, usually from the disabled list, who controls the car with a simple up-and-down control device. The car generally travels up and down the shaft continuously, either stopping for observed passengers or being called by vocal shouts. The

FIGURE 20.43. Shaft enclosed with wire fencing material. (Courtesy of Turner Construction Company.)

construction manager employs the operators and accepts liability for running the temporary elevators.

Cars can also be released to the construction manager for his use when almost complete with cab and hall doors operating by buttons inside the car. A common practice is to line the entire finished cab with a soft protective material such as cane board over which a ⅝-in. plywood protection is carefully fitted. Marble doorways, bronze door frames, hall doors, and car doors must also be protected with either heavy paper or wood corner pieces. The construction manager should release the elevator back to the elevator company as soon as he can operate without that car so that final adjusting, testing, and acceptance by the architect can take place.

21

Noise and Vibration Control

21.1 INTRODUCTION   Most of the disagreeable noise and vibration inside a building emanates from the mechanical and electrical systems. Fans, plumbing and chilled water pumps, motors, vibrating pipes, rattling ductwork, and rush of air in ductwork are some of the worst offenders. Figure 21.1 shows typical sound sources and transmission paths. Conveyor and pneumatic tube systems in office buildings and banks, and elevator machines, motor generator sets, and elevator door mechanisms in apartment houses, also present problems. Cooling towers probably offend neighboring building occupants more than your own. Electrically transformers hum, relays click, starters groan, and circuit breakers snap.

In office buildings and apartment houses most complaints come from the area immediately below an upper-level machine room. Either the designer failed to properly identify and acoustically provide proper treatment or something went wrong in execution, or some of each. Usually, the machine room is operating before the floor immediately below reaches completion. The prudent construction manager and designer should inspect this area in the relative quiet of an evening or a weekend during the construction period to determine if an acoustical problem exists. The solution will be quicker and cheaper to achieve if identified before the occupants arrive.

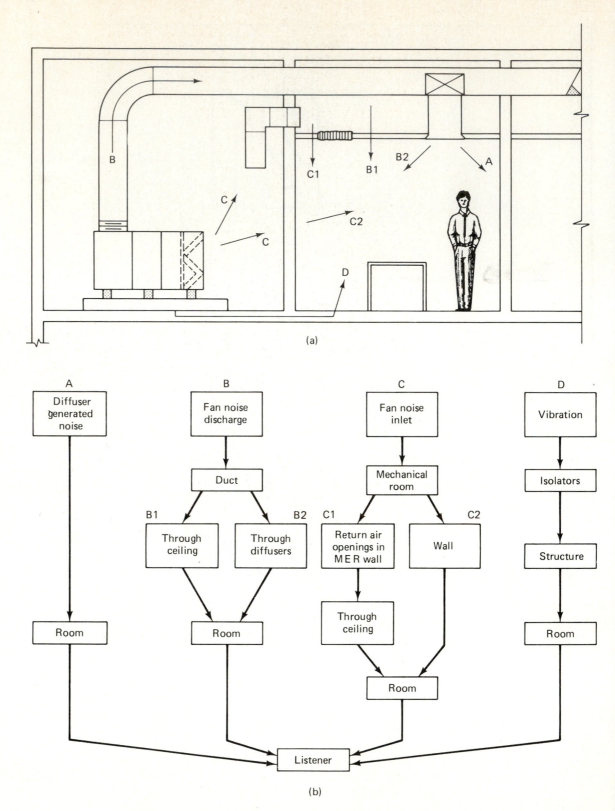

FIGURE 21.1. (a) Typical sound sources; (b) transmission paths. (Courtesy of ASHRAE.)

21.2 AIRBORNE NOISE CONTROL: THE FLOATING FLOOR

By preventing *airborne noise* from leaving or entering a space, an acoustical solution can be achieved (Fig. 21.1). Figure 21.2 shows a typical machine room with floor, wall, and ceiling acoustical treatment. The solution is architectural, not mechanical–electrical. The solution is curative rather than preventive in that the noise exists but

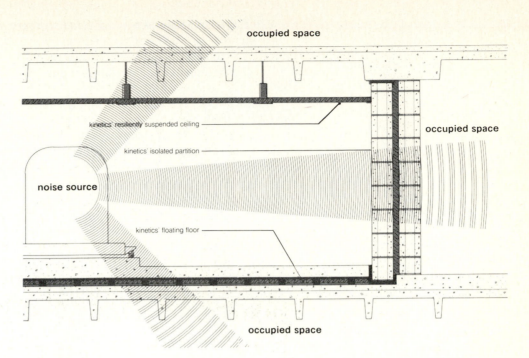

FIGURE 21.2. Machine room acoustical treatment. (Courtesy of Peabody Noise Control.)

is controlled. The use of a *floating floor* is very effective. Most designers use a waterproof membrane over the structural slab to protect the occupied space below.

"There are two common methods for constructing floating floors:

1. A system where *individual pads* are spaced on 0.3 to 0.6 m (1 to 2 ft) centers each way and covered with plywood or sheetmetal. Low density absorption material can be provided between the pads to reduce the effect of tunneling. Reinforced concrete is placed directly on the waterproofed panels and cured. Then, the equipment is set [Fig. 21.3].

FIGURE 21.3. Floating floor construction. (Courtesy of ASHRAE.)

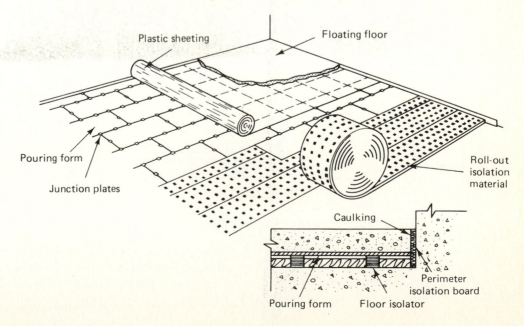

2. A system where *cast-in-place canisters* are placed on 0.6 to 1.2 m (2 to 4 ft) centers each way on the structural floor. Reinforced concrete is poured over the canisters; after curing, the entire slab is raised into operating position, the canister access holes are grouted, and the equipment is set [Fig. 21.4].

In either system, mechanical equipment may be placed on the floating slab unless it operates near the resonance frequency of the floating slab. Mechanical equipment may also be located on extensions of the structural slab which penetrate through the floating floor. A qualified professional should evaluate the system since short circuiting, flanking, and lack of attention to details could materially reduce the performance of the floating slab. Note that floating floors are designed and installed specifically to control airborne sound transmission. They are not intended to be used in lieu of vibration isolators."*

In all acoustical work much *attention should be directed to details* so that structural, architectural, and mechanical items do not contact each other. Openings must be gasketed and tightly sealed. This sealing process serves a dual purpose in preventing the spread of fire and smoke around pipes and ductwork. Some important details are shown in Figs. 21.5 to 21.9.

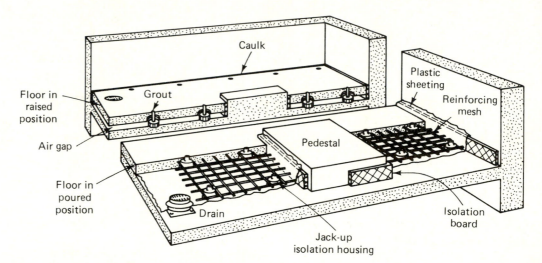

FIGURE 21.4. Typical floating floor construction. (Courtesy of ASHRAE.)

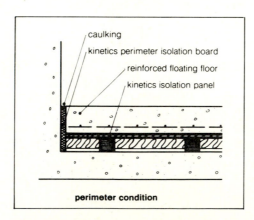

FIGURE 21.5. Perimeter condition. (Courtesy of Peabody Noise Control.)

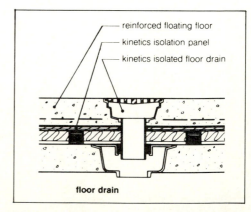

FIGURE 21.6. Floor drain. (Courtesy of Peabody Noise Control.)

*ASHRAE, *Sound and Vibration Control,* 1980 Systems Volume, Chap. 35.

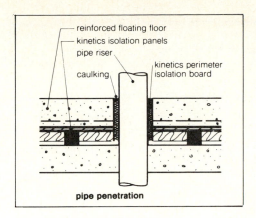

FIGURE 21.7. Pipe penetration. (Courtesy of Peabody Noise Control.)

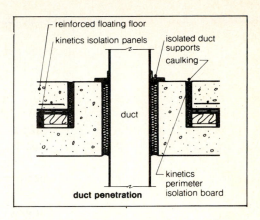

FIGURE 21.8. Duct penetration. (Courtesy of Peabody Noise Control.)

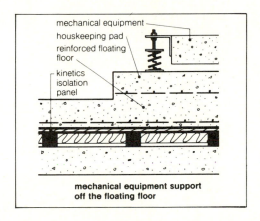

FIGURE 21.9. Mechanical equipment support off the floating floor. (Courtesy of Peabody Noise Control.)

21.3   WALL PARTITION SYSTEMS

In acoustical treatment the unwanted sound must not be permitted to run around or flank other acoustic treatment. Since mass solves acoustical conditions, cement masonry partitions effectively combat noise problems (Fig. 21.10). Isolating the partition from the rest of the structure prevents sound transmission outside the limits of the wall (Fig. 21.11). Partition isolating brackets restrain the wall at the head

FIGURE 21.10. Resiliently supported partition detail. (Courtesy of Peabody Noise Control.)

FIGURE 21.11. Plan view of a vertical resilient partition connection. (Courtesy of Peabody Noise Control.)

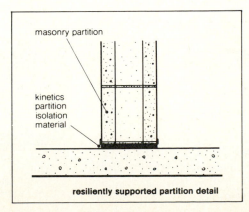

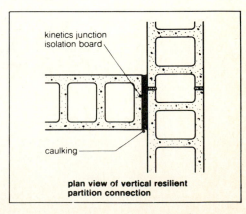

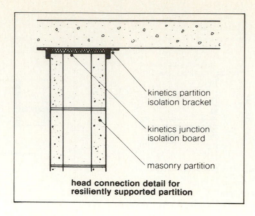

FIGURE 21.12. Head connection detail for a resiliently supported partition. (Courtesy of Peabody Noise Control.)

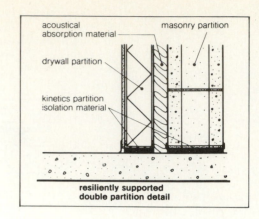

FIGURE 21.13. Resiliently supported double partition detail. (Courtesy of Peabody Noise Control.)

connection (Fig. 21.12). Double partitions are highly effective and there are many different types from which the architect and acoustical consultant can choose. Figure 21.13 is one example.

21.4 SUSPENDED CEILINGS

Despite the fact that a *suspended ceiling* performs better acoustically than acoustical material applied directly to the structural slab above a machine room, most designers specify the applied method. A large machine room has hundreds of pipe, duct, conduit, and equipment hangers, each one of which would make the suspended ceiling and mechanical and electrical installation more complicated and costly. Second, a suspended ceiling would require additional headroom, which would also add to the building cost. Suspended ceilings are best suited for architectural areas. Figures 21.14 to 21.17 show details of isolation techniques for suspended ceilings.

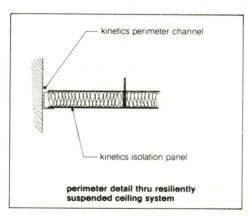

FIGURE 21.14. Perimeter detail through a resiliently suspended ceiling system. (Courtesy of Peabody Noise Control.)

FIGURE 21.15. Alternate methods of ceiling support using spring hangers. (Courtesy of Peabody Noise Control.)

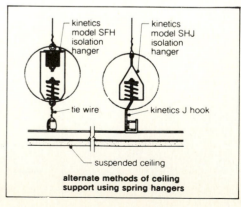

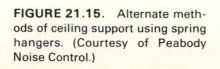

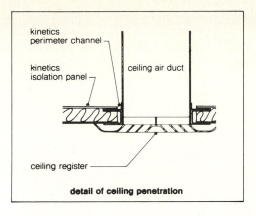

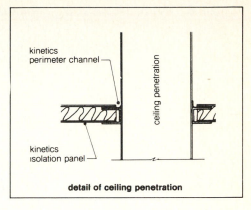

FIGURE 21.16. Detail of ceiling penetration. (Courtesy of Peabody Noise Control.)

FIGURE 21.17. Detail of ceiling penetration. (Courtesy of Peabody Noise Control.)

21.5 BAFFLES   In industrial and recreational areas high noise levels can be diminished by the use of *noise control baffles*. One scheme utilizes 2 ft × 4 ft × 1½ in. of fiberglass sealed in a fire-retardant vinyl film cover hung in an acoustical control pattern. Figure 21.18 shows an industrial application, and Fig. 21.19 shows two patterns for hanging the baffles.

FIGURE 21.18. Use of noise control baffles. Noise control baffles are designed to effectively reduce overall ambient noise levels in industrial, recreational, or other high-noise areas. (Courtesy of Peabody Noise Control.)

FIGURE 21.19. Acoustical baffle patterns. (Courtesy of Peabody Noise Control.)

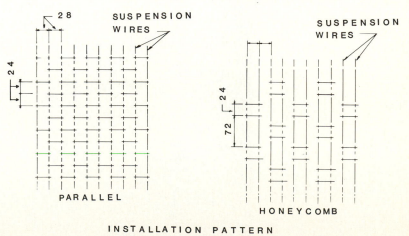

PARALLEL

HONEYCOMB

INSTALLATION PATTERN

21.6   INERTIA BASES   Mounting plumbing and chilled water pumps, air compressors, centrifugal fans, and chillers on *inertia bases* lowers the center of gravity and greatly reduces vibration. The inertia bases come with sturdy brackets designed to transfer the machine and inertia base load through the suspension system to the structural slab (Figs. 21.20 and 21.21). Figure 21.22 shows a steel inertia base with reinforcing steel and anchor bolts ready for concreting. For inertia bases carrying heavy equipment with large unbalanced forces, the isolation spring shown in Fig. 21.23, by *hanging* the load from the spring, markedly reduces lateral movement in the inertia base. In those cases where the

FIGURE 21.20. Adjusting a spring suspension system. (Courtesy of Peabody Noise Control.)

FIGURE 21.21. Centrifugal fan installation. (Courtesy of Peabody Noise Control.)

designer decides that an inertia base can be eliminated, compressors, pumps, fans, and air conditioning units directly mounted on spring-vibration isolators produce satisfactory results. The spring assembly rests on a noise isolation pad (Figs. 21.24 and 21.25). In Fig. 21.26 the pump and motor are on a spring-mounted inertia base, the electric motor has a flexible cable connection, two different-size rubber pipe expansion joints connect to the piping, and in the background the wall is built with acoustical cement masonry units—a good example of controlling airborne noise and vibration.

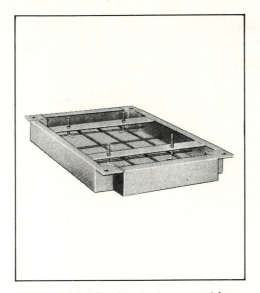

FIGURE 21.22.   Inertia-base steel form. (Courtesy of Peabody Noise Control.)

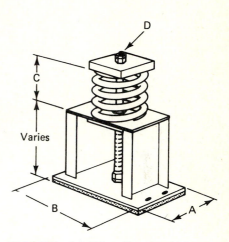

FIGURE 21.23. Suspended-type mounting. (Courtesy of Peabody Noise Control.)

FIGURE 21.24.   Vibration isolator. (Courtesy of Peabody Noise Control.)

FIGURE 21.25.   Vibration isolator. (Courtesy of Peabody Noise Control.)

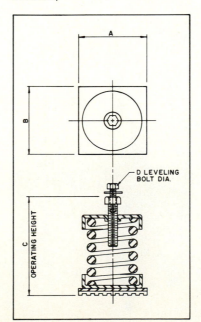

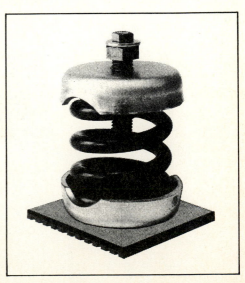

FIGURE 21.26.   Well-executed pump installation.

21.7   ISOLATION HANGERS

Ductwork, transformers, conduit, piping, fans, and other mechanical equipment are supported on hangers from the structural slab. The vibration produced in these parts of the mechanical and electrical system are readily dissipated by the use of *vibration isolation hangers.* Figure 21.27 shows a type that utilizes a molded fiberglass insert that transfers the load to the lower spring section. Neoprene is also used extensively for this purpose. Figure 21.28 shows a variety of vibration isolation and plain hangers fastened to beam and trapeze ready for pipe installation.

FIGURE 21.27.   Suspended-type isolation hanger. (Courtesy of Peabody Noise Control.)

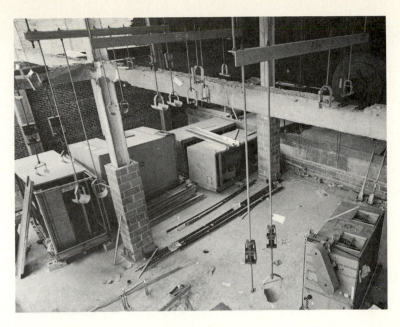

FIGURE 21.28.   Hangers for a machine room. (Photo by Gil Amiaga, N.Y.C.)

**21.8   ROOFTOP
EQUIPMENT ISOLATION**

Isolation rails are specifically designed and engineered to isolate most packaged rooftop equipment from the roof structure, thereby minimizing the transmission of vibration to the structure. Rails provide support and a continuous air and water seal while effectively isolating the transmission of vibration. The rails consist of extruded or roll-formed shapes and a neoprene or vinyl water seal. Corners are factory fabricated, requiring only minimum field assembly and attachment to the conventional roof curb. Spring isolators are placed according to actual weight distribution to maintain equipment level. Optional restraints are available to minimize motion due to wind loads, and other factors (see Fig. 21.29).

FIGURE 21.29.   Sectional view of a rooftop isolation rail. (Courtesy of Peabody Noise Control.)

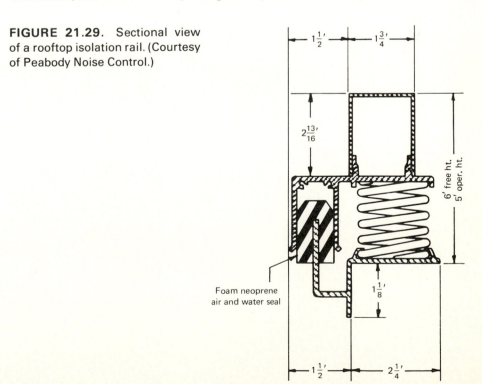

Foam neoprene
air and water seal

21.9   ACHIEVING ACOUSTICAL GOALS

We have probably all reacted at one time or another to the untimely flushing of a noisy toilet. Water closets need not create undesirable noise. Equipment with superior acoustic performance can be specified by the designer, but usually at an added cost. Fans, diffusers, through-wall air conditioners, induction units, pumps, light switches, transformers, and many other kinds of equipment can be purchased with superior acoustical qualities. Initially, the acoustical engineer will prefer to design around this type of equipment. He then has available for acoustically treating the best in mechanical and electrical systems, architectural, and structural wall, ceiling, and floor materials and systems. He can utilize acoustical venting devices (Fig. 21.30) and the various sound control devices just discussed, including the duct silencers and liners mentioned in Sections 15.9 and 15.11. Above all else, the *acoustical engineer should participate* in the architectural, structural, mechanical, and electrical *design* rather than being consulted after the structure proves acoustically faulty, as happens all too often.

The construction people have the responsibility to correctly install the architectural, structural, mechanical, and electrical systems to meet the acoustical plans and specifications. This means that the construction supervisor *must* pay attention to detail. Quite often, proper adjustment of the vibration spring devices at the inertia bases is all that need be done. Care must be taken to make sure that *every* trade works toward the same acoustical goal. All work must be inspected to prevent unwanted contact of mechanical systems and walls, ceilings, or floors. It is very important that acoustically constructed walls have acoustic doors and frames. Once again, an on-site inspection with the machinery running and outside conditions quiet will detect troubles while they are still easy to solve. If you do have a noise-producing pipe duct or pneumatic tube already installed, you might suggest enclosing it in a cocoon of metal lath and vermiculite plaster. It works.

FIGURE 21.30.   Acoustical venting devices permit a flow of air between spaces without the objectionable transmission of sound. (Courtesy of Industrial Acoustics Company, New York, New York.)

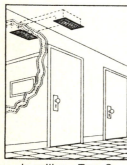
In ceiling—Type C

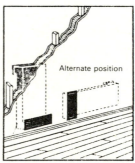
Alternate position

In wall—Type W

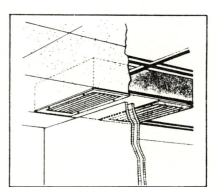

22

Comfort Control

22.1 FUNCTIONS In our private residences we soon become familiar with how to control our comfort. The thermostat, usually electric, sends the signal that starts and stops our oil burner, gas furnace, or air conditioning unit. In a single home usually one thermostat controls one piece of apparatus for a single system that heats or cools the entire house. The unit operates at a constant speed and temperature until the thermostat shuts it down. In commercial buildings we can have *numerous* systems, each one containing *controllers* which activate *controlled devices* such as valves and dampers. By means of controls the flow of heated and chilled water and steam is regulated. We seek to control both flow and temperature. In air handling, dampers regulate the amount of fresh air received and stale air exhausted. At the terminals we control the final temperature of air entering the occupied space and the rate of flow. To summarize, we control the water, steam, and air that we *supply* to a system and then control their *use* to produce the required comfort condition.

22.2 THE CONTROL CONTRACTOR HVAC contractors sublet their control work to nationally recognized companies that specialize in manufacturing, installing, and tuning control systems. These systems, though they run all through the building, occupy infinitesimal space and can hardly be noticed. The control contractor has the responsibility of making the HVAC systems perform as the various parts of the building are completed. At times only 10% of a large building are occupied initially. Whether it is summer, spring, fall, or winter, the controls must operate to give the initial occupants the comfort they deserve. The control company must as expeditiously as possible satisfy these changing conditions until the entire building is complete, occupied, and functioning. Working with the HVAC contractor they must meet the specifications, improvise on operation during interim periods, train the owner's personnel to take over the operation of the HVAC systems, and through technology satisfy occupants' complaints. Since they deal in human comfort they must respond to complaints about conditions being too hot, too cold, too dry, too humid, too drafty, and too stagnant. Basically, they concentrate on delivering the designed quantity of air at the proper temperature and humidity while diplomatically emphasizing that they have met design criteria. From a construction supervisor's standpoint the control contractor is a sub-subcontractor and a highly specialized manufacturing-oriented one at that. However, from a customer satisfaction viewpoint his importance to the project should not be underestimated, since through good performance and customer relations he can be very helpful in convincing an owner that his contractors have performed well.

22.3 PNEUMATIC CONTROL

As shown in Fig. 22.1, a small compressor supplies the compressed air at 15 to 25 psig to the *pneumatic control system*. Reacting to conditions, the controller regulates the air becomes drier. The biwood element uses yew and cedar with the grains at right accordingly. The control can be two-position, the "on–off" type, or proportional, the type that moves the controlled device to a position consistent with the demand conditions.

The control system should do the following:

1. Meet the required conditions of temperature, humidity, and airflow
2. Operate to control safety by controlling temperature liquid levels, pressure, and so on
3. Operate economically

It should be sensitive enough to prevent wide fluctuations of conditions while avoiding cycling every few minutes.

FIGURE 22.1. Air compressors and receiver tank for a pneumatic control system.

**22.4 SENSING
ELEMENTS**

The control system utilizes devices that react to temperature, humidity, and pressure.

Temperature. Bimetal elements have two dissimilar metals, such as brass and nickel steel alloy, which differ greatly in expansion (Fig. 22.2). *Remote bulb elements* are used extensively in ductwork and also in tanks and piping. They have a bulb filled with heat-sensitive liquid or gas, a capillary tube, and a diaphragm which indicates temperature through movement (Fig. 22.3).

Relative Humidity. Relative humidity can be measured by wood, membrane, or hair and CAB (cellulose acetate butyrate). In Fig. 22.4 the membrane shortens as the air becomes drier. The biwood element uses yew and cedar with the grains at right angles. Humidity changes cause the element to bend. The CAB type utilizes a loop that shortens with a drop in relative humidity, causing bending (Fig. 22.5a), or as a link assembly that shortens (Fig. 22.5b).

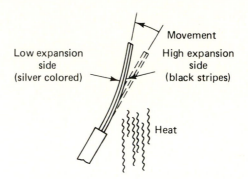

FIGURE 22.2. Bimetal temperature-sensing element. (Courtesy of Johnson Controls Inc.)

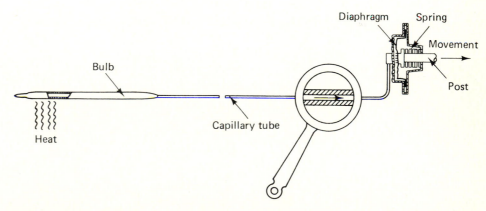

FIGURE 22.3. Remote bulb temperature-sensing element. (Courtesy of Johnson Controls Inc.)

FIGURE 22.4. Relative humidity-sensing elements. (Courtesy of Johnson Controls Inc.)

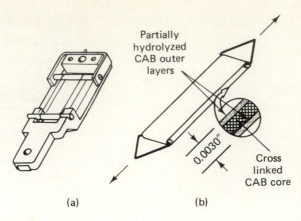

FIGURE 22.5. CAB relative humidity-sensing elements. (Courtesy of Johnson Controls Inc.)

Pressure. Pressure is measured by three types of elements: *the diaphragm, the bellows,* and the *bourden spring tube.* In the diaphragm type increased pressure makes the diaphragm bulge, producing recordable motion (Fig. 22.6a). In the bellows type a flattened tube bent into a circle or spiral elongates to produce motion (Fig. 22.6b). In the bourden tube type a flattened tube bent into a partial circle tends to straighten under pressure (Fig. 22.6c). Figure 22.7 shows sensing devices mounted on the side of an air conditioning casing unit.

FIGURE 22.6. Pressure-sensing element: (a) diaphragm type; (b) bellows type; (c) Bourden spring tube type. (Courtesy of Johnson Controls Inc.)

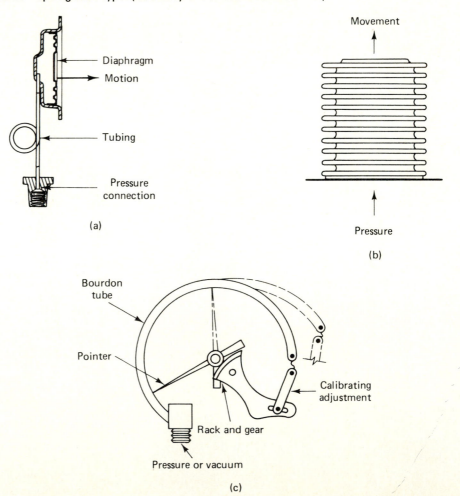

FIGURE 22.7. Sensing devices on the side of an air conditioning casing.

22.5 CONTROLLED DEVICES* "A controlled device is the final piece of equipment in a control system, which regulates the controlled variable in accordance with the demands of the controller. This device may be a valve controlling the flow of water or steam, or a damper operator regulating a damper to control air flow" [Figs. 22.8 and 22.9].

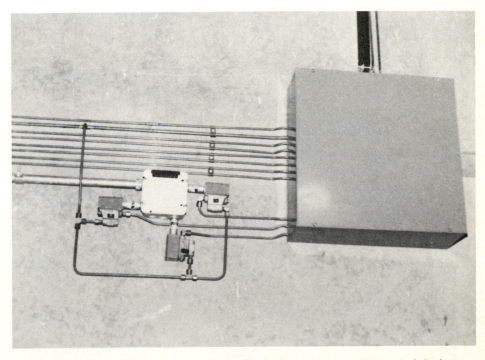

FIGURE 22.8. Small copper and/or plastic tubing connects the control devices.

---

*Section 22.5 is taken from Johnson Controls Inc., *Fundamentals of Pneumatic Control* (Johnson Controls Training Manual), pp. 53, 54.

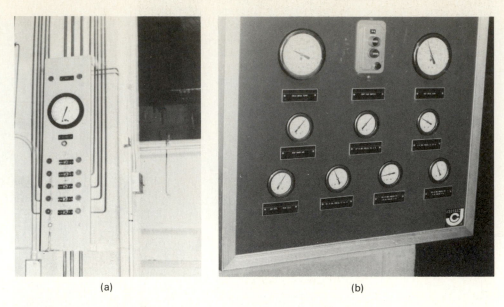

(a)   (b)

FIGURE 22.9.   (a) Local control station; (b) instrument panel with a pneumatic system.

Damper Operators

"Pneumatic damper operators of the piston type have a long, powerful straight stroke which requires no lever arrangements. Air from the controller is applied to the molded diaphragm which has a positive seal to prevent air leakage [Fig. 22.10]. This air pressure expands the diaphragm forcing the piston and stem outward against the force of the spring. The movement of the piston varies proportionally with the air pressure applied to the diaphragm. This air pressure, from the controller, varies over the full pressure range.

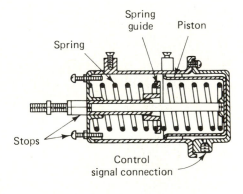

FIGURE 22.10.   Components of the piston damper operator: (a) normal position; (b) full stroke. (Courtesy of Johnson Controls Inc.)

(a)

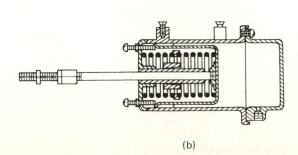

(b)

The spring returns the operator to its normal position when the air pressure is removed from the diaphragm. Full movement of the operator can be restricted to set limits by using various spring ranges. Assume a spring range of 5 to 10 psig. With this spring the operator is in its normal position when the air pressure applied to the diaphragm is 5 psig or less. Between 5 and 10 psig the stroke will be proportional to the air pressure on the diaphragm. Above 10 psig the operator will be at its maximum stroke.

Damper operators can be mounted on the damper frame and coupled directly to the damper blades. In some cases the operator is mounted on the ductwork and coupled to the damper blade axis through a crank arm and linkage arrangement [Fig. 22.11].

Air flow control applications include throttling, face-and-bypass [Fig. 22.12] and mixing of outside and recirculated air [Fig. 22.13].

FIGURE 22.11. Damper operator fastened to ductwork.

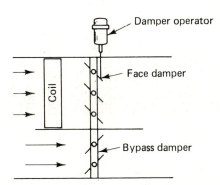

FIGURE 22.12. Face-and-bypass damper application. (Courtesy of Johnson Controls Inc.)

FIGURE 22.13. Dampers used to mix outside air and recirculated air. (Courtesy of Johnson Controls Inc.)

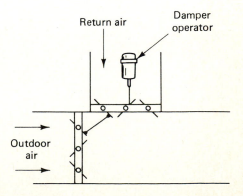

Valves

Valves control the flow of steam, water or other fluid media, and are available in a variety of body patterns to meet various piping arrangements, and to suit all applications and installations. Also there are a number of inner valves available, each capable of producing a different flow characteristic. However, there are only three basic body styles and all patterns fall into one of these styles; normally open, normally closed and three-way.

Among the pneumatic operators furnished with valves are the *piston top* [Fig. 22.14] and the *rubber diaphragm* [Fig. 22.15]. The former is used mainly on small valves controlling radiators, convectors and terminal air conditioning units, while the latter is used to control large valves regulating flow through heating and cooling coils on central or zoned systems.

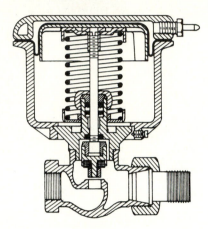

FIGURE 22.14. Piston top valve operator on a normally open valve. (Courtesy of Johnson Controls Inc.)

FIGURE 22.15. Rubber diaphragm operator on a normally open valve. (Courtesy of Johnson Controls Inc.)

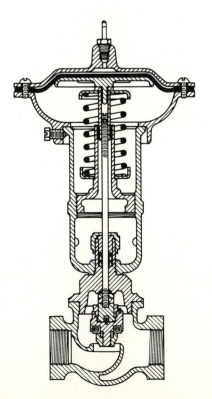

Valves function in a manner similar to damper operators. Air pressure or hydraulic fluid is applied to the rubber diaphragm which expands and opposes the force of the spring. This causes the inner valve (stem and disc or plug) to move toward the seat of a normally open valve [Figs. 22.14 and 22.15] stopping the flow or away from the seat of a normally closed valve [Fig. 22.16] allowing flow. When air pressure is removed from the diaphragm, the spring will return the inner valve to its normal position.

The *three-way valve* may be a *mixing valve* [Fig. 22.17] or a *bypass valve*. There are three connections on this type, common, normally open and normally closed.

The *mixing valve* has two inlets and one common outlet. The controller positions the inner valve so that the entire flow is from either one of the two inlets or a portion from both.

The *bypass valve* has one inlet and two outlets. The controller positions the inner valve so that the entire flow is directed to either one of the two outlets or a portion to both."

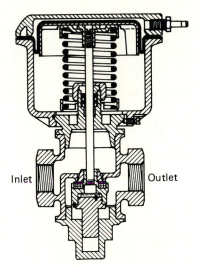

Inlet   Outlet

FIGURE 22.16. Normally closed valve with a piston top operator. (Courtesy of Johnson Controls Inc.)

FIGURE 22.17. Three-way mixing valve with a piston top operator. (Courtesy of Johnson Controls Inc.)

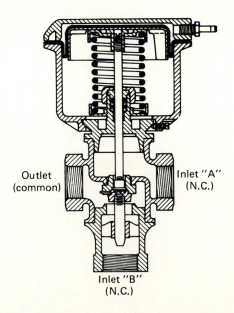

Outlet (common)   Inlet "A" (N.C.)

Inlet "B" (N.C.)

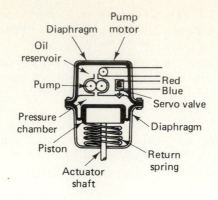

FIGURE 22.18. Electrohydraulic actuator. (Courtesy of Johnson Controls Inc.)

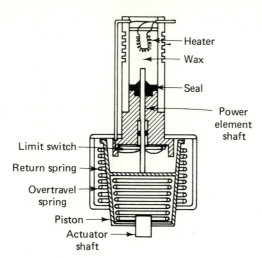

FIGURE 22.19. Thermal-type actuator. (Courtesy of Johnson Controls Inc.)

22.6 ELECTRONIC CONTROLS

Mechanical systems are also controlled electrically by electronic systems. Electrical control signals are sent through a low-voltage direct-current system based on electric sensing devices. The system utilizes electrohydraulic actuators in which the servo valve is operated by low-voltage 0- to 16-volt dc and the oil pump applies hydraulic pressure on the diaphragm (Fig. 22.18).

Another kind of actuator is the thermal type, in which wax melts and as a liquid applies pressure to the power element shaft and the piston (Fig. 22.19).

22.7 TESTING, ADJUSTING, AND BALANCING

The purchase and installation of all the elements of a building's environmental systems does not ensure that the specified design objectives and the comfort of the occupants will automatically be met. A transitional stage between actual construction and owners' operation takes place with testing, adjusting, and balancing. Along with concern for meeting the design objectives, building owners want the mechanical system to perform with optimum efficiency. The process also has use in resolving those unfortunate problem cases where the specified equipment piping, ductwork, and terminals still do not produce the required comfort.

Balancers are special technicians working directly for or as subcontractors to the HVAC contractor. Their job consists of (1) balancing the air and water distribution, (2) adjusting the total system to provide design quantities, (3) measuring electrical usage, (4) verifying the performance of equipment and automatic controls, and (5) checking for sound and vibration. In carrying out this assignment they report the following:*

*ASHRAE, *1976 Systems Handbook: Testing, Adjusting and Balancing,* p. 40.6.

"1. Design

   a. Cfm to be delivered
   b. Fan static pressure
   c. Motor horsepower
   d. Percent of outside air under minimum conditions
   e. Rpm of the fan
   f. The brake horsepower required to obtain this cfm at the design static pressure

2. Installed

   a. The equipment manufacturer
   b. The size of unit installed
   c. The arrangement of the air handling unit
   d. The class fan
   e. The nameplate information voltage, and so on
   f. Tested voltage
   g. No-load amperes on three-phase and single-phase ½ horsepower and up

3. Field tests

   a. Fan rpm
   b. Power readings
   c. Calculated approximate brake horsepower
   d. Total suction pressure
   e. Static discharge pressure
   f. Fan static pressure

4. Terminal outlets

   a. Locate outlet by room and position
   b. Outlet manufacturer and type
   c. Outlet size
   d. Indicate neck area for diffusers and core area for grilles
   e. Insert manufacturer's outlet factor
   f. Indicate the design cfm and the required velocity in fpm to obtain this cfm
   g. Indicate the test velocities and cfm
   h. Indicate adjustment pattern and cfm
   i. Induction unit information when used."

Air Balancing

The exact procedure is too complex for our purposes, but is thoroughly explained in Chapter 40 of the ASHRAE 1976 Systems Handbook. However, some concept of the procedure can be realized from the following. The initial step is the simple testing of the system for leaks. Balancing requires all related equipment in operation and duct dampers fully opened. One method is to start at the supply fan taking readings and ending at the terminals. Readings are taken at each location where the ducts divide, and the dampers are set so that each duct gets the designed cubic feet per minute of flow. Minimum and maximum outdoor air tests must also be run. The air supply is divided and subdivided so that eventually at each terminal there is sufficient air available for making the final adjustment to meet the design criteria. The process of testing, adjusting, and balancing of complex mechanical systems requires skill, diligence, and patience. The process must be repeated until the systems have been tuned to achieve optimum performance.

Water Balancing

Water balancing can be very accurately attained by direct flow measurement. The system components themselves can be used for this purpose, such as control valves, terminal units, chillers, and so on. In addition, venturi, orifice plate, and pitot tube flow meters, together with pumps, flow limiting devices, and presettable balance valves, will provide flow information. In certain circumstances a combination of flow and temperature readings can also be utilized. The overall objective is system balance together with minimum operating cost. One example would be the reduction of excess pump head by trimming the pump impeller rather than allowing the excess head to be absorbed by throttle valves, which would impose a life-long operating cost penalty to system operation. The flushing and cleaning of all systems is an important initial step.

22.8 PROBLEMS   A leaky roof, a cost overrun, a delay in occupancy, and a comfort problem are four fairly certain ways of infuriating an owner. Not all comfort problems can be solved with tinkering and tuning. Where insufficient air is the problem, often the sheaves on the supply fan are changed to speed up the fan and supply more air. Fan efficiency suffers somewhat, but this often solves the comfort problem. Occasionally, the ductwork has to be enlarged and restrictions eliminated to overcome excessive duct resistance. Sometimes water temperatures have to be raised or lowered to meet proper terminal readings. The project will benefit if fair treatment, cooperation, and good judgment are exercised by the owner, architect, engineers, and the contractors in solving these problems.

appendix

Construction Management Guidelines*

"The General Contractor has traditionally been the manager of the construction process in whatever form that process has taken and will continue to fill this role. However, the complexities of today's construction techniques and the difficulties in managing the ever-increasing number of highly-specialized subcontractors occasionally require new approaches to the methods historically used by the General Contractor in his management function.

The Associated General Contractors of America, in keeping with its recognized position of leadership in the construction industry, has established a fundamental policy with regard to construction management.

This policy is intended to act as a guide to all parties contemplating the use of construction management in constructing any type of project.

In establishing guidelines for this contracting method, AGC does not endorse it as a substitute for any other successful contracting method. The Association maintains its long-standing position relative to competitive bidding and the single contract system. However, the construction management type contract does present certain distinct advantages on some projects. Therefore, AGC strongly recommends to owners that the guidelines for construction management as adopted by it be followed on all projects on which construction management is to be used.

*Reprinted with the permission of the Associated General Contractors of America; copyright 1979.

Far from being an innovation, construction management has been used successfully for many years by owners and contractors, especially in the private sector. This method allows the owner and the architect-engineer to have available to them the services of the General Contractor not only during the construction phase but during the design phase as well. Therefore, the AGC policy formalizes for those unfamiliar with construction management the procedures and methods which can best produce the product desired.

The following questions and answers are provided in order to outline clearly the AGC definition of Construction Management.

1. What is Construction Management?

It is one effective method of satisfying an Owner's building needs. It treats the project planning, design and construction phases as integrated tasks within a construction system. These tasks are assigned to a Construction Team consisting of the Owner, the Construction Manager and the Architect-Engineer. Members of the Construction Team ideally work together from project inception to project completion, with the common objective of best serving the Owner's interests. Interactions between construction cost, quality, and completion schedule are carefully examined by the Team so that a project of maximum value to the Owner is realized in the most economic time frame.

2. What is a Construction Manager?

The Construction Manager is the qualified general contracting organization which performs the Construction Management under a professional services contract with the Owner. The Construction Manager, as the construction professional on the Construction Team, will work with the Owner and the Architect-Engineer from the beginning of design through construction completion, and provide leadership to the Team.

Construction Management may take different forms depending upon the requirements of the Owner, the requirements of various administrative agencies that may be involved with the Project, and, the requirements of the law in the jurisdiction of the Project. The Construction Manager should provide the complete services noted in this booklet during the planning and design phases and he should perform complete management services during the construction phase. Depending on the Owner's requirements noted above, the contractors may either have subcontracts with the Construction Manager or prime contract directly with the Owner. There may or may not be a guaranteed maximum price, and the Construction Manager may or may not be permitted to perform work with his own forces. The AGC documents referred to in this booklet can be used under any of these forms of Construction Management.

3. What is a Construction Team?

The Construction Team is the group responsible for the planning, design and construction of the Project. Its members are the Construction Manager, the Architect-Engineer, and the Owner. It is important to the success of the Project that the Owner assign to the Construction Team competent personnel with the authority to make timely decisions concerning budget and program. The Architect-Engineer should be willing to serve as a cooperating member of the Construction Team using the Construction Management approach. The responsibilities and duties of the Construction Manager are described below.

4. How is the Construction Team formed?

The Owner selects the Construction Manager and the Architect-Engineer.

5.   *When is the Construction Team formed?*

The Owner forms the team very early in the planning stage of a project.

6.   *Why is the Construction Team formed?*

The Owner believes that his interests will best be served if he has available to him, from the very conception of a project, the services of a Construction Manager in addition to the services of a competent Architect-Engineer. These persons will then work together, under the Owner's direction to develop the best and most economical construction program. The Construction Manager's knowledge and experience are of particular value to the Owner in the following areas:

a.   Involvement of the Construction Manager during the planning and design provides the Owner with reliable current information about probable costs and schedules.

b.   The Construction Manager can start construction and order long-delivery material items before the total design is completed, thus allowing the Owner beneficial use of the Project at the earliest possible date, and protecting him against rising costs in an inflationary market.

c.   The Construction Manager and the Architect-Engineer can engage in value analyses of alternative design and construction procedures from the early stages of design development. These analyses will enable the Architect-Engineer to make early major design decisions based upon accurate information relative to cost and time as well as functional and aesthetic considerations.

7.   *How is the Architect-Engineer member of the Construction Team selected?*

The Architect-Engineer may be selected in the conventional manner. He will work closely with the Construction Manager and the Owner on the Construction Team and must be willing to support the procedure.

8.   *What does the Construction Manager do during the Planning and Design Phases?*

He provides a wide range of professional services during both these phases. Specific assignments include the following:

a.   He will consult with, advise, assist and make recommendations to the Owner and Architect-Engineer on all aspects of planning for the Project construction.

b.   He will review the Architectural, Civil, Mechanical, Electrical and Structural plans and specifications as they are being developed, and advise and make recommendations with respect to such factors as construction feasibility, possible economies, availability of materials and labor, time requirements for procurement and construction, and projected costs. He will provide input for life cycle cost studies and energy conservation requirements. He will assist in the coordination of all sections of the drawings and specifications, without, however, assuming any of the Architect-Engineer's normal responsibilities for design.

c.   He will make budget estimates based on the Owner's program and other available information. The first estimate may be a parameter type and subsequent estimates will be in increasing detail as quantity surveys are developed based on developing plans and specifications. He will continue to review and refine these estimates as the development of the plans and specifications proceeds, and will advise the Owner and the Architect-Engineer if it appears that the budgeted targets for the Project cost and/or completion will not be met. He will prepare a final cost estimate when plans and specifications are complete. On most projects, the general contractor-Construction Manager, because of his financial responsibility, will be able to provide a guaranteed maximum price at the time the Construction Team has developed the drawings and specifications to a point where the scope of the Project is defined.

d. He will recommend for purchase and expedite the procurement of long-lead items to ensure their delivery by the required dates.

e. He will make recommendations to the Owner and the Architect-Engineer regarding the division of work in the plans and specifications to facilitate the bidding and awarding of contracts, taking into consideration such factors as time of performance, availability of labor, overlapping trade jurisdictions, and provisions for temporary facilities. He will make a market survey and solicit the interest of capable contractors.

f. He will review plans and specifications with the Architect-Engineer to eliminate areas of conflict and overlapping in the work to be performed by the various contractors.

g. As working drawings and specifications are completed, he will take competitive bids on the work. After analyzing the bids thus received, he will either award contracts or recommend to the Owner that such contracts be awarded. The exact procedure will depend upon his contract with the Owner.

h. At an early stage in the Project, he will prepare a progress schedule for all project activities by the Owner, Architect-Engineer, contractors, and himself. He will closely monitor the schedule during both the design and construction phases of the Project and be responsible for providing all parties with periodic reports as to the status of each activity with respect to the Project schedule.

9. *What does the Construction Manager do during the construction phase?*

The Construction Manager assumes responsibility for managing the Project during construction in much the same way as the general contractor traditionally has done.

a. He will maintain competent supervisory staff to coordinate and provide general direction of the work and progress of the contractors on the Project.

b. He will observe the work as it is being performed, until final completion and acceptance by the Owner, to assure that the materials furnished and work performed are in accordance with working drawings and specifications.

c. He will establish an organization and lines of authority in order to carry out the overall plans of the Construction Team.

d. He will establish procedures for coordination among the Owner, Architect-Engineer, contractors and Construction Manager with respect to all aspects of the Project and implement such procedures. He will maintain job site records and make appropriate progress reports.

e. In cooperation with the Architect-Engineer, he will establish and implement procedures to be followed for expediting and processing all shop drawings, samples, catalogs, and other Project documents.

f.   He will implement an effective labor policy in conformance with local, state, and national labor laws. He will review the safety and EEO programs of each contractor and make appropriate recommendations.

g.   He will review and process all applications for payment by involved contractors and material suppliers in accordance with the terms of the contract.

h.   He will make recommendations for and process requests for changes in the work and maintain records of change orders.

i.   He will furnish either with his own forces or others all General Conditions items as required.

j.   He will perform portions of the work with his own forces if requested by the Owner to do so.

k.   He will schedule and conduct job meetings to ensure the orderly progress of the work.

l.   When the Project is of sufficient size and complexity, the Construction Manager will provide data processing services as may be appropriate.

m.   He will refer all questions relative to interpretation of design intent to the Architect-Engineer.

n.   He will continue the close monitoring of the Project progress schedule, coordinating and expediting the work of all of the contractors and his own forces, and provide periodic status reports to the Team.

o.   He will establish and maintain an effective cost control system, monitoring all Project costs. He will schedule and conduct appropriate meetings to review costs and be responsible for providing periodic reports to the Team on cost status.

10.   *What qualifications should an Owner consider when selecting a Construction Manager?*

The Construction Manager will be selected on the basis of an objective analysis of his professional and general contracting qualifications, and major considerations will be given to:

a.   Demonstration of his ability to perform projects comparable in design, scope and complexity.

b.   The recommendation of Owners for whom the contractor has performed Construction Management.

c.   His financial strength, bonding capacity, and ability to assume a financial risk if the Owner requires it.

d.   The qualifications of in-house staff personnel who will manage the Project.

e.   His demonstrated ability to work cooperatively with the Owner and the Architect-Engineer throughout the Project, and to display leadership and initiative in performing his tasks as a member of the Construction Team.

f.   The demonstration of successful management systems which have been employed by the Construction Manager for the purpose of conceptual estimating, budgeting, scheduling and cost controls.

g.   The Construction Manager's knowledge of and ability to implement the most effective overall insurance program for the Project.

h.   The Construction Manager's resources to provide the requisite services in the design and construction phases for the technical portions of the Project as well as the architectural and structural portions.

i.   The Construction Manager's capability to perform work on the Project with his own forces if it is advantageous to the Project.

11.   How is the Construction Manager of the Construction Team selected?

The selection of the Construction Manager should be based upon the ability of the prospective Construction Manager to perform the services required of him, and his ability to meet procedural requirements of administrative agencies that may be involved with the Project. Most importantly the selection should be based on an objective appraisal of the prospective Construction Manager's professional and general contracting qualifications to perform the Construction Manager functions as a member of the Construction Team. The fee should be one of the factors considered in making the selection, but it should not be the only one.

12.   How is the Construction Manager paid?

For his services the Construction Manager is paid a professional fee commensurate with the complexity of the Project, time of construction, the degree of financial risk, required level of staffing, and as otherwise negotiated with the Owner.

In addition to the fee the Owner will normally be responsible for paying:

a. The actual net cost of the contracts, usually awarded on a competitive basis.
b. The actual net cost of such general conditions work required to be done by the Construction Manager, and the actual costs of Project staff, both direct and indirect.
c. The actual net cost of any construction work that may be performed by the Construction Manager's own forces.

All of the costs outlined above can be included under a Guaranteed Maximum Price at the time the Construction Team has developed the drawings and specifications to a point where the scope of the Project is sufficiently defined.

13.   What form of Contract should be used on a Construction Management Contract?

The form of contract will depend upon whether or not a Guaranteed Maximum Price is required and whether the contractors have subcontracts with the Construction Manager or have direct contracts with the Owner. These features of the contract may be mandated by the procedural requirements of administrative agencies that may be involved with the Project or by law. AGC Document No. 8b - "General Conditions for Owner-Construction Manager Agreement" may be used in any of these cases. For the agreement between the Owner and Construction Manager, AGC Document No. 8— "Standard Form of Agreement Between Owner and Construction Manager" is recommended except when all the contractors have contracts directly with the Owner. In that case, AGC Document No. 8d is recommended.

The foregoing questions and answers are designed to clarify use of the Construction Management concept as applied by members of the Associated General Contractors of America. They are offered as guidelines, with the understanding— which every general contractor possesses—that each construction project is different from any other and that special problems are necessarily met by special solutions. Skill, integrity, and responsibility will continue to prevail in arriving at such solutions."

Bibliography

AGC No. 8 Standard Form of Agreement Between Owner and Construction Manager.

AGC No. 8a Amendment to Owner–Construction Manager Contract.

AGC No. 8b General Conditions for Owner–Construction Manager Contract.

AGC No. 8d Standard Form of Agreement Between Owner and Construction Manager [Owner Awards All Trade Contracts].

AGC No. 62a Construction Management Control Process.

AGC No. 62b Owner Guidelines for Selection of a Construction Manager.

AGC No. 70 An Owner's Guide to Building Construction Contracting Methods.

AGC No. 76 Construction Management for the General Contractor.

AGC No. 113 Construction Management Slide/Cassette Presentation.

appendix

Temporary Plumbing*

A. The General Trades Contractor shall provide the following:

1. Furnish and install enclosures for temporary toilets.
2. Furnish and maintain supplies for temporary toilets and permanent facilities towards end of job.
3. Maintain all of the above facilities in a state of cleanliness throughout life of job.
4. Remove all temporary enclosures when no longer required as directed by the Manager of Construction.
5. Provide and maintain temporary portable toilets prior to installation of temporary toilets by plumber. Note that temporary toilets will be provided by the Special Foundations Contractor and the Structural Steel Contractor for their own operation.
6. There shall be one toilet for every 40 men on the job.

B. The Plumbing Contractor shall provide the following:

1. Pay the costs of water for construction services.
2. Legal Requirements: The plumbing work for construction purposes shall comply with all Federal, State and City requirements. This subcontractor shall obtain and pay for any required permit or inspections related to this work.
3. *Scope of Work:* Contractor shall furnish, install, operate and maintain all temporary plumbing work for the entire project. The work shall include, but is not limited to, the following:

* Appendix B is taken from Vincent G. Bush, *Construction Management: A Handbook for Contractors, Architects and Students,* 1973, pp. 112, 113.

a. *Standpipes:* Permanent 6" standpipes included under the basic plumbing work located adjacent to Stairs A and B shall be installed as closely as possible following erection of the structural steel, and in no case more than five floors behind erected steel. Properly valved hose connection shall be provided at each floor, threaded in conformity with Columbus Fire Department requirements. Permanent standpipe riser work shall be installed such that at least one standpipe riser is in service and follows the steel erection as required by the Fire Department. Install cross connections, supplemented by temporary piping where necessary. Furnish and maintain at the highest operational outlet on each standpipe, a proper box containing sufficient hose to reach all parts of the floor, a 1–1/8" nozzle, spanner wrenches, and hose straps. Temporary booster pumps are to be furnished and installed by the Plumbing Contractor to maintain water pressure as required for the upper floors. Install proper Fire Department connections at the ground floor so that standpipes are made available to the Fire Department immediately upon installation and maintain them available at all times. The entire installation is to be in conformity with Chapter 8, Article 81, of N.B.F.U. Pamphlet No. 14 and Paragraph BB25-14 of the Ohio Building Code.

b. *Water Service Tower:* Provide a temporary water service meter, and distribution piping for the initial phase of construction. As soon as possible install permanent water service at riser east of Stair A at columns 3C and 4C, from which temporary water shall be installed for service throughout the building. Provide proper valves on the risers and run-outs to localize shut down when additional piping is added. Temporary 3/4" valve connections for various construction trades shall be provided at each floor complete with movable water barrel, overflow basin and drain line. Temporary booster pumps are to be furnished and installed by the Plumbing Contractor to maintain water pressure as required for the upper floors. This Contractor shall furnish all controls and starters but not including disconnect switches for complete automatic operation of systems. Costs of water for construction will be paid by the Plumbing Contractor.

c. *Toilet Facilities:* As the construction of the building progresses temporary toilet rooms are to be provided in the North Core for use of construction personnel at the 2nd, 4th, 6th, 10th, 14th, 18th, 22nd, 26th, 30th, 34th, 38th, and 41st floors. Toilet rooms will consist of two water closets and a 5 foot galvanized sheet metal urinal, except that the toilet rooms at the 6th and 41st floors shall have one water closet. Two water closets and urinals shall be provided in the East Core at 1st level down as soon as permanent piping is completed. Water is to be provided from the temporary water riser mentioned above. Temporary soil lines are to be run above floor and connected to permanent soil stacks at west column P3 (F-G, 2-3) location of temporary toilet in Bay F-2 G-2 to F-3 G-3. Portable toilets will be used in Tower prior to temporary toilet facilities above mentioned. Enclosures, supplies and cleaning will be provided by the General Trades Contractor.

d. Four temporary drains shall be installed at approximately the 18th floor and connected to permanent leaders. Remove when directed by the Manager of Construction.

4. *Material:* Minimum cost is the basic requirement consistent with material and workmanship which will satisfactorily meet conditions of the job.

C

appendix

Temporary Heat*

A. General:

1. Before portions of the building are enclosed temporary heat is the responsibility of each contractor as required for his work. After portions of the building are enclosed the temporary heat becomes the responsibility of the heating contractor.

2. Each Contractor shall advise the Heating Contractor daily as to temperatures required in various parts of the building. It shall be each Contractor's responsibility that temperature as recommended by the manufacturer of the material concerned and as required for proper installation are maintained while materials are stored in the building, or being installed and for the length of time recommended following installation.

B. Enclosure:

1. The Manager of Construction will determine at which time the building becomes enclosed and in general the building will be deemed enclosed as each floor or group of floors is permanently enclosed from the weather.

2. All Contractors shall bring the entire project to satisfactory enclosure as soon as possible, concentrating on any area which can be brought to satisfactory enclosure without inhibiting progress on any other area.

* Appendix C is taken from Vincent G. Bush, *Construction Management: A Handbook for Contractors, Architects and Students,* 1973, pp. 109, 110, 111. Reprinted with permission of Reston Publishing Co., a Prentice-Hall Co., 11480 Sunset Hills Road, Reston, VA 22090.

3. Satisfactory enclosure is defined as the condition when exterior wall enclosures and floor above is enclosed and/or roofing is completed, exterior window and door openings are installed and closed with either temporary or permanent doors in position and glass or an acceptable temporary substitute in sash. Conditions of heat loss shall approximate those of final building conditions.

4. Prior to satisfactory enclosure, cold weather protection shall be provided by each Contractor as required for his work to allow work to proceed at a pace required to meet the construction schedule. Protection shall be provided by contractor responsible for the particular phase of construction needing protection and for the benefit of all trades encompassed by his contract.

5. Methods of providing cold weather protection are the Contractor's responsibility. However, methods consisting of heavy tarpaulins, heavy sheet plastic, or other comparable means are recommended. Actual methods employed shall be adequate to fully enclose required portion of new construction so as to maintain heat and afford protection from weather while work is under way, and shall be subject to approval of the Manager of Construction.

C. General Trades Contractor's Responsibilities:

1. The General Trades Contractor shall provide temporary closures at windows and floors where penetrations are required for material movement, hoists, etc., as are required to close building. The General Trades Contractor shall remove these closures as directed by the Manager of Construction.

2. Well before plastering begins, and continuous throughout settling and drying periods, temperatures of 65° F. at days and 55° F. at nights shall be maintained. During this period, no finish woodwork, wood finish flooring, resilient flooring, or flexible wall coverings shall be installed or stored in the building, and no finish painting or applying of finish wall coatings shall be undertaken.

D. Heating Contractor's Responsibilities:

1. The Heating Contractor shall bear all cost for fuel in the operation of the temporary heating system and the permanent heating system when used as a source of temporary heat.

2. *Legal Requirements:* The heating work for construction purposes shall comply with all Federal, State and City requirements. This Contractor shall obtain and pay for any required permits or inspections pertaining to this work.

3. *Scope of Work:* Heating will in general be provided by gas fire unit heaters. These unit heaters will be installed throughout the building as required by construction progress, and maintained for the full 1972-1973 heating season. Temporary heat for concrete operation will be maintained by the General Trades Contractor.

4. *Maintenance:* Contractor is to include in his proposal any maintenance required to keep the system in operation.

5. *Metering and Fuel Payments:* The permanent meter will be installed as part of the plumbing contract. All fuel payments will be paid for by the Heating Contractor.

6. Minimum cost is the basic requirement consistent with material and workmanship which will satisfactorily meet conditions of the job, but it must meet all safety requirements, standards and codes, including all Federal, State and City requirements.

7. *Removal and Salvage:* The Heating Contractor shall disassemble and move from the property all temporary material and equipment used for temporary heat when its use is no longer required and when the removal of such material will not cause damage or removal of permanent work then installed. Temporary connections in permanent 5" gas riser will be plugged and sealed. Surplus will be salvaged and become the property of the Heating Contractor.

8. *Unit Prices:* The Heating Contractor shall submit unit prices in Form of Proposal for additions to or deductions from number of typical floors requiring temporary heat for the 1972-1973 heating season.

9. *Use of Permanent Heating System as a Source of Temporary Heat:*

   a. The Heating Contractor shall expedite his work so that the completed or partially completed permanent heating installation may be used as the source for temporary heat at the earliest possible date.

b. The permanent duct system shall not be used for supplying temporary heat until the filter units are installed, filters shall be frequently changed (or cleaned) and new filters shall be installed at the completion of the contract.

c. The permanent convectors or other permanent space heaters may be temporarily set and used for temporary heating, but before they are permanently set, they must be thoroughly cleaned, with all cement and plaster removed, and the shop coat restored to its original finish. Deflect direct heat away from glass.

d. During this period the Heating Contractor shall be responsible for the maintenance, operation and supervision of the plant, and shall arrange to be on call at all times, including nights, Sundays and holidays, to service the plant and keep it in operation; he shall also cooperate with other Contractors in every way to the end that heat will be maintained continuously in the building and no damage done to the heating plant.

e. All costs for gas fuel required to operate the heating plant during this period shall be borne by the Heating Contractor. See paragraphs below for responsibilities of Electric and Plumbing Contractors. All other costs attributable to providing temporary heat as herein described shall be borne by this Heating Contractor, including cost of providing and installing all valves, traps and space heating units used temporarily, also the costs of temporarily setting any permanent space heating units, re-conditioning same before they are finally installed as permanent units, etc., including wages of all required operating personnel and all costs for any temporary electrical controls and connections required, including costs for maintenance, operation and supervision of same.

f. If the permanent heating or cooling plant is used for temporary heating or cooling, it is understood that this use in no way affects the guarantee which shall become effective at the time of completion and acceptance of the building.

E. Plumbing Contractor's Responsibilities:

The Plumbing Contractor is required to install permanent gas piping, permanent gas meters and regulating valves so that they are available for use of the Heating Contractor to provide temporary heat. Any cost for water consumption to provide temporary heat will be the responsibility of the Plumbing Contractor.

F. Electrical Contractor's Responsibilities:

The Electrical Contractor shall be required to pay all electric power costs in connection with operation of temporary heating as hereinbefore described. The Electrical Contractor is also required to pay all power costs in connection with operation of induction units containing electric heating when permanent heating system is in operation for temporary heat.

TEMPORARY LIGHT & POWER FOR CONSTRUCTION PURPOSES:

A. Legal Requirements: The electrical work for construction purposes shall conform to all Federal, State and Municipal requirements. The Electrical Contractor shall obtain and pay for any required applications, permits and inspections pertaining to this work.

B. Insurance Requirements: The requirements are the same as for the permanent installation. The work shall conform to the requirements of the "National Electric Code" and the "National Safety Code".

C. General: Temporary work shall be installed by the Electrical Contractor in such a manner as not to interfere with the permanent construction. If such interference does occur it will be the responsibility of the Electrical Contractor to make such changes as may be required to overcome the interference. The cost of these changes will be included as part of the Electrical Contractor's base price. All costs of electricity will be paid by the Electrical Contractor, including cost of electricity for lighting on sidewalk barricade and project construction signing on face of barricade.

D. Material: As the life of this installation is limited, and as this installation will not form a part of the finished building, minimum cost is the basic requirement consistent with material and workmanship which will satisfactorily meet job conditions.

E. Maintenance: All temporary facilities are to be maintained and kept in good operating condition. Maintenance men necessary to perform this work shall be provided in accordance with requirements. Maintenance time will include allowance for normal working hours for all trades including ½ hour start up and ½ hour shut down overtime as required. Regular overtime during week days, Saturday, Sunday, and holidays is not to be included. Quote unit price for overtime beyond these requirements.

The Electrical Contractor shall be responsible for installing and maintaining a reasonably balanced system and shall take current readings on the feeders at regular intervals as required by the Manager of Construction.

F. Transfer to Permanent Service: When power becomes available from the permanent building service it will be necessary to transfer temporary requirements to this source. Transfers are to be made when directed by the Manager of Construction. Cost of such work is to be included in the electrical contract.

G. Removal and Salvage: The Electrical Contractor shall disassemble and remove from the property all temporary electrical wiring and equipment when its use is no longer required. Surplus materials will be salvaged and the salvage value reflected in the lump sum price.

H. Protection: The Electrical Contractor shall protect his installation against weather damage and the normal operations of other trades.

I. Work by Others: See Specification Section 2E Special Requirements for work by Special Foundations Contractor and Section 5F Special Requirements for work by Structural Steel Contractor.

J. Scope of the Work: The Electrical Contractor shall provide labor and material for the installation and maintenance of temporary light and power as may be required during the period of construction.

Index